ANNUAL EDITIONS

Environment 10/11

Twenty-Ninth Edition

EDITOR

Zachary Sharp
University of New Mexico

Zachary Sharp is a Regents Professor in the Department of Earth and Planetary Sciences at the University of New Mexico. He received a BS in Geology from U.C. Berkeley and a Masters and PhD from the University of Michigan. He was a postdoctoral fellow at the Carnegie Institution of Washington and then spent eight years at the Université de Lausanne in Switzerland, before moving to New Mexico. He has over 100 publications, including a book on stable isotope geochemistry, and has been teaching a course on environmental science for the last Eight years. His passion is educating students about environmental issues.

The McGraw·Hill Companies

ANNUAL EDITIONS: ENVIRONMENT, TWENTY-NINTH EDITION

Published by McGraw-Hill, a business unit of The McGraw-Hill Companies, Inc., 1221 Avenue of the Americas, New York, NY 10020.
Annual Editions is published by the **Contemporary Learning Series** group within The McGraw-Hill Higher Education division.

1 2 3 4 5 6 7 8 9 0 WDQ/WDQ 1 0 9 8 7 6 5 4 3 2 1 0

ISBN 978–0–07–351556–4
MHID 0–07–351556–6
ISSN 0272–9008

Managing Editor: *Larry Loeppke*
Director Specialized Production: *Faye Schilling*
Developmental Editor: *Dave Welsh*
Editorial Coordinator: *Mary Foust*
Editorial Assistant: *Cindy Hedley*
Production Service Assistant: *Rita Hingtgen*
Permissions Coordinator: *DeAnna Dausener*
Senior Marketing Manager: *Julie Keck*
Marketing Communications Specialist: *Mary Klein*
Marketing Coordinator: *Alice Link*
Senior Project Manager: *Joyce Watters*
Production Supervisor: *Sue Culbertson*
Cover Graphics: *Kristine Jubeck*

Compositor: Laserwords Private Limited
Cover Image: U.S. Department of Energy (inset); © Getty Images/RF (background)

Library in Congress Cataloging-in-Publication Data
Main entry under title: Annual Editions: Environment. 2010/2011.
1. Environment—Periodicals. I. Sharp, Zachary, *comp.* II. Title: Environment.
658'.05

www.mhhe.com

Editors/Academic Advisory Board

Members of the Academic Advisory Board are instrumental in the final selection of articles for each edition of ANNUAL EDITIONS. Their review of articles for content, level, and appropriateness provides critical direction to the editors and staff. We think that you will find their careful consideration well reflected in this volume.

ANNUAL EDITIONS: Environment 10/11
29th Edition

EDITOR

ACADEMIC ADVISORY BOARD MEMBERS

Preface

In publishing ANNUAL EDITIONS we recognize the enormous role played by the magazines, newspapers, and journals of the public press in providing current, first-rate educational information in a broad spectrum of interest areas. Many of these articles are appropriate for students, researchers, and professionals seeking accurate, current material to help bridge the gap between principles and theories and the real world. These articles, however, become more useful for study when those of lasting value are carefully collected, organized, indexed, and reproduced in a low-cost format, which provides easy and permanent access when the material is needed. That is the role played by ANNUAL EDITIONS.

With the year 2010 come new possibilities for environmental progress. Four decades have passed since the first recognition that anthropogenic climate change poses a potential damage to our planet's health. As more data are collected and more research is conducted, the threat appears ever more serious. Ocean temperatures are the highest on record, glaciers are melting at alarming rates, and as the magnitude of climate change effects is revisited, estimates are invariably adjusted upward. At this point, all industrialized countries have signed the Kyoto protocol (an amendment to the United Nations Convention on Climate Change aimed at reducing greenhouse gas emissions) except one—The United States of America. After taking office, President Bush announced that he had no intention of supporting the Kyoto accords, pointing out that the science behind global warming was inexact and that other major polluters, such as China, were exempt from the first round of reductions. Following eight years of reluctance to address climate change, the United States now has a new president who is committed to reducing our greenhouse gas emissions. President Obama signed an executive order mandating emission reductions for government agencies and is working with Congress to implement a set of cap-and-trade regulations aimed at reducing greenhouse gas emissions. One might expect that real change is on its way, except for the new 300-pound gorilla in the room, the economy.

President Obama inherited what is arguably the most severe recession since the Great Depression. Unemployment has skyrocketed, banks foreclose every week, the housing market is weak (but recovering), and federal deficits have ballooned as Congress attempts to lessen the effects and intensity of the recession. How do hard economic times affect our ability and desire to combat environmental issues? Can one possibly argue that this is the time for our government to impose further environmental restrictions on our industries? When people are losing their jobs and homes, can they be asked to make further sacrifices in the name of the environment?

The predictable (and "safe") reaction of our political leaders is that this country cannot afford to tackle environmental problems in the present economy. The consensus is that people cannot afford to pay higher gas prices, more for their energy, or more for food and other necessities. Furthermore, manufacturers cannot afford the higher costs of production and in a global market would have a harder time competing with countries that do not share strict environmental regulation. These are powerful arguments for acting prudently when formulating environmental regulations—at least for the near term. But what are the costs of inaction? Stronger and more frequent hurricanes. Intense megafires across the Western United States. Increased and prolonged droughts in the nation's heartland. Rising sea levels and flooded coastal cities. Melting of glaciers and billions of people without fresh water. Massive human displacement. Loss of species and habitat. The cost of inaction is far too high to ignore. A simple analogy is to consider a homeowner who just lost his job or suffered a large pay cut. Then, to compound his woes, there appears a leak in his roof. In spite of the large costs of repairs, the homeowner recognizes that delaying repair will only result in far higher costs down the line. And so, the repairs are made, painful as the expense might be, and his house is saved. Do we have a similar vision for our planet's well-being?

While there has been a steady increase in the public awareness of climate change issues over the past decade, the latest poll on global warming shows a sharp reversal in the long-term trend. A recent Pew Research Center poll finds that the number of Americans who believe the earth is warming is at a three-year low, with only 57 percent believing that there is strong evidence that the Earth has gotten hotter over the past decades, and only 36 percent believing that the increased temperatures are related to human activity. Why, when the scientific evidence is stronger than ever, when there is nearly universal agreement amongst climate scientists that anthropogenic climate change is real, and when floods, hurricanes, droughts, and fires are headline news, is the public so out of synch with the science?

Perhaps the attention span of the American public has been overextended. People wait and watch for the apocalyptic predictions of climate change to materialize, and yet year after year they see nothing—or at most a "gradual change". Maybe it's because the industries that have the most to lose from increased greenhouse gas regulations have waged a successful publicity campaign sowing doubt on the scientific findings? Maybe the scientists are not doing a good job of communicating with the public? Perhaps it is the perceived hopelessness of the entire situation. Given the near impossible task of seriously reducing the greenhouse gas emissions so integral in keeping this nation functioning, people may prefer to just throw up their hands and ignore the whole issue entirely, convincing themselves that it's "probably not real anyway". Without hope, inaction is inevitable. What's the cause of our changing perceptions? Most likely, the real answer is a combination of all the above.

There is no silver bullet. There's no single thing that we can do to completely turn this situation around. But there are many little things that make a difference, and while each by itself may sound trivial or even silly, the combination of them can result in significant reductions in greenhouse gas emissions. Little things like adding insulation to our homes, turning down the thermostat, driving our cars less, recycling, taking the bus, turning out the lights, eating less "energy-intensive" foods, drinking tap water instead of bottled water, turning off "vampire" appliances. But while our individual actions are necessary, they are not enough to correct the underlying causes of global warming. Governments must get involved. And rather than viewing greenhouse gas reductions as a painful but necessary step, intervention must be viewed in a positive light. We create jobs by building wind farms. A recent installation in New Mexico provides local jobs, funds local schools, creates permanent revenue to the county and local landowners, and supports the nascent U.S. turbine manufacturers (the New Mexico turbines were made in California). The federal and state governments provide generous tax subsidies for installation of home-based photovoltaic system and insulation upgrades. A $100 million grant from the Department of Energy is being used to create the infrastructure for all-electric vehicle recharging stations in five states. Government resources have been provided for solar research, both in the private and public sector.

The above examples of government support benefit our country. They provide jobs, they reduce our dependence on foreign oil, and ultimately they reduce our global carbon footprint. Like the example of the hole in the roof, the costs of change are real and immediate. But the long-term benefits and savings are also real. They are an investment in our future. It is beholden on all of us—students, professors, scientists, politicians, and the general public—to foster an awareness and proactive vision of our energy future. The articles in this book provide a framework for further discussion about what we all can do to make "green" the new color for our country.

Zachary Sharp
Editor

Contents

Preface iv
Correlation Guide xii
Topic Guide xiii
Internet References xv

UNIT 1
The Global Environment: An Emerging World View

Unit Overview xviii

1. **IPCC, 2007: Summary for Policymakers,** Susan Solomon et al. (eds.), In: *Climate Change 2007: The Physical Science Basis*. Contribution of Working group I to the Fourth Assessment Report of the Intergovernmental Panel on Climate Change, 2007
This article summarizes the results of the 2007 Nobel Peace Prize winning effort by the Intergovernmental Panel on Climate Change ***(IPCC)*** report on our "understanding of the human and natural drivers of climate change." Importantly, it presents estimates of how ***global warming*** will affect future climates. The degree of confidence of the assessments is indicated. Even the cautious consensus reached by this large group of scientists from around the world is troubling. 2

2. **Global Warming Battlefields: How Climate Change Threatens Security,** Michael T. Klare, *Current History,* **106,** 2007
What are the ***societal effects*** of climate change identified by the IPCC report on "Climate Change Impacts, Adaptation and Vulnerability"? They are not limited to humanitarian disasters. It is likely that ethnic conflict, insurgencies, and civil violence due to diminished supplies of vital resources. Diminished rainfall and river flow, rising sea level, and more frequent and are also causes severe storms will affect the ability of underdeveloped societies to meet basic ***sustainability levels.*** Water scarcity, limited food availability, and coastal inundation are all effects that will be felt across the entire planet. 16

3. **China Needs Help with Climate Change,** Kelly Sims Gallagher, *Current History,* 2007
Until recently, the U.S. was the "bad boy" of climate change, emitting 25% of CO_2 while having only 5% of the population. While the U.S. remains a high *per capita* emitter, overall, ***China*** has overtaken the U.S. as the biggest CO_2 emitter. China's growth rate is breathtaking, doubling every three and a half years. In this article, the author argues that the U.S. and China need to work together to develop technologies and strategies to lower CO_2 emissions. 23

4. **Where Oil and Water Do Mix: Environmental Scarcity and Future Conflict in the Middle East and North Africa,** Jason J. Morrissette and Douglas A. Borer, *Parameters,* Winter 2004–2005
Much of the past history of conflict in the North African/Southwest Asian culture realm has been based in religion, ideology, and territory. Future conflict in this area is more likely to be based in ***environmental scarcity***—such as too little oil and not enough water—to support the ***population growth*** that is far outpacing economic growth. 28

The concepts in bold italics are developed in the article. For further expansion, please refer to the Topic Guide.

UNIT 2
Population, Policy, and Economy

Unit Overview 36

5. **Do Global Attitudes and Behaviors Support Sustainable Development?,** Anthony A. Leiserowitz, Robert W. Kates, and Thomas M. Parris, *Environment,* November 2005
   Sustainable development is necessary in order to meet human needs and maintain the life-support systems of the planet. Advocates of achieving sustainability on a global scale recognize that significant change in values, attitudes, and behavior in ***human societies*** will be necessary before sustainability can be achieved. 38

6. **Paying for Climate Change,** Benjamin Jones, Michael Keen, and Jon Strand, *Finance & Development,* March 2008
   How much should be spent on climate change ***remediation?*** To address this question, Jones *et al.* argue that ***economists*** are needed to assess the costs of the threat and the remedies. Individuals, firms, and governments do not have monetary ***incentives*** to limit carbon emissions, and any benefits from voluntary reductions benefit the entire community rather than those making the sacrifice. This article assesses the cost-benefit analysis of taxing emissions, predicting R&D breakthroughs, and the very-long term effects baffle economic theory. And if that's not enough, who sacrifices, who pays? 54

7. **High-Tech Trash: Will Your Discarded TV or Computer End up in a Ditch in Ghana?,** Chris Carroll, *National Geographic,* January 2008
   Over the next few years, 30 to 40 million PCs will become obsolete. How does one dispose of a PC, and what is its fate? More than 70 percent go to ***landfill,*** toxic components be damned. Of the 20 percent that is ***recycled,*** a large amount ends up in third-world countries, where valuable components, such as silver and gold, are extracted. Unfortunately, the environmental conditions of such recycling are appalling. 58

8. **Down with Carbon: Scientists Work to Put the Greenhouse Gas in Its Place,** Sid Perkins, *Science News,* May 10, 2008
   If we can't stop burning it, how about getting rid of it? ***Down with Carbon*** discusses possible ***carbon sequestration*** (or ***carbon storage***) possibilities—Ideas such as fertilization of the oceans (to increase algal blooms), extracting CO_2 directly from smoke stacks and pumping it back into the ground, and burying dead wood are all considered. Each idea has pros and cons. 62

UNIT 3
Energy: Present and Future Problems

Unit Overview 66

9. **Gassing up with Hydrogen,** Sunita Stayapal, John Petrovic, and George Thomas, *Scientific American,* April 2007
   Hydrogen is a super-clean fuel that has been touted as the ideal substitute for gasoline. In this *Scientific American* article, one of the formidable obstacles to a ***hydrogen economy*** is discussed, namely hydrogen containment or storage. Hydrogen is a very powerful fuel, but it occurs normally as a non-compressible gas, making it very difficult to store on a car. Numerous futuristic options for storage are discussed. 69

The concepts in bold italics are developed in the article. For further expansion, please refer to the Topic Guide.

10. Wind Power: Obstacles and Opportunities, Martin J. Pasqualetti, *Environment,* September 2004

Wind power is one of the oldest energy sources, used to power mills and water pumps for thousands of years. It is now one of the most promising of the ***alternative energy*** strategies. But in spite of its environmental attributes, wind power meets with considerable local resistance because of aesthetics, noise, and potential damage to bird populations. The proper strategy is to develop wind power in sites where it meets the least resistance. **73**

11. A Solar Grand Plan, Ken Zweibel, James Mason, and Vasilis Fthenakis, *Scientific American,* January 2008

"By 2050 ***solar power*** could end U.S. dependence on foreign oil and slash greenhouse gas emissions." Sound too good to be true? This article gives an overview of the state of the art and future possibilities in relation to how a significant portion of our electricity could be generated by this clean, ***renewable resource.*** **86**

12. The Rise of Renewable Energy, Daniel M. Kammen, *Scientific American,* September 2006

Solar cells, wind turbines, and biofuels are discussed as possible clean, renewable energy sources for the future. Daniel Kammen is hopeful that "renewables" can replace fossil fuel as as our major energy source. If the United States puts more money into research and development and has the political will to change, a major switch is possible. This energy change will be facilitated by the dramatically increasing prices of petroleum and natural gas. **93**

13. What Nuclear Renaissance?, Christian Parenti, *The Nation,* May 2008

Where are the new ***nuclear power*** plants? There has been a great deal of hype for this ***carbon-free power supply.*** And yet, construction of the most recent nuclear power plant in the United States began in 1973. Why haven't more been built? The reason, according to Parenti, is that they cost too much and carry too much risk. Investors simply aren't willing to put up the cash for new construction, even with huge government subsidies. Costs and terrorist concerns are simply too high. **100**

14. The Biofuel Future: Scientists Seek Ways to Make Green Energy Pay Off, Rachel Ehrenberg, *Science News,* August 1, 2009

Both fossil fuels and ***biofuels*** can be burned to generate electricity. The difference is that biofuel burning is ***carbon neutral;*** the CO_2 generated from biofuel burning was only recently drawn out of the atmosphere during plant photosynthesis. But biofuels can only work if they produce more energy than it takes to make them. And they cannot displace large amounts of valuable agricultural land that is needed for our food supply. Ehrenberg discusses the pros and cons of our biofuel future, describing the hurdles and promise. **104**

15. Putting Your Home on an Energy Diet, Marianne Lavelle, *U.S. News and World Report,* October 19, 2008

This is a short article with helpful suggestions for reducing ***home energy budgets.*** Realizing that "The cleanest and cheapest kilowatt-hour is the one we do not have to produce," the author points out that a 40 percent savings on electricity could be realized by building "green" and by simple ***energy savers*** such as turning off "vampire electronics". **107**

The concepts in bold italics are developed in the article. For further expansion, please refer to the Topic Guide.

UNIT 4
Biosphere: Endangered Species

Unit Overview 110

16. Forest Invades Tundra . . . and the New Tenants Could Aggravate Global Warming, Janet Raloff, *Science News,* July 5, 2008

Temperatures in the ***Arctic*** are climbing at a rate of twice the global average. Scientists are now finding that the boundary between northern forests and the Arctic ***tundra*** is slowly creeping northward as temperatures rise. The snow covered tundra typically reflects light, keeping temperatures low. As trees invade the tundra, they absorb sunlight raising temperatures. Drought and fire have hurt both the tundra and forests. Throw in advancing shrubs, active microbes, and you have a seriously disturbed ***ecosystem.*** 112

17. America's Coral Reefs: Awash with Problems, Tundi Agardy, *Issues in Science and Technology,* Winter 2004

America's coral reefs—the rain forests of the oceans—are in trouble. This delicate ecosystem has declined by 80 percent over the last three decades. ***Overfishing,*** fertilizers, sediment input, and ocean warming all contribute to the dramtic decline in this invaluable resource, both a thing of intrinsic beauty and economic value. The public is generally unaware of the destruction going on beneath the waves, but as author Jeffrey McNeely explains, there is hope. Public and private efforts can ultimately reverse the fate of these national treasures. 114

18. Seabird Signals, Doreen Cubie, *National Wildlife,* August/September 2008

Cassin's auklets are the "canary in the coalmine" of the ocean. On a remote island, the nesting pairs have decreased by a factor of five. The ecosystem is changing, the food web is unraveling and the auklets' population is plunging. Possible explanations include ***warming ocean temperature*** or changing ocean circulation patterns. 120

19. Taming the Blue Frontier, Sarah Simpson, *Conservation Magazine,* April/June 2009

Fish farms or ***aquaculture*** reduce overharvesting of the ocean's wild stocks. But large fish farms also generate huge amounts of fish waste and spread disease. Author Sarah Simpson discusses cutting-edge technologies that drastically reduce the environmental damage and create sustainable aquaculture practices. Some remarkable ideas are being tested. 123

20. Nature's Revenge, Donovan Webster, *Best Life,* November 2008

What happens when a city surrounded by forests is abandoned virtually overnight? This experiment, though not planned or anticipated, is playing out following the tragic ***nuclear accident*** at ***Chernobyl.*** The radiation-blanketed city is uninhabitable, creating one of the largest wildlife preserves in Europe. The forest is taking back the city. Donovan Webster is given a creepy tour of the highly ***radioactive*** ghost town. 127

The concepts in bold italics are developed in the article. For further expansion, please refer to the Topic Guide.

UNIT 5
Resources: Land and Water

Unit Overview 134

21. Tracking U.S. Groundwater: Reserves for the Future?, William M. Alley, *Environment,* April 2006

The term ***groundwater*** reserves implies that the supply of groundwater, like other limited natural resources, can be depleted. Current rates of extraction for irrigation and other uses far exceed the rates of natural replacement, placing this precious water resource in jeopardy. The depletion of groundwater reserves has gone from being a local problem to a national- and probably even global one. 136

22. How Much Is Clean Water Worth?, Jim Morrison, *National Wildlife,* February/March 2005

When the value of a clean water resource is calculated in monetary terms, it becomes increasingly clear that ***conservation*** methods make both economic and ecologic sense. The tricky part is manipulating the economic system that drives our behavior so that it makes sense to invest in and protect natural assets—like ***clean water.*** 147

23. Searching for Sustainability: Forest Policies, Smallholders, and the Trans-Amazon Highway, Eirivelthon Lima et al., *Environment,* January/February 2006

Commercial ***logging*** in the ***Amazon*** has traditionally been an ecologically destructive process, as cleared areas were occupied by farmers who extended the clearing process. The development of the major economic corridor of the Trans-Amazon Highway illustrates how logging can be converted from a destructive force to one that promotes ***sustainable development.*** 150

24. Diet, Energy, and Global Warming, Gidon Eshel and Pamela A. Martin, *Earth Interactions,* December 2006

Want to save the planet? Forget driving a Prius. Become a ***vegetarian*** instead! Using a thorough set of facts, statistics, and calculations, the authors conclude that your ***carbon footprint*** can be drastically reduced if you change your diet to a less energy-intensive one. Animal-based foods require far more energy for their production, compared with vegetarian foods, and there appear to be no negative health effects associated with a balanced vegetarian diet. 158

25 Landfill-on-Sea, Daisy Dumas, *Ecologist,* February 7, 2008

In the middle of the Pacific Ocean, in an area covering twice the size of France, there is a non-degradable ***plastic garbage*** dump. 100 million tons of plastic are used each year, and discarded plastic entering the Pacific Ocean will ultimately end up in the ***Central Pacific Gyre,*** contributing to the ***Great Pacific Garbage Patch.*** And the garbage patch is growing. The toxic side-effects of the non-biodegradable plastic are very hazardous to animals at all trophic levels. 167

The concepts in bold italics are developed in the article. For further expansion, please refer to the Topic Guide.

UNIT 6
The Politics of Climate Change

Unit Overview 170

26 The Truth about Denial, Sharon Begley, *Newsweek,* August 13, 2007

In this remarkable exposé, Sharon Begley describes the sophisticated efforts by the ***oil industry and lobby*** to dispel the science behind global warming. Much as the tobacco industry did a generation ago, big oil is working hard to sow doubt about global warming and to even question whether it is such a bad thing in the first place. Although the overwhelming majority of climate scientists agree that global warming is a serious threat to humanity, Begley explains how the powerful oil-funded "denial machine" is succeeding at spreading confusion and thereby preventing any action to combat this global threat. 172

27. Swift Boating, Stealth Budgeting, and Unitary Executives, James Hansen, *World Watch,* November/December 2006

The scientific thesis that ***anthropogenic climate change*** is altering the global environmental system has reached the point of consensus among scientists. The few scientists who do not accept global climate change are referred to as ***contrarians.*** Unfortunately, they approach the global warming problem as if they were lawyers, not scientists, whose job it is to defend a client rather than seek the truth. 177

28. The Myth of the 1970s Global Cooling Scientific Consensus, Thomas C. Peterson, William M. Connolly, and John Fleck, *Bulletin of the American Meteorological Society,* September 2008

One of the most common arguments used by climate change skeptics is that scientists don't really know what's going on. One has to only look back to the 1970s, when they thought the world was cooling, not warming. Or so the argument goes. Peterson *et al.* explore this common perception and find that the myth is not true. The overwhelming majority of peer-reviewed articles from the time predicted that temperatures would rise, not fall. The ***global cooling*** argument used so commonly by ***climate skeptics*** is simply not supported by the facts. 181

29. How to Stop Climate Change: The Easy Way, Mark Lynas, *New Statesman,* November 8, 2007

We become further convinced that climate change is real and that something must be done. Are we doing anything? Mark Lynas suggests three straightforward ways to reduce our carbon consumption: (1) Stop debating, start doing; (2) focus on big wins; (3) use technology. By looking at the overwhelming problems in pieces, it becomes clear that we can move forward with environmental protection. 193

30. Environmental Justice for All, Leyla Kokmen, *Utne,* March/April 2008

Low-income communities suffer from ***urban pollution*** to a far greater extent than other groups. The reasons are obvious: Property values are low, the financial resources of a low-income community to fight the incursion of a polluting factory are limited, as is its ability to litigate against health hazards. A new factory located in a depressed area brings not only pollution risks, but jobs. And so a delicate balancing act must be performed. The trick is to fight poverty (bring in the jobs) and pollution at the same time. 196

Test-Your-Knowledge Form 200

Article Rating Form 201

The concepts in bold italics are developed in the article. For further expansion, please refer to the Topic Guide.

Correlation Guide

The *Annual Editions* series provides students with convenient, inexpensive access to current, carefully selected articles from the public press. **Annual Editions: Environment 10/11** is an easy-to-use reader that presents articles on important topics such as *the global environment, population policy, energy use and policy,* and many more. For more information on *Annual Editions* and other *McGraw-Hill Contemporary Learning Series* titles, visit www.mhhe.com/cls.

This convenient guide matches the units in **Annual Editions: Environment 10/11** with the corresponding chapters in two of our best-selling McGraw-Hill Environmental Science textbooks by Enger/Smith and Cunningham/Cunningham.

Annual Editions: Environment 10/11	Environmental Science: A Study of Interrelationships, 12/e by Enger/Smith	Environmental Science: A Global Concern, 11/e, by Cunningham/ Cunningham
Unit 1: The Global Environment	**Chapter 1:** Environmental Interrelationships **Chapter 2:** Environmental Ethics **Chapter 19:** Environmental Policy and Decision Making	**Chapter 1:** Understanding Our Environment **Chapter 2:** Principles of Science and Systems **Chapter 15:** Air, Weather, and Climate **Chapter 24:** Environmental Policy, Law, and Planning
Unit 2: Population, Policy, and Economy	**Chapter 3:** Environmental Risk: Economics, Assessment, and Management **Chapter 7:** Populations: Characteristics and Issues **Chapter 8:** Energy and Civilization **Chapter 9:** Energy Sources **Chapter 19:** Environmental Policy and Decision Making	**Chapter 6:** Population Biology **Chapter 7:** Human Populations **Chapter 23:** Ecological Economics **Chapter 24:** Environmental Policy, Law, and Planning
Unit 3: Energy: Present and Future Problems	**Chapter 8:** Energy and Civilization **Chapter 9:** Energy Sources **Chapter 10:** Nuclear Energy	**Chapter 19:** Conventional Energy **Chapter 20:** Sustainable Energy
Unit 4: Biosphere: Endangered Species	**Chapter 5:** Interactions: Environments and Organisms **Chapter 6:** Kinds of Ecosystems and Communities **Chapter 11:** Biodiversity Issues	**Chapter 5:** Biomes: Global Patterns of Life **Chapter 11:** Biodiversity: Preserving Species
Unit 5: Resources: Land and Water	**Chapter 12:** Land-Use Planning **Chapter 13:** Soil and Its Uses **Chapter 14:** Agricultural Methods and Pest Management **Chapter 15:** Water Management	**Chapter 5:** Biomes: Global Patterns of Life **Chapter 10:** Farming: Conventional and Sustainable Practices **Chapter 12:** Biodiversity: Preserving Landscapes **Chapter 14:** Geology and Earth Resources **Chapter 17:** Water Use and Management **Chapter 18:** Water Pollution
Unit 6: The Politics of Climate Change	**Chapter 8:** Energy and Civilization **Chapter 19:** Environmental Policy and Decision Making	**Chapter 7:** Human Populations **Chapter 24:** Environmental Policy, Law, and Planning

Topic Guide

This topic guide suggests how the selections in this book relate to the subjects covered in your course. You may want to use the topics listed on these pages to search the Web more easily.

On the following pages a number of websites have been gathered specifically for this book. They are arranged to reflect the units of this Annual Editions reader. You can link to these sites by going to *http://www.mhhe.com/cls.*

All the articles that relate to each topic are listed below the bold-faced term.

Alternative energy

8. Down with Carbon: Scientists Work to Put the Greenhouse Gas in Its Place
9. Gassing up with Hydrogen
10. Wind Power: Obstacles and Opportunities
11. A Solar Grand Plan
12. The Rise of Renewable Energy
13. What Nuclear Renaissance?
15. Putting Your Home on an Energy Diet

Arctic

16. Forest Invades Tundra . . . and the New Tenants Could Aggravate Global Warming

Biofuels

14. The Biofuel Future: Scientists Seek Ways to Make Green Energy Pay Off

Biosphere

16. Forest Invades Tundra . . . and the New Tenants Could Aggravate Global Warming
17. America's Coral Reefs: Awash with Problems
18. Seabird Signals
20. Nature's Revenge

Carbon sequestration

8. Down with Carbon: Scientists Work to Put the Greenhouse Gas in Its Place

China

3. China Needs Help with Climate Change

Conservation

15. Putting Your Home on an Energy Diet
19. Taming the Blue Frontier
22. How Much Is Clean Water Worth?

Economics

6. Paying for Climate Change
7. High-Tech Trash: Will Your Discarded TV or Computer End up in a Ditch in Ghana?
24. Diet, Energy, and Global Warming
29. How to Stop Climate Change: The Easy Way

Environmental scarcity

4. Where Oil and Water Do Mix: Environmental Scarcity and Future Conflict in the Middle East and North Africa
19. Taming the Blue Frontier
24. Diet, Energy, and Global Warming

Global warming & Climate change

1. IPCC, 2007: Summary for Policymakers
2. Global Warming Battlefields: How Climate Change Threatens Security
3. China Needs Help with Climate Change
8. Down with Carbon: Scientists Work to Put the Greenhouse Gas in Its Place
16. Forest Invades Tundra . . . and the New Tenants Could Aggravate Global Warming
24. Diet, Energy, and Global Warming
27. Swift Boating, Stealth Budgeting, and Unitary Executives
28. The Myth of the 1970s Global Cooling Scientific Consensus

Groundwater

21. Tracking U.S. Groundwater: Reserves for the Future?
22. How Much Is Clean Water Worth?

Hydrogen

9. Gassing up with Hydrogen

Nuclear energy

13. What Nuclear Renaissance?
20. Nature's Revenge

Oceans

17. America's Coral Reefs: Awash with Problems
18. Seabird Signals
19. Taming the Blue Frontier
25. Landfill-on-Sea

Politics and policy

26. The Truth about Denial
27. Swift Boating, Stealth Budgeting, and Unitary Executives
28. The Myth of the 1970s Global Cooling Scientific Consensus
29. How to Stop Climate Change: The Easy Way
30. Environmental Justice for All

Pollution

7. High-Tech Trash: Will Your Discarded TV or Computer End up in a Ditch in Ghana?
19. Taming the Blue Frontier
20. Nature's Revenge
25. Landfill-on-Sea
30. Environmental Justice for All

Population

4. Where Oil and Water Do Mix: Environmental Scarcity and Future Conflict in the Middle East and North Africa

Recycling

7. High-Tech Trash: Will Your Discarded TV or Computer End up in a Ditch in Ghana?
19. Taming the Blue Frontier

Renewable energy

8. Down with Carbon: Scientists Work to Put the Greenhouse Gas in Its Place
9. Gassing up with Hydrogen
10. Wind Power: Obstacles and Opportunities
11. A Solar Grand Plan
12. The Rise of Renewable Energy
14. The Biofuel Future: Scientists Seek Ways to Make Green Energy Pay Off

Resources

21. Tracking U.S. Groundwater: Reserves for the Future?
23. Searching for Sustainability: Forest Policies, Smallholders, and the Trans-Amazon Highway

Solar energy

11. A Solar Grand Plan

Sustainability

2. Global Warming Battlefields: How Climate Change Threatens Security
5. Do Global Attitudes and Behaviors Support Sustainable Development?
19. Taming the Blue Frontier
23. Searching for Sustainability: Forest Policies, Smallholders, and the Trans-Amazon Highway
24. Diet, Energy, and Global Warming

Wind power

10. Wind Power: Obstacles and Opportunities

Internet References

The following Internet sites have been selected to support the articles found in this reader. These sites were available at the time of publication. However, because websites often change their structure and content, the information listed may no longer be available. We invite you to visit *http://www.mhhe.com/cls* for easy access to these sites.

Annual Editions: Environment 10/11

General Sources

Britannica's Internet Guide
http://www.britannica.com
This site presents extensive links to material on world geography and culture, encompassing material on wildlife, human lifestyles, and the environment.

CIA Factbook
https://www.cia.gov/library/publications/the-world-factbook/index.html
This site is the United States government's official source for data on the population, production, resources, geography, political systems, and other important characteristics of each of the world's countries.

CO_2 Calculator
http://actonco2.direct.gov.uk/index.html
This site is a fun link to illustrate where we use energy in our daily lives and how we can reduce our CO_2 footprint. Graphical interface.

EnviroLink
http://www.envirolink.org
One of the world's largest environmental information clearinghouses, EnviroLink is a grassroots nonprofit organization that unites organizations and volunteers around the world and provides up-to-date information and resources.

Global Climate Change, NASA's Eyes on the Earth
http://climate.nasa.gov
A remarkably informative and graphical discussion about climate change.

IPPC (Internation Panel on Climate Change)
http://www.ipcc-wg2.org
Link to the Nobel Prize–winning report on climate change. An incredible resource. Very detailed, but with a short informative Summary for Policy Makers.

Library of Congress
http://www.loc.gov
Examine this extensive website to learn about resource tools, library services/resources, exhibitions, and databases in many different subfields of environmental studies.

The New York Times
http://www.nytimes.com
Browsing through the archives of the New York Times will provide a wide array of articles and information related to the different subfields of the environment.

SocioSite: Sociological Subject Areas
http://www.pscw.uva.nl/sociosite/TOPICS
This huge sociological site from the University of Amsterdam provides many discussions and references of interest to students of the environment, such as the links to information on ecology and consumerism.

U.S. Geological Survey
http://www.usgs.gov
This site and its many links are replete with information and resources in environmental studies, from explanations of El Niño to discussion of concerns about water resources.

Wikipedia
http://www.wikipedia.org
A free encyclopedia with millions of articles contributed collaboratively using Wiki software. Remarkable resource created by users the world over.

UNIT 1: The Global Environment: An Emerging World View

Alternative Energy Institute (AEI)
http://www.altenergy.org
The AEI will continue to monitor the transition from today's energy forms to the future in a "surprising journey of twists and turns." This site is the beginning of an incredible journey.

Arctic Climate Impact Assessment
http://www.acia.uaf.edu
An international project of the Arctic Council and the International Arctic Science Committee (IASC), to evaluate and synthesize knowledge on climate variability, climate change, and increased ultraviolet radiation and their consequences.

Earth Science Enterprise
http://www.earth.nasa.gov
Information about NASA's Mission to Planet Earth program and its Science of the Earth System can be found here. Surf to learn about satellites, El Niño, and even "strategic visions" of interest to environmentalists.

Global Climate Change
http://climate.jpl.nasa.gov
JPL's graphic-rich site discussing all aspects of climate change. A must-see.

IISDnet
http://www.iisd.org
The International Institute for Sustainable Development, a Canadian organization, presents information through gateways entitled Business, Climate Change, Measurement and Assessment, and Natural Resources. IISD Linkages is its multimedia resource for environment and development policy makers.

National Geographic Society
http://www.nationalgeographic.com
Links to National Geographic's huge archive are provided here. There is a great deal of material related to the atmosphere, the oceans, and other environmental topics.

Internet References

Research and Reference (Library of Congress)
http://lcweb.loc.gov/rr

This research and reference site of the Library of Congress will lead to invaluable information on different countries. It provides links to numerous publications, bibliographies, and guides in area studies that can be of great help to environmentalists.

Solstice: Documents and Databases
http://solstice.crest.org/index.html

In this online source for sustainable energy information, the Center for Renewable Energy and Sustainable Technology (CREST) offers documents and databases on renewable energy, energy efficiency, and sustainable living. The site also offers related websites, case studies, and policy issues.

United Nations
http://www.unsystem.org

Visit this official website locator for the United Nations System of Organizations to get a sense of the scope of international environmental inquiry today. Various UN organizations concern themselves with everything from maritime law to habitat protection to agriculture.

United Nations Environment Programme (UNEP)
http://www.unep.ch

Consult this home page of UNEP for links to critical topics of concern to environmentalists, including desertification, migratory species, and the impact of trade on the environment. The site will direct you to useful databases and global resource information.

World Resources Institute (WRI)
http://www.wri.org

The World Resources Institute is committed to change for a sustainable world and believes that change in human behavior is urgently needed to halt the accelerating rate of environmental deterioration in some areas. It sponsors not only the general website above but also The Environmental Information Portal (www.earthtrends.wri.org) that provides a rich database on the interaction between human disease, pollution, and large-scale environmental, development, and demographic issues.

UNIT 2: Population, Policy, and Economy

The Hunger Project
http://www.thp.org

Browse through this nonprofit organization's site to explore the ways in which it attempts to achieve its goal: the sustainable end to global hunger through leadership at all levels of society. The Hunger Project contends that the persistence of hunger is at the heart of the major security issues that are threatening our planet.

Poverty Mapping
http://www.povertymap.net

Poverty maps can quickly provide information on the spatial distribution of poverty. This site provides maps, graphics, data, publications, news, and links that provide the public with poverty mapping from the global to the subnational level.

World Health Organization
http://www.who.int

The home page of the World Health Organization provides links to a wealth of statistical and analytical information about health and the environment in the developing world.

World Population and Demographic Data
http://geography.about.com/cs/worldpopulation

On this site, information about world population and additional demographic data for all the countries of the world are provided.

WWW Virtual Library: Demography & Population Studies
http://demography.anu.edu.au/VirtualLibrary

This is a definitive guide to demography and population studies. A multitude of important links to information about global poverty and hunger can be found here.

UNIT 3: Energy: Present and Future Problems

Alliance for Global Sustainability (AGS)
http://globalsustainability.org

The AGS is a cooperative venture seeking solutions to today's urgent and complex environmental problems. Research teams from four universities study large-scale, multidisciplinary environmental problems that are faced by the world's ecosystems, economies, and societies.

Alternative Energy Institute, Inc.
http://www.altenergy.org

On this site created by a nonprofit organization, discover how the use of conventional fuels affects the environment. Also learn about research work on new forms of energy.

Department of Energy–Energy Efficiency and Renewable Energy
http://www.eere.energy.gov

The U.S. government's official site regarding energy, its use and government policy.

Energy and the Environment: Resources for a Networked World
http://zebu.uoregon.edu/energy.html

An extensive array of materials having to do with energy sources—both renewable and nonrenewable—as well as other topics of interest to students of the environment is found on this site.

Fuel economy (Department of Energy)
http://www.fueleconomy.gov

Gas mileage (MPG), greenhouse gas emissions, air pollution ratings, and safety information for new and used cars and trucks. Also discusses alternative fuel vehicles.

Institute for Global Communication/EcoNet
http://www.igc.org

This environmentally friendly site provides links to dozens of governmental, organizational, and commercial sites having to do with energy sources. Resources address energy efficiency, renewable generating sources, global warming, and more.

Nuclear Power Introduction
http://library.thinkquest.org/17658/pdfs/nucintro.pdf

Information regarding alternative energy forms can be accessed here. There is a brief introduction to nuclear power and a link to maps that show where nuclear power plants exist.

U.S. Department of Energy
http://www.energy.gov

Scrolling through the links provided by this Department of Energy home page will lead to information about fossil fuels and a variety of sustainable/renewable energy sources.

UNIT 4: Biosphere: Endangered Species

Endangered Species
http://www.endangeredspecie.com

This site provides a wealth of information on endangered species anywhere in the world. Links providing data on the causes, interesting facts, law issues, case studies, and other issues on endangered species are available.

Internet References

Friends of the Earth
http://www.foe.co.uk/index.html

Friends of the Earth, a nonprofit organization based in the United Kingdom, pursues a number of campaigns to protect the Earth and its living creatures. This site has links to many important environmental sites, covering such broad topics as ozone depletion, soil erosion, and biodiversity.

Natural Resources Defense Council
http://nrdc.org

The Natural Resources Defense Council (NRDC) uses law, science, and the support of more than 1 million members and activists to protect the planet's wildlife, plants, water, soils, and other resources. The site provides abundant information on global issues and political responses.

Smithsonian Institution Website
http://www.si.edu

Looking through this site, which will provide access to many of the enormous resources of the Smithsonian, offers a sense of the biological diversity that is threatened by humans' unsound environmental policies and practices.

World Wildlife Federation (WWF)
http://www.wwf.org

This home page of the WWF leads to an extensive array of information links about endangered species, wildlife management and preservation, and more. It provides many suggestions for how to take an active part in protecting the biosphere.

UNIT 5: Resources: Land and Water

Global Climate Change
http://www.puc.state.oh.us/consumer/gcc/index.html

The goal of this PUCO (Public Utilities Commission of Ohio) site is to serve as a clearinghouse of information related to global climate change. Its extensive links provide an explanation of the science and chronology of global climate change, acronyms, definitions, and more.

National Oceanic and Atmospheric Administration (NOAA)
http://www.noaa.gov

Through this home page of NOAA, you can find information about coastal issues, fisheries, climate, and more.

National Operational Hydrologic Remote Sensing Center (NOHRSC)
http://www.nohrsc.nws.gov

Flood images are available at this site of the NOHRSC, which works with the U.S. National Weather Service to track weatherrelated information.

Terrestrial Sciences
http://www.cgd.ucar.edu/tss

The Terrestrial Sciences Section (TSS) is part of the Climate and Global Dynamics (CGD) Division at the National Center for Atmospheric Research (NCAR) in Boulder, Colorado. Scientists in the section study land-atmosphere interactions, in particular surface forcing of the atmosphere, through model development, application, and observational analyses. Here, you'll find a link to VEMAP, the Vegetation/Ecosystem Modeling and Analysis Project.

UNIT 6: The Politics of Climate Change

Energy Justice Network
http://www.energyjustice.net

Energy Justice is the grassroots energy agenda, supporting communities threatened by polluting energy and waste technologies. Taking direction from our grassroots base and the Principles of Environmental Justice, we advocate a clean energy, zero-emission, zero-waste future for all.

Persistent Organic Pollutants (POP)
http://www.chem.unep.ch/pops

Visit this site to learn more about persistent organic pollutants (POPs) and the issues and concerns surrounding them.

RealClimate
http://www.realclimate.org

A blog for climate science with information from climate scientists. This is a one-stop link for resources that people can use to get up to speed on the issue of climate change

School of Labor and Industrial Relations (SLIR): Hot Links
http://www.lir.msu.edu/hotlinks

Michigan State University's SLIR page connects to industrial relations sites throughout the world. It has links to U.S. government statistics, newspapers and libraries, international intergovernmental organizations, and more.

Space Research Institute
http://arc.iki.rssi.ru/eng/index.htm

For a change of pace, browse through this home page of Russia's Space Research Institute for information on its Environment Monitoring Information Systems, the IKI Satellite Situation Center, and its Data Archive.

Worldwatch Institute
http://www.worldwatch.org

The Worldwatch Institute, dedicated to fostering the evolution of an environmentally sustainable society, presents this site with access to World Watch Magazine and State of the World 2000. Click on In the News and Press Releases for discussions of current problems.

UNIT 1

The Global Environment: An Emerging World View

Unit Selections

1. **IPCC, 2007: Summary for Policymakers,** Susan Solomon et al. (eds.)
2. **Global Warming Battlefields: How Climate Change Threatens Security,** Michael T. Klare
3. **China Needs Help with Climate Change,** Kelly Sims Gallagher
4. **Where Oil and Water Do Mix: Environmental Scarcity and Future Conflict in the Middle East and North Africa,** Jason J. Morrissette and Douglas A. Borer

Key Points to Consider

- What are the global ramification of climate change? Why is the public so at odds with the overwhelming concensus of the scientific community? What do we know with certainty about climate change and what questions are still debated? How far should we be willing to go to mitigate its effects?
- What are the interconnections between different nations? How should first-world versus developing nations interact to improve our environmental concerns?
- How can human societies and cultures adapt to such trends in order to prevent significant environmental disruption?
- What is the interplay between climate change, environmental degradation, and population growth?
- What is the relationship between human attitudes and behavior and the attempts to develop systems of economic development that can be environmentally sustainable?

Student Website

www.mhhe.com/cls

Internet References

Alternative Energy Institute (AEI)
http://www.altenergy.org

Arctic Climate Impact Assessment
http://www.acia.uaf.edu

Earth Science Enterprise
http://www.earth.nasa.gov

Global Climate Change
http://climate.jpl.nasa.gov

IISDnet
http://www.iisd.org

National Geographic Society
http://www.nationalgeographic.com

Research and Reference (Library of Congress)
http://lcweb.loc.gov/rr

Solstice: Documents and Databases
http://solstice.crest.org/index.html

United Nations
http://www.unsystem.org

United Nations Environment Programme (UNEP)
http://www.unep.ch

World Resources Institute (WRI)
http://www.wri.org/

Without doubt, the most important environmental issue we face today is climate change. The Intergovernmental Panel on Climate Change (IPCC) released its Fourth Assessment Report on Climate Change in 2007. For this work, the members of the panel were awarded the Nobel Peace Prize. A summary of their assessment of the predicted effects of increased greenhouse gas levels is presented in the first issue (Intergovernmental Panel on Climate Change—IPCC Summary for Policymakers) (note that this report is followed by another IPCC report on expected outcomes that can be found at http://www.ipcc.ch/pdf/assessment-report/ar4/syr/ar4_syr_spm.pdf). In order to achieve consensus amongst all members, the conclusions are very conservative, with most climate scientists expecting even more dramatic climate changes to occur in the coming decades—in our lifetimes and our children's lifetimes. Even so, the results are sobering, to say the least. The summary states, "*Warming of the climate system is unequivocal, as is now evident from observations of increases in global average air and ocean temperatures, widespread melting of snow and ice, and rising global mean sea level*," with 11 of the last 12 years being the warmest since moderns records began in 1850. And "*Anthropogenic warming and sea level rise would continue for centuries due to the timescales associated with climate processes and feedbacks, even if greenhouse gas concentrations were to be stabilized.*" The IPCC report is alarming and will hopefully awaken us to action.

In Article 2, "Global Warming Battlefields" the widespread hardships that are predicted will lead to violent conflict and unprecedented humanitarian trauma. The poorest countries will be hardest hit, with increasing drought, decreasing food supply, inundated coastal cities, and massive migration. First-world nations will have no choice but to deal with the conflicts, diminished resources, and mass migration.

Article 3, "China Needs Help with Climate Change," illustrates the depth of the problem. Only a few years ago, we assumed that if we collectively could reduce the greenhouse gas emissions of the most developed nations, we would return to a manageable situation. Now we recognize that developing nations, notably China and India must be part of the equation. The article argues that developed and developing nations must work together to forge consensus and find solutions to reducing greenhouse gas emissions and confront the challenges of impending climate change.

© Dr. Parvinder Sethi

The last article deals with the prediction of future conflicts in North Africa/Southwest Asia. Tying back into the problems of climate change and global warming, this region will suffer from water shortages, loss of revenues from falling oil production and exportation and a rapid population growth. U.S. military analyses have suggested that one of the greatest threats to our security is the displacement and disenfranchisement that is expected as the climate changes.

Climate Change 2007

The Physical Science Basis, Summary for Policymakers

INTERGOVERNMENTAL PANEL ON CLIMATE CHANGE

Introduction

The Working Group I contribution to the IPCC Fourth Assessment Report describes progress in understanding of the human and natural drivers of climate change[1], observed climate change, climate processes and attribution, and estimates of projected future climate change. It builds upon past IPCC assessments and incorporates new findings from the past six years of research. Scientific progress since the TAR is based upon large amounts of new and more comprehensive data, more sophisticated analyses of data, improvements in understanding of processes and their simulation in models, and more extensive exploration of uncertainty ranges.

The basis for substantive paragraphs in this Summary for Policymakers can be found in the chapter sections specified in curly brackets.

Human and Natural Drivers of Climate Change

Changes in the atmospheric abundance of greenhouse gases and aerosols, in solar radiation and in land surface properties alter the energy balance of the climate system. These changes are expressed in terms of radiative forcing[2], which is used to compare how a range of human and natural factors drive warming or cooling influences on global climate. Since the Third Assessment Report (TAR), new observations and related modelling of greenhouse gases, solar activity, land surface properties and some aspects of aerosols have led to improvements in the quantitative estimates of radiative forcing.

> Global atmospheric concentrations of carbon dioxide, methane and nitrous oxide have increased markedly as a result of human activities since 1750 and now far exceed pre-industrial values determined from ice cores spanning many thousands of years (see Figure SPM-1). The global increases in carbon dioxide concentration are due primarily to fossil fuel use and land-use change, while those of methane and nitrous oxide are primarily due to agriculture. {2.3, 6.4, 7.3}

- Carbon dioxide is the most important anthropogenic greenhouse gas (see Figure SPM-2). The global atmospheric concentration of carbon dioxide has increased from a pre-industrial value of about 280 ppm to 379 ppm[3] in 2005. The atmospheric concentration of carbon dioxide in 2005 exceeds by far the natural range over the last 650,000 years (180 to 300 ppm) as determined from ice cores. The annual carbon dioxide concentration growth-rate was larger during the last 10 years (1995–2005 average: 1.9 ppm per year), than it has been since the beginning of continuous direct atmospheric measurements (1960–2005 average: 1.4 ppm per year) although there is year-to-year variability in growth rates. {2.3, 7.3}
- The primary source of the increased atmospheric concentration of carbon dioxide since the pre-industrial period results from fossil fuel use, with land use change providing another significant but smaller contribution. Annual fossil carbon dioxide emissions[4] increased from an average of 6.4 [6.0 to 6.8][5] GtC (23.5 [22.0 to 25.0] $GtCO_2$) per year in the 1990s, to 7.2 [6.9 to 7.5] GtC (26.4 [25.3 to 27.5] $GtCO_2$) per year in 2000–2005 (2004 and 2005 data are interim estimates). Carbon dioxide emissions associated with land-use change are estimated to be 1.6 [0.5 to 2.7] GtC (5.9 [1.8 to 9.9] $GtCO_2$) per year over the 1990s, although these estimates have a large uncertainty. {7.3}
- The global atmospheric concentration of methane has increased from a pre-industrial value of about 715 ppb to 1732 ppb in the early 1990s, and is 1774 ppb in 2005. The atmospheric concentration of methane in 2005 exceeds by far the natural range of the last 650,000 years (320 to 790 ppb) as determined from ice cores. Growth rates have declined since the early 1990s, consistent with total emissions (sum of anthropogenic and natural sources) being nearly constant during this period. It is *very likely*[6] that the observed increase in methane concentration is due to anthropogenic activities, predominantly agriculture and fossil fuel use, but relative contributions from different source types are not well determined. {2.3, 7.4}
- The global atmospheric nitrous oxide concentration increased from a pre-industrial value of about 270 ppb to 319 ppb in 2005. The growth rate has been approximately

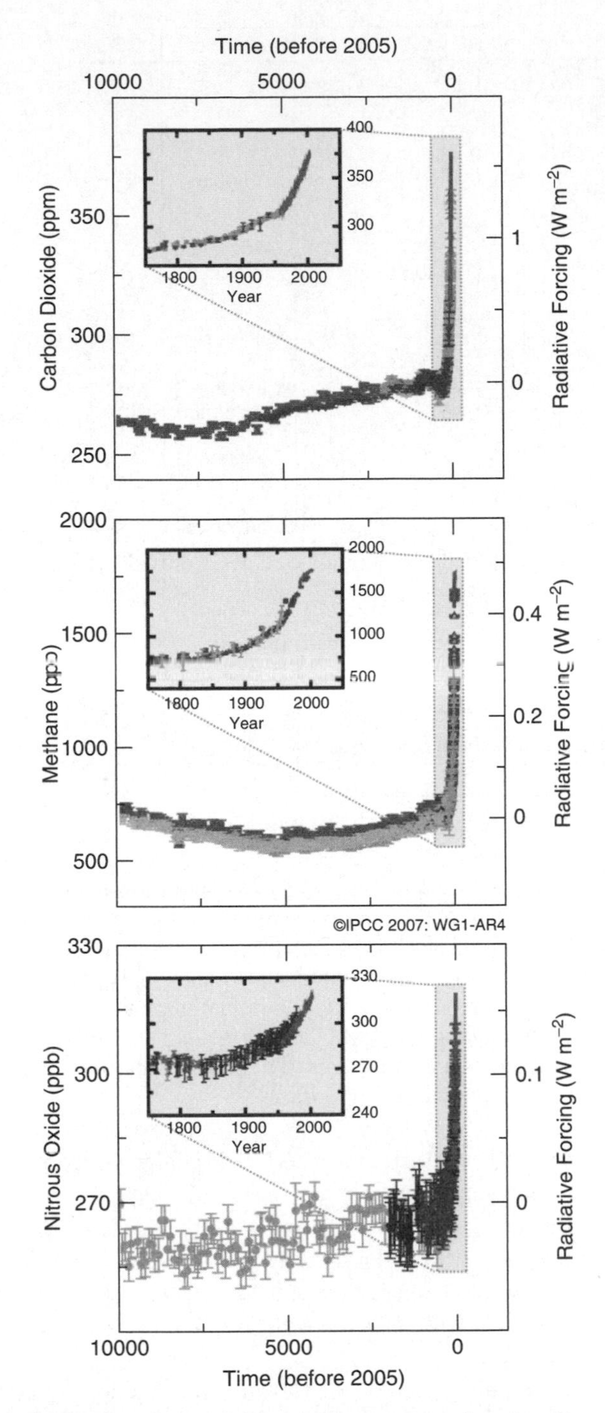

Figure SPM-1 Changes in greenhouse gases from ice-Core and modern data. Atmospheric concentrations of carbon dioxide, methane and nitrous oxide over the last 10,000 years (large panels) and since 1750 (inset panels). {Figure 6.4}

constant since 1980. More than a third of all nitrous oxide emissions are anthropogenic and are primarily due to agriculture. {2.3, 7.4}

- The combined radiative forcing due to increases in carbon dioxide, methane, and nitrous oxide is +2.30 [+2.07 to +2.53] W m^{-2}, and its rate of increase during the industrial era is *very likely* to have been unprecedented in more than 10,000 years (see Figures SPM-1 and SPM-2). The carbon dioxide radiative forcing increased by 20% from 1995 to 2005, the largest change for any decade in at least the last 200 years. {2.3, 6.4}

> The understanding of anthropogenic warming and cooling influences on climate has improved since the Third Assessment Report (TAR), leading to *very high confidence*[7] that the globally averaged net effect of human activities since 1750 has been one of warming, with a radiative forcing of +1.6 [+0.6 to +2.4] W m^{-2}. (see Figure SPM-2). {2.3. 6.5, 2.9}

- Anthropogenic contributions to aerosols (primarily sulphate, organic carbon, black carbon, nitrate and dust) together produce a cooling effect, with a total direct radiative forcing of −0.5 [−0.9 to −0.1] W m^{-2} and an indirect cloud albedo forcing of −0.7 [−1.8 to −0.3] W m^{-2}. These forcings are now better understood than at the time of the TAR due to improved *in situ,* satellite and ground-based measurements and more comprehensive modelling, but remain the dominant uncertainty in radiative forcing. Aerosols also influence cloud lifetime and precipitation. {2.4, 2.9, 7.5}
- Significant anthropogenic contributions to radiative forcing come from several other sources. Tropospheric ozone changes due to emissions of ozone-forming chemicals (nitrogen oxides, carbon monoxide, and hydrocarbons) contribute +0.35 [+0.25 to +0.65] W m^{-2}. The direct radiative forcing due to changes in halocarbons[8] is +0.34 [+0.31 to +0.37] W m^{-2}. Changes in surface albedo, due to land-cover changes and deposition of black carbon aerosols on snow, exert respective forcings of −0.2 [−0.4 to 0.0] and +0.1 [0.0 to +0.2] W m^{-2}. Additional terms smaller than ∓0.1 W m^{-2} are shown in Figure SPM-2. {2.3, 2.5, 7.2}
- Changes in solar irradiance since 1750 are estimated to cause a radiative forcing of +0.12 [+0.06 to +0.30] W m^{-2}, which is less than half the estimate given in the TAR. {2.7}

Direct Observations of Recent Climate Change

Since the TAR, progress in understanding how climate is changing in space and in time has been gained through improvements and extensions of numerous datasets and data analyses, broader geographical coverage, better understanding of uncertainties, and a wider variety of measurements. Increasingly comprehensive observations are available for glaciers and snow cover since the 1960s, and for sea level and ice sheets since about the past decade. However, data coverage remains limited in some regions.

- Eleven of the last twelve years (1995–2006) rank among the 12 warmest years in the instrumental record of global surface temperature[9] (since 1850).

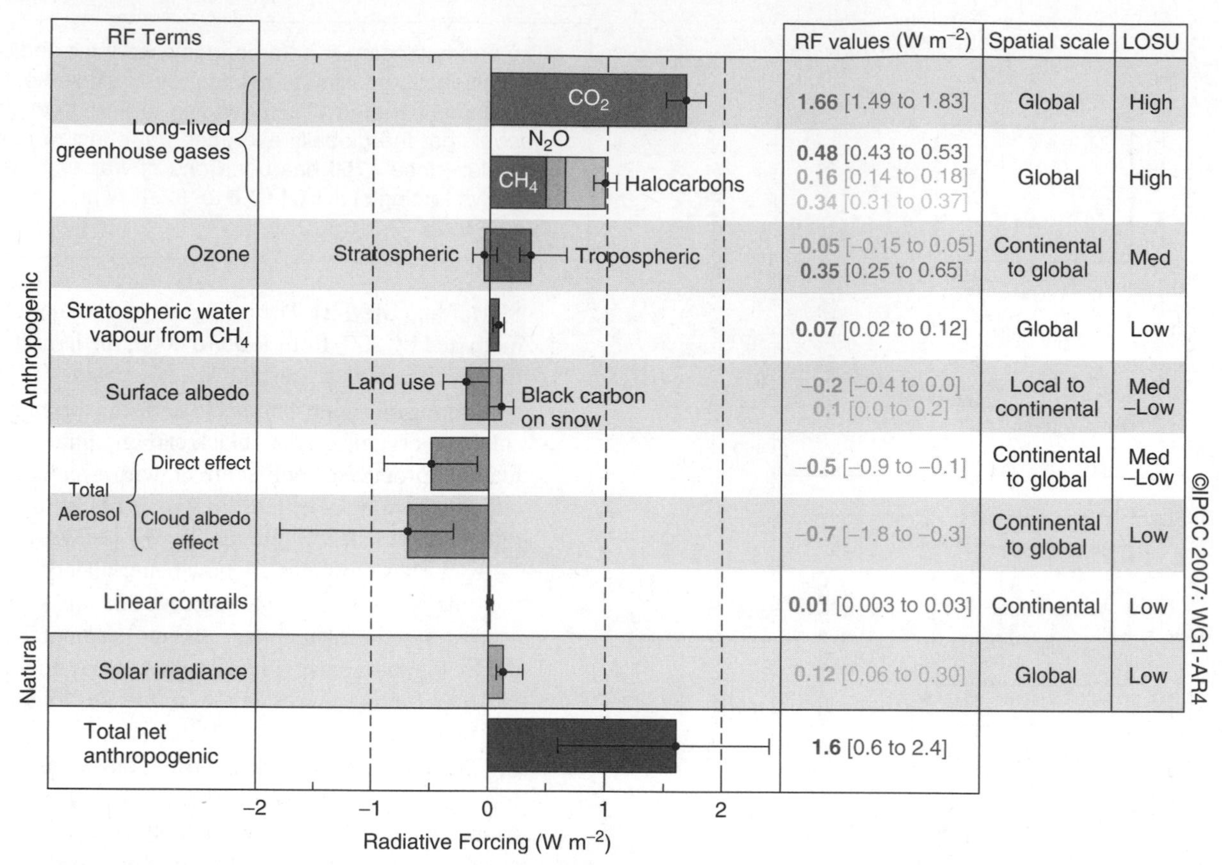

Figure SPM-2 Radiative forcing components. Global-average radiative forcing (RF) estimates and ranges in 2005 for anthropogenic carbon dioxide (CO_2), methane (CH_4), nitrous oxide (N_2O) and other important agents and mechanisms, together with the typical geographical extent (spatial scale) of the forcing and the assessed level of scientific understanding (LOSU). The net anthropogenic radiative forcing and its range are also shown. These require summing asymmetric uncertainty estimates from the component terms, and cannot be obtained by simple addition. Additional forcing factors not included here are considered to have a very low LOSU. Volcanic aerosols contribute an additional natural forcing but are not included in this figure due to their episodic nature. Range for linear contrails does not include other possible effects of aviation on cloudiness. {2.9, Figure 2.20}

Warming of the climate system is unequivocal, as is now evident from observations of increases in global average air and ocean temperatures, widespread melting of snow and ice, and rising global average sea level (see Figure SPM-3). {3.2, 4.2, 5.5}

The updated 100-year linear trend (1906–2005) of 0.74 [0.56 to 0.92]°C is therefore larger than the corresponding trend for 1901–2000 given in the TAR of 0.6 [0.4 to 0.8] °C. The linear warming trend over the last 50 years (0.13 [0.10 to 0.16] °C per decade) is nearly twice that for the last 100 years. The total temperature increase from 1850–1899 to 2001–2005 is 0.76 [0.57 to 0.95] °C. Urban heat island effects are real but local, and have a negligible influence (less than 0.006 °C per decade over land and zero over the oceans) on these values. {3.2}

- New analyses of balloon-borne and satellite measurements of lower- and mid-tropospheric temperature show warming rates that are similar to those of the surface temperature record and are consistent within their respective uncertainties, largely reconciling a discrepancy noted in the TAR. {3.2, 3.4}
- The average atmospheric water vapour content has increased since at least the 1980s over land and ocean as well as in the upper troposphere. The increase is broadly consistent with the extra water vapour that warmer air can hold. {3.4}
- Observations since 1961 show that the average temperature of the global ocean has increased to depths of at least 3000 m and that the ocean has been absorbing more than 80% of the heat added to the climate system. Such warming causes seawater to expand, contributing to sea level rise (see Table SPM-1). {5.2, 5.5}
- Mountain glaciers and snow cover have declined on average in both hemispheres. Widespread decreases in glaciers and ice caps have contributed to sea level rise (ice caps do not include contributions from the Greenland and Antarctic ice sheets). (See Table SPM-1.) {4.6, 4.7, 4.8, 5.5}
- New data since the TAR now show that losses from the ice sheets of Greenland and Antarctica have *very likely* contributed to sea level rise over 1993 to 2003

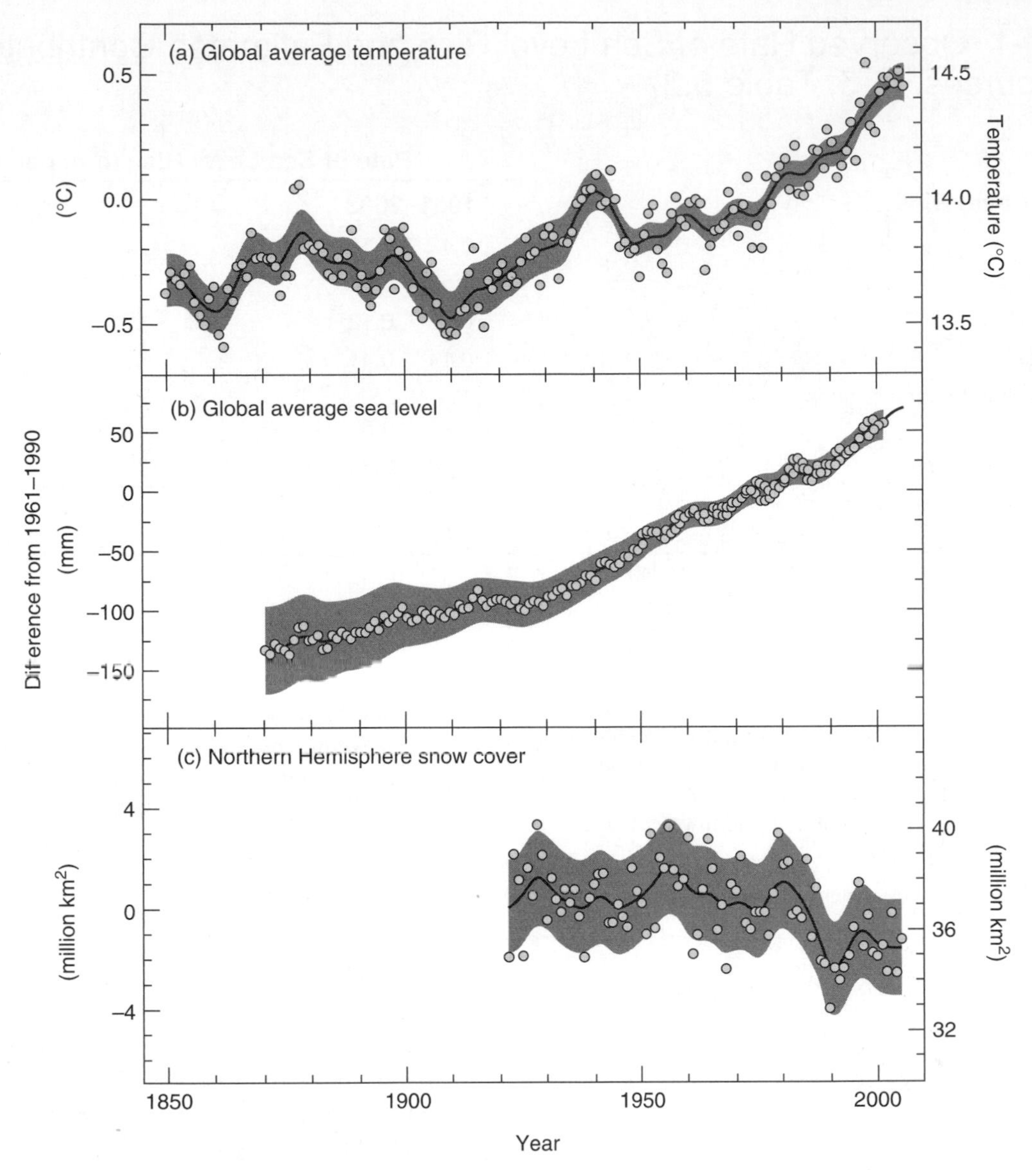

Figure SPM-3 Changes in temperature, sea level and northern hemisphere snow cover. The shaded areas are the uncertainty intervals estimated from a comprehensive analysis of known uncertainties (a and b) and from the time series (c). {FAQ 3.1, Figure 1, Figure 4.2 and Figure 5.13}

(see Table SPM-1). Flow speed has increased for some Greenland and Antarctic outlet glaciers, which drain ice from the interior of the ice sheets. The corresponding increased ice sheet mass loss has often followed thinning, reduction or loss of ice shelves or loss of floating glacier tongues. Such dynamical ice loss is sufficient to explain most of the Antarctic net mass loss and approximately half of the Greenland net mass loss. The remainder of the ice loss from Greenland has occurred because losses due to melting have exceeded accumulation due to snowfall. {4.6, 4.8, 5.5}

- Global average sea level rose at an average rate of 1.8 [1.3 to 2.3] mm per year over 1961 to 2003. The rate was faster over 1993 to 2003, about 3.1 [2.4 to 3.8] mm per year. Whether the faster rate for 1993 to 2003 reflects decadal variability or an increase in the longer-term trend is unclear. There is *high confidence* that the rate of observed sea level rise increased from the 19th to the 20th century. The total 20th century rise is estimated to be 0.17 [0.12 to 0.22] m. {5.5}
- For 1993–2003, the sum of the climate contributions is consistent within uncertainties with the total sea level rise that is directly observed (see Table SPM-1). These estimates are based on improved satellite and *in-situ* data now available. For the period of 1961 to 2003, the sum of climate contributions is estimated to be smaller than the observed sea level rise. The TAR reported a similar discrepancy for 1910 to 1990. {5.5}
- Average Arctic temperatures increased at almost twice the global average rate in the past 100 years. Arctic temperatures have high decadal variability, and a warm period was also observed from 1925 to 1945. {3.2}

Table SPM-1 Observed Rate of Sea Level Rise and Estimated Contributions from Different Sources. {5.5, Table 5.3}

Source of Sea Level Rise	Rate of Sea Level Rise (mm per Year)	
	1961–2003	1993–2003
Thermal expansion	0.42 ∓ 0.12	1.6 ∓ 0.5
Glaciers and ice caps	0.50 ∓ 0.18	0.77 ∓ 0.22
Greenland ice sheet	0.05 ∓ 0.12	0.21 ∓ 0.07
Antarctic ice sheet	0.14 ∓ 0.41	0.21 ∓ 0.35
Sum of individual climate contributions to sea level rise	1.1 ∓ 0.5	2.8 ∓ 0.7
Observed total sea level rise	1.8 ∓ 0.5[a]	3.1 ∓ 0.7[a]
Difference (Observed minus sum of estimated climate contributions)	0.7 ∓ 0.7	0.3 ∓ 1.0

[a]Data prior to 1993 are from tide gauges and after 1993 are from satellite altimetry.

At continental, regional, and ocean basin scales, numerous long-term changes in climate have been observed. These include changes in Arctic temperatures and ice, widespread changes in precipitation amounts, ocean salinity, wind patterns and aspects of extreme weather including droughts, heavy precipitation, heat waves and the intensity of tropical cyclones[10]. {3.2, 3.3, 3.4, 3.5, 3.6, 5.2}

- Satellite data since 1978 show that annual average Arctic sea ice extent has shrunk by 2.7 [2.1 to 3.3]% per decade, with larger decreases in summer of 7.4 [5.0 to 9.8]% per decade. These values are consistent with those reported in the TAR. {4.4}
- Temperatures at the top of the permafrost layer have generally increased since the 1980s in the Arctic (by up to 3 °C). The maximum area covered by seasonally frozen ground has decreased by about 7% in the Northern Hemisphere since 1900, with a decrease in spring of up to 15% {4.7}
- Long-term trends from 1900 to 2005 have been observed in precipitation amount over many large regions[11]. Significantly increased precipitation has been observed in eastern parts of North and South America, northern Europe and northern and central Asia. Drying has been observed in the Sahel, the Mediterranean, southern Africa and parts of southern Asia. Precipitation is highly variable spatially and temporally, and data are limited in some regions. Long-term trends have not been observed for the other large regions assessed[11]. {3.3, 3.9}
- Changes in precipitation and evaporation over the oceans are suggested by freshening of mid and high latitude waters together with increased salinity in low latitude waters. {5.2}
- Mid-latitude westerly winds have strengthened in both hemispheres since the 1960s. {3.5}
- More intense and longer droughts have been observed over wider areas since the 1970s, particularly in the tropics and subtropics. Increased drying linked with higher temperatures and decreased precipitation have contributed to changes in drought. Changes in sea surface temperatures (SST), wind patterns, and decreased snowpack and snow cover have also been linked to droughts. {3.3}
- The frequency of heavy precipitation events has increased over most land areas, consistent with warming and observed increases of atmospheric water vapour. {3.8, 3.9}
- Widespread changes in extreme temperatures have been observed over the last 50 years. Cold days, cold nights and frost have become less frequent, while hot days, hot nights, and heat waves have become more frequent (see Table SPM-2). {3.8}
- There is observational evidence for an increase of intense tropical cyclone activity in the North Atlantic since about 1970, correlated with increases of tropical sea surface temperatures. There are also suggestions of increased intense tropical cyclone activity in some other regions where concerns over data quality are greater. Multi-decadal variability and the quality of the tropical cyclone records prior to routine satellite observations in about 1970 complicate the detection of long-term trends in tropical cyclone activity. There is no clear trend in the annual numbers of tropical cyclones. {3.8}
- A decrease in diurnal temperature range (DTR) was reported in the TAR, but the data available then extended only from 1950 to 1993. Updated observations reveal that DTR has not changed from 1979 to 2004 as both day- and night-time temperature have risen at about the same rate. The trends are highly variable from one region to another. {3.2}
- Antarctic sea ice extent continues to show inter-annual variability and localized changes but no statistically

Table SPM-2 Recent Trends, Assessment of Human Influence on the Trend, and Projections for Extreme Weather Events for Which There Is an Observed Late 20th Century Trend. {Tables 3.7, 3.8, 9.4, Sections 3.8, 5.5, 9.7, 11.2–11.9}

Phenomenon [a] and Direction of Trend	Likelihood That Trend Occurred in Late 20th Century (Typically Post 1960)	Likelihood of a Human Contribution to Observed Trend [b]	Likelihood of Future Trends Based on Projections for 21st Century Using SRES Scenarios
Warmer and fewer cold days and nights over most land areas	*Very likely* [c]	*Likely* [d]	*Virtually certain* [d]
Warmer and more frequent hot days and nights over most land areas	*Very likely* [e]	*Likely (nights)* [d]	*Virtually certain* [d]
Warm spells / heat waves. Frequency increases over most land areas	*Likely*	*More likely than not* [f]	*Very likely*
Heavy precipitation events. Frequency (or proportion of total rainfall from heavy falls) increases over most areas	*Likely*	*More likely than not* [f]	*Very likely*
Area affected by droughts increases	*Likely* in many regions since 1970s	*More likely than not*	*Likely*
Intense tropical cyclone activity increases	*Likely* in some regions since 1970	*More likely than not* [f]	*Likely*
Increased incidence of extreme high sea level (excludes tsunamis) [g]	*Likely*	*More likely than not* [f,h]	*Likely* [i]

[a] See Table 3.7 for further details regarding definitions.

[b] See Table TS-4, Box TS.3.4 and Table 9.4.

[c] Decreased frequency of cold days and nights (coldest 10%).

[d] Warming of the most extreme days and nights each year.

[e] Increased frequency of hot days and nights (hottest 10%).

[f] Magnitude of anthropogenic contributions not assessed. Attribution for these phenomena based on expert judgement rather than formal attribution studies.

[g] Extreme high sea level depends on average sea level and on regional weather systems. It is defined here as the highest 1% of hourly values of observed sea level at a station for a given reference period.

[h] Changes in observed extreme high sea level closely follow the changes in average sea level {5.5.2.6}. It is *very likely* that anthropogenic activity contributed to a rise in average sea level. {9.5.2}

[i] In all scenarios, the projected global average sea level at 2100 is higher than in the reference period {10.6}. The effect of changes in regional weather systems on sea level extremes has not been assessed.

> Some aspects of climate have not been observed to change. {3.2, 3.8, 4.4, 5.3}

significant average trends, consistent with the lack of warming reflected in atmospheric temperatures averaged across the region. {3.2, 4.4}

- There is insufficient evidence to determine whether trends exist in the meridional overturning circulation of the global ocean or in small scale phenomena such as tornadoes, hail, lightning and dust-storms. {3.8, 5.3}

A Paleoclimatic Perspective

Paleoclimatic studies use changes in climatically sensitive indicators to infer past changes in global climate on time scales ranging from decades to millions of years. Such proxy data (e.g., tree ring width) may be influenced by both local temperature and other factors such as precipitation, and are often representative of particular seasons rather than full years. Studies since the TAR draw increased confidence from additional data showing coherent behaviour across multiple indicators in different parts of the world. However, uncertainties generally increase with time into the past due to increasingly limited spatial coverage.

Paleoclimate information supports the interpretation that the warmth of the last half century is unusual in at least the previous 1300 years. The last time the polar regions were significantly warmer than present for an extended period (about 125,000 years ago), reductions in polar ice volume led to 4 to 6 metres of sea level rise. {6.4, 6.6}

- Average Northern Hemisphere temperatures during the second half of the 20th century were *very likely* higher than during any other 50-year period in the last 500 years and *likely* the highest in at least the past 1300 years. Some recent studies indicate greater variability in Northern Hemisphere temperatures than suggested in the TAR, particularly finding that cooler periods existed in the 12 to 14th, 17th, and 19th centuries. Warmer periods prior to the 20th century are within the uncertainty range given in the TAR. {6.6}
- Global average sea level in the last interglacial period (about 125,000 years ago) was *likely* 4 to 6 m higher than during the 20th century, mainly due to the retreat of polar ice. Ice core data indicate that average polar temperatures at that time were 3 to 5 °C higher than present, because of differences in the Earth's orbit. The Greenland ice sheet and other Arctic ice fields *likely* contributed no more than 4 m of the observed sea level rise. There may also have been a contribution from Antarctica. {6.4}

Understanding and Attributing Climate Change

This assessment considers longer and improved records, an expanded range of observations, and improvements in the simulation of many aspects of climate and its variability based on studies since the TAR. It also considers the results of new attribution studies that have evaluated whether observed changes are quantitatively consistent with the expected response to external forcings and inconsistent with alternative physically plausible explanations.

Most of the observed increase in globally averaged temperatures since the mid-20th century is *very likely* due to the observed increase in anthropogenic greenhouse gas concentrations[12]. This is an advance since the TAR's conclusion that "most of the observed warming over the last 50 years is *likely* to have been due to the increase in greenhouse gas concentrations". Discernible human influences now extend to other aspects of climate, including ocean warming, continental-average temperatures, temperature extremes and wind patterns (see Figure SPM-4 and Table SPM-2). {9.4, 9.5}

- It is *likely* that increases in greenhouse gas concentrations alone would have caused more warming than observed because volcanic and anthropogenic aerosols have offset some warming that would otherwise have taken place. {2.9, 7.5, 9.4}
- The observed widespread warming of the atmosphere and ocean, together with ice mass loss, support the conclusion that it is *extremely unlikely* that global climate change of the past fifty years can be explained without external forcing, and *very likely* that it is not due to known natural causes alone. {4.8, 5.2, 9.4, 9.5, 9.7}
- Warming of the climate system has been detected in changes of surface and atmospheric temperatures, temperatures in the upper several hundred metres of the ocean and in contributions to sea level rise. Attribution studies have established anthropogenic contributions to all of these changes. The observed pattern of tropospheric warming and stratospheric cooling is *very likely* due to the combined influences of greenhouse gas increases and stratospheric ozone depletion. {3.2, 3.4, 9.4, 9.5}
- It is *likely* that there has been significant anthropogenic warming over the past 50 years averaged over each continent except Antarctica (see Figure SPM-4). The observed patterns of warming, including greater warming over land than over the ocean, and their changes over time, are only simulated by models that include anthropogenic forcing. The ability of coupled climate models to simulate the observed temperature evolution on each of six continents provides stronger evidence of human influence on climate than was available in the TAR. {3.2, 9.4}
- Difficulties remain in reliably simulating and attributing observed temperature changes at smaller scales. On these scales, natural climate variability is relatively larger making it harder to distinguish changes expected due to external forcings. Uncertainties in local forcings and feedbacks also make it difficult to estimate the contribution of greenhouse gas increases to observed small-scale temperature changes. {8.3, 9.4}
- Anthropogenic forcing is *likely* to have contributed to changes in wind patterns[13], affecting extra-tropical storm tracks and temperature patterns in both hemispheres. However, the observed changes in the Northern Hemisphere circulation are larger than simulated in response to 20th century forcing change. {3.5, 3.6, 9.5, 10.3}
- Temperatures of the most extreme hot nights, cold nights and cold days are *likely* to have increased due to anthropogenic forcing. It is *more likely than not* that anthropogenic forcing has increased the risk of heat waves (see Table SPM-2). {9.4}
- The equilibrium climate sensitivity is a measure of the climate system response to sustained radiative forcing. It is not a projection but is defined as the global average surface warming following a doubling of carbon dioxide concentrations. It is *likely* to be in the range 2 to 4.5 °C with a best estimate of about 3 °C, and is *very unlikely*

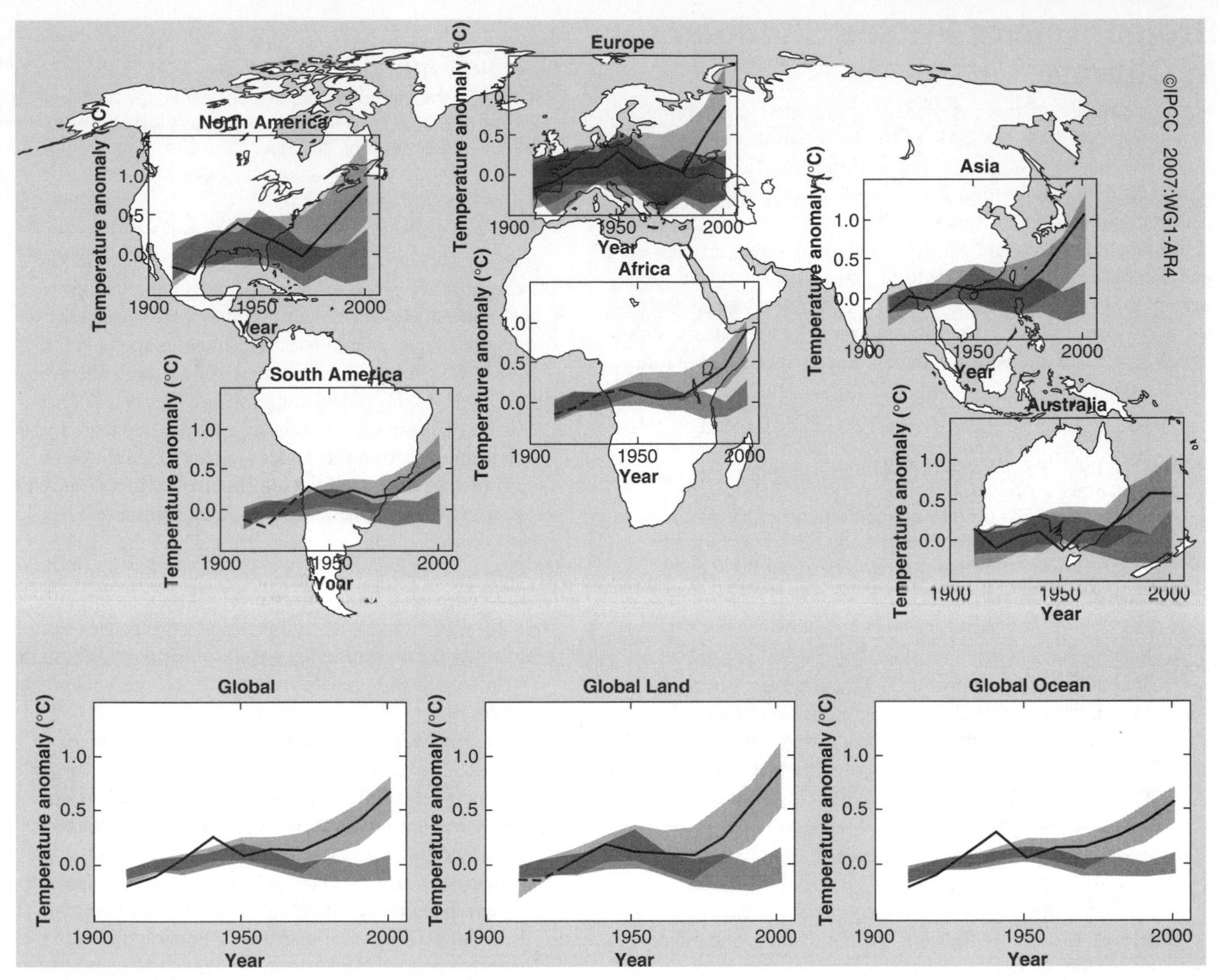

Figure SPM-4. Global and continental temperature change. Comparison of observed continental- and global-scale changes in surface temperature with results simulated by climate models using natural and anthropogenic forcings. Decadal averages of observations are shown for the period 1906–2005 (black line) plotted against the centre of the decade and relative to the corresponding average for 1901–1950. Lines are dashed where spatial coverage is less than 50%. Light gray shaded bands show the 5–95% range for 19 simulations from 5 climate models using only the natural forcings due to solar activity and volcanoes. Medium gray shaded bands show the 5–95% range for 58 simulations from 14 climate models using both natural and anthropogenic forcings. {FAQ 9.2, Figure 1}

Analysis of climate models together with constraints from observations enables an assessed *likely* range to be given for climate sensitivity for the first time and provides increased confidence in the understanding of the climate system response to radiative forcing. {6.6, 8.6, 9.6, Box 10.2}

to be less than 1.5 °C. Values substantially higher than 4.5 °C cannot be excluded, but agreement of models with observations is not as good for those values. Water vapour changes represent the largest feedback affecting climate sensitivity and are now better understood than in the TAR. Cloud feedbacks remain the largest source of uncertainty. {8.6, 9.6, Box 10.2}

- It is *very unlikely* that climate changes of at least the seven centuries prior to 1950 were due to variability generated within the climate system alone. A significant fraction of the reconstructed Northern Hemisphere interdecadal temperature variability over those centuries is *very likely* attributable to volcanic eruptions and changes in solar irradiance, and it is *likely* that anthropogenic forcing contributed to the early 20th century warming evident in these records. {2.7, 2.8, 6.6, 9.3}

Projections of Future Changes in Climate

A major advance of this assessment of climate change projections compared with the TAR is the large number of simulations available from a broader range of models. Taken together with additional information from observations, these provide a quantitative basis for estimating likelihoods for many aspects of future climate change. Model simulations cover a range of possible futures including idealised emission or concentration assumptions. These include SRES[14] illustrative marker scenarios for the 2000–2100 period and model experiments with greenhouse gases and aerosol concentrations held constant after year 2000 or 2100.

For the next two decades a warming of about 0.2 °C per decade is projected for a range of SRES emission scenarios. Even if the concentrations of all greenhouse gases and aerosols had been kept constant at year 2000 levels, a further warming of about 0.1 °C per decade would be expected. {10.3, 10.7}

- Since IPCC's first report in 1990, assessed projections have suggested global averaged temperature increases between about 0.15 and 0.3 °C per decade for 1990 to 2005. This can now be compared with observed values of about 0.2 °C per decade, strengthening confidence in near-term projections. {1.2, 3.2}
- Model experiments show that even if all radiative forcing agents are held constant at year 2000 levels, a further warming trend would occur in the next two decades at a rate of about 0.1 °C per decade, due mainly to the slow response of the oceans. About twice as much warming (0.2 °C per decade) would be expected if emissions are within the range of the SRES scenarios. Best-estimate projections from models indicate that decadal-average warming over each inhabited continent by 2030 is insensitive to the choice among SRES scenarios and is *very likely* to be at least twice as large as the corresponding model-estimated natural variability during the 20th century. {9.4, 10.3, 10.5, 11.2–11.7, Figure TS-29}
- Advances in climate change modelling now enable best estimates and *likely* assessed uncertainty ranges to be given for projected warming for different emission scenarios. Results for different emission scenarios are provided explicitly in this report to avoid loss of this policy-relevant information. Projected globally-averaged surface warmings for the end of the 21st century (2090–2099) relative to 1980–1999 are shown in Table SPM-3. These illustrate the differences between lower to higher SRES emission scenarios and the projected warming uncertainty associated with these scenarios. {10.5}

Continued greenhouse gas emissions at or above current rates would cause further warming and induce many changes in the global climate system during the 21st century that would *very likely* be larger than those observed during the 20th century. {10.3}

- Best estimates and *likely* ranges for globally average surface air warming for six SRES emissions marker scenarios are given in this assessment and are shown in Table SPM-3. For example, the best estimate for the low scenario (B1) is 1.8°C (*likely* range is 1.1 °C to 2.9 °C), and the best estimate for the high scenario (A1FI) is 4.0 °C (*likely* range is 2.4 °C to 6.4 °C). Although these projections are broadly consistent with the span quoted in the TAR (1.4 to 5.8 °C), they are not directly comparable (see Figure SPM-5). The AR4 is more advanced as it provides best estimates and an assessed likelihood range for each of the marker scenarios. The new assessment of the *likely* ranges now relies on a larger number of climate models of increasing complexity and realism, as well as new information regarding the nature of feedbacks from the carbon cycle and constraints on climate response from observations. {10.5}
- Warming tends to reduce land and ocean uptake of atmospheric carbon dioxide, increasing the fraction of anthropogenic emissions that remains in the atmosphere. For the A2 scenario, for example, the climate-carbon cycle feedback increases the corresponding global average warming at 2100 by more than 1 °C. Assessed upper ranges for temperature projections are larger than in the TAR (see Table SPM-3) mainly because the broader range of models now available suggests stronger climate-carbon cycle feedbacks. {7.3, 10.5}
- Model-based projections of global average sea level rise at the end of the 21st century (2090–2099) are shown in Table SPM-3. For each scenario, the midpoint of the range in Table SPM-3 is within 10% of the TAR model average for 2090–2099. The ranges are narrower than in the TAR mainly because of improved information about some uncertainties in the projected contributions[15]. {10.6}
- Models used to date do not include uncertainties in climate-carbon cycle feedback nor do they include the full effects of changes in ice sheet flow, because a basis in published literature is lacking. The projections include a contribution due to increased ice flow from Greenland and Antarctica at the rates observed for 1993–2003, but these flow rates could increase or decrease in the future. For example, if this contribution were to grow linearly with global average temperature change, the upper ranges of sea level rise for SRES scenarios shown in Table SPM-3 would increase by 0.1 m to 0.2 m. Larger values cannot be excluded, but understanding of these effects is too limited to assess

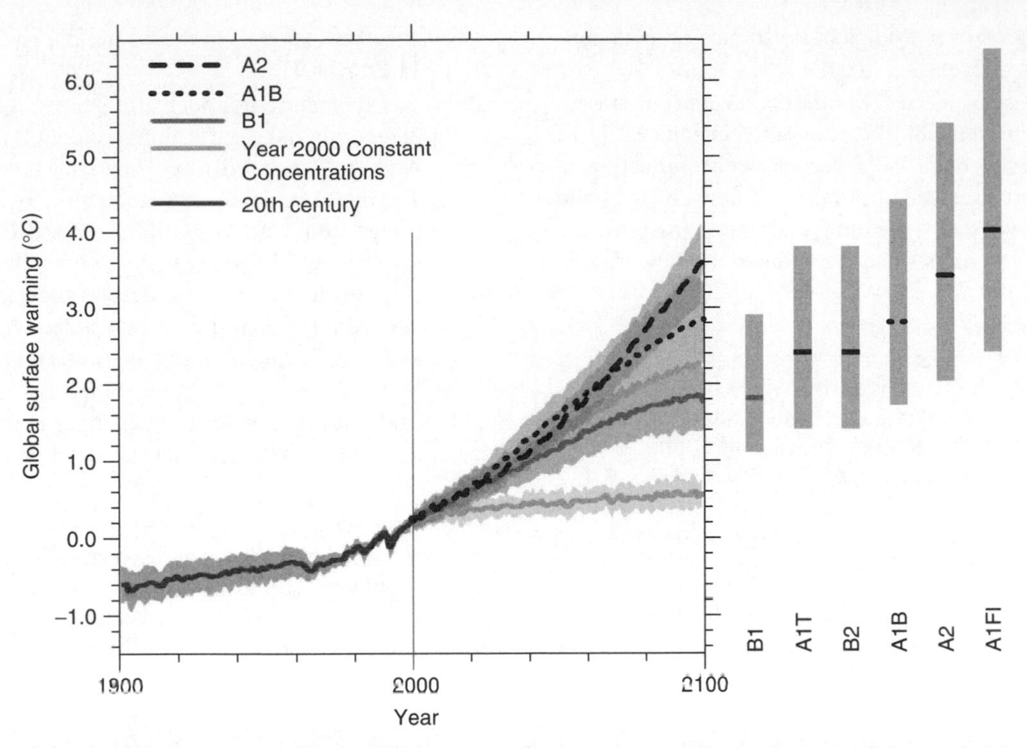

Figure SPM-5 Multi-model averages and assessed ranges for surface warming. Solid lines are multi-model global averages of surface warming (relative to 1980–99) for the scenarios A2, A1B and B1, shown as continuations of the 20th century simulations. Shading denotes the plus/minus one standard deviation range of individual model annual averages. The orange line is for the experiment where concentrations were held constant at year 2000 values. The gray bars at right indicate the best estimate (solid line within each bar) and the *likely* range assessed for the six SRES marker scenarios. The assessment of the best estimate and *likely* ranges in the gray bars includes the AOGCMs in the left part of the figure, as well as results from a hierarchy of independent models and observational constraints. {Figures 10.4 and 10.29}

Table SPM-3 Projected Globally Averaged Surface Warming and Sea Level Rise at the End of the 21st Century. {10.5, 10.6, Table 10.7}

	Temperature Change (°C at 2090–2099 Relative to 1980–1999) [a]		**Sea Level Rise (m at 2090–2099 Relative to 1980–1999)**
Case	**Best Estimate**	***Likely* Range**	**Model-Based Range Excluding Future Rapid Dynamical Changes in Ice Flow**
Constant Year 2000 concentrations [b]	0.6	0.3–0.9	NA
B1 scenario	1.8	1.1–2.9	0.18–0.38
A1T scenario	2.4	1.4–3.8	0.20–0.45
B2 scenario	2.4	1.4–3.8	0.20–0.43
A1B scenario	2.8	1.7–4.4	0.21–0.48
A2 scenario	3.4	2.0–5.4	0.23–0.51
A1FI scenario	4.0	2.4–6.4	0.26–0.59

[a]These estimates are assessed from a hierarchy of models that encompass a simple climate model, several Earth Models of Intermediate Complexity (EMICs), and a large number of Atmosphere-Ocean Global Circulation Models (AOGCMs).

[b]Year 2000 constant composition is derived from AOGCMs only.

their likelihood or provide a best estimate or an upper bound for sea level rise. {10.6}

- Increasing atmospheric carbon dioxide concentrations leads to increasing acidification of the ocean. Projections based on SRES scenarios give reductions in average global surface ocean pH[16] of between 0.14 and 0.35 units over the 21st century, adding to the present decrease of 0.1 units since pre-industrial times. {5.4, Box 7.3, 10.4}

There is now higher confidence in projected patterns of warming and other regional-scale features, including changes in wind patterns, precipitation, and some aspects of extremes and of ice. {8.2, 8.3, 8.4, 8.5, 9.4, 9.5, 10.3, 11.1}

- Projected warming in the 21st century shows scenario-independent geographical patterns similar to those observed over the past several decades. Warming is expected to be greatest over land and at most high northern latitudes, and least over the Southern Ocean and parts of the North Atlantic ocean (see Figure SPM-6). {10.3}
- Snow cover is projected to contract. Widespread increases in thaw depth are projected over most permafrost regions. {10.3, 10.6}
- Sea ice is projected to shrink in both the Arctic and Antarctic under all SRES scenarios. In some projections, Arctic late-summer sea ice disappears almost entirely by the latter part of the 21st century. {10.3}
- It is *very likely* that hot extremes, heat waves, and heavy precipitation events will continue to become more frequent. {10.3}
- Based on a range of models, it is *likely* that future tropical cyclones (typhoons and hurricanes) will become more intense, with larger peak wind speeds and more heavy precipitation associated with ongoing increases of tropical SSTs. There is less confidence in projections of a global decrease in numbers of tropical cyclones. The apparent increase in the proportion of very intense storms since 1970 in some regions is much larger than simulated by current models for that period. {9.5, 10.3, 3.8}
- Extra-tropical storm tracks are projected to move poleward, with consequent changes in wind, precipitation, and temperature patterns, continuing the broad pattern of observed trends over the last half-century. {3.6, 10.3}
- Since the TAR there is an improving understanding of projected patterns of precipitation. Increases in the amount of precipitation are *very likely* in high-latitudes, while decreases are *likely* in most subtropical land regions (by as much as about 20% in the A1B scenario in 2100, see Figure SPM-7), continuing observed patterns in recent trends. {3.3, 8.3, 9.5, 10.3, 11.2 to 11.9}
- Based on current model simulations, it is *very likely* that the meridional overturning circulation (MOC) of the Atlantic Ocean will slow down during the 21st century. The multi-model average reduction by 2100 is 25% (range from zero to about 50%) for SRES emission scenario A1B. Temperatures in the Atlantic region are projected to increase despite such changes due to the much larger warming associated with projected increases of greenhouse gases. It is *very unlikely* that the MOC will undergo a large abrupt transition during the 21st century. Longer-term changes in the MOC cannot be assessed with confidence. {10.3, 10.7}

Anthropogenic warming and sea level rise would continue for centuries due to the timescales associated with climate processes and feedbacks, even if greenhouse gas concentrations were to be stabilized. {10.4, 10.5, 10.7}

- Climate carbon cycle coupling is expected to add carbon dioxide to the atmosphere as the climate system warms, but the magnitude of this feedback is uncertain. This increases the uncertainty in the trajectory of carbon dioxide emissions required to achieve a particular stabilisation level of atmospheric carbon dioxide concentration. Based on current understanding of climate carbon cycle feedback, model studies suggest that to stabilise at 450 ppm carbon dioxide, could require that cumulative emissions over the 21st century be reduced from an average of approximately 670 [630 to 710] GtC (2460 [2310 to 2600] $GtCO_2$) to approximately 490 [375 to 600] GtC (1800 [1370 to 2200] $GtCO_2$). Similarly, to stabilise at 1000 ppm this feedback could require that cumulative emissions be reduced from a model average of approximately 1415 [1340 to 1490] GtC (5190 [4910 to 5460] $GtCO_2$) to approximately 1100 [980 to 1250] GtC (4030 [3590 to 4580] $GtCO_2$). {7.3, 10.4}
- If radiative forcing were to be stabilized in 2100 at B1 or A1B levels[11] a further increase in global average temperature of about 0.5 °C would still be expected, mostly by 2200. {10.7}
- If radiative forcing were to be stabilized in 2100 at A1B levels[11], thermal expansion alone would lead to 0.3 to 0.8 m of sea level rise by 2300 (relative to 1980–1999). Thermal expansion would continue for many centuries, due to the time required to transport heat into the deep ocean. {10.7}
- Contraction of the Greenland ice sheet is projected to continue to contribute to sea level rise after 2100. Current models suggest ice mass losses increase with temperature more rapidly than gains due to precipitation and that the surface mass balance becomes negative

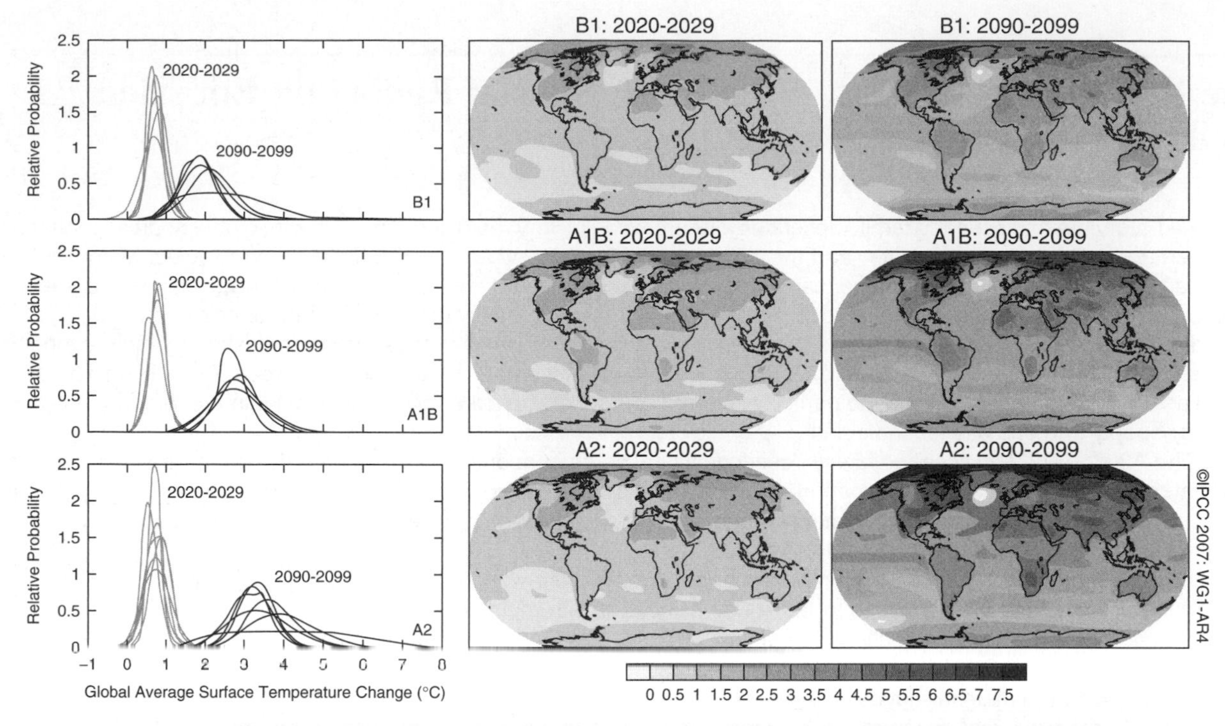

Figure SPM-6 AOGCM projections of surface temperatures. Projected surface temperature changes for the early and late 21st century relative to the period 1980–1999. The central and right panels show the Atmosphere-Ocean General Circulation multi-Model average projections for the B1 (top), A1B (middle) and A2 (bottom) SRES scenarios averaged over decades 2020–2029 (center) and 2090–2099 (right). The left panel shows corresponding uncertainties as the relative probabilities of estimated global average warming from several different AOGCM and EMICs studies for the same periods. Some studies present results only for a subset of the SRES scenarios, or for various model versions. Therefore the difference in the number of curves, shown in the left-hand panels, is due only to differences in the availability of results. {Figures 10.8 and 10.28}

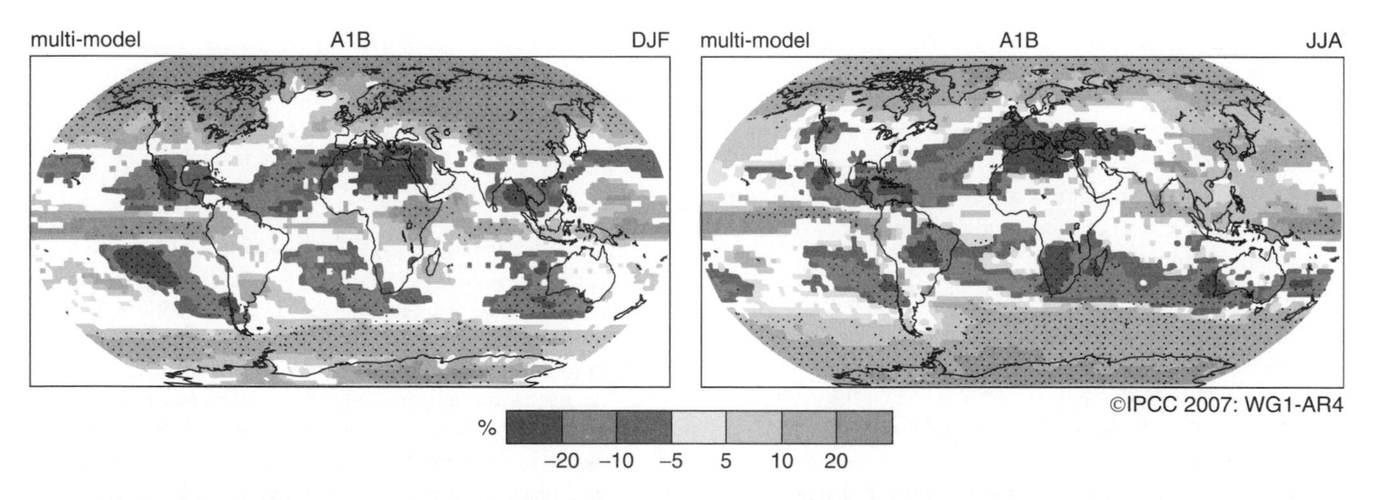

Figure SPM-7 Projected patterns of precipitation changes. Relative changes in precipitation (in percent) for the period 2090–2099, relative to 1980–1999. Values are multi-model averages based on the SRES A1B scenario for December to February (left) and June to August (right). White areas are where less than 66% of the models agree in the sign of the change and stippled areas are where more than 90% of the models agree in the sign of the change. {Figure 10.9}

at a global average warming (relative to pre-industrial values) in excess of 1.9 to 4.6 °C. If a negative surface mass balance were sustained for millennia, that would lead to virtually complete elimination of the Greenland ice sheet and a resulting contribution to sea level rise of about 7 m. The corresponding future temperatures in Greenland are comparable to those inferred for the last interglacial period 125,000 years ago, when paleoclimatic information suggests reductions of polar land ice extent and 4 to 6 m of sea level rise. {6.4, 10.7}

The Emission Scenarios of the IPCC Special Report on Emission Scenarios (SRES)[17]

A1. The A1 storyline and scenario family describes a future world of very rapid economic growth, global population that peaks in mid-century and declines thereafter, and the rapid introduction of new and more efficient technologies. Major underlying themes are convergence among regions, capacity building and increased cultural and social interactions, with a substantial reduction in regional differences in per capita income. The A1 scenario family develops into three groups that describe alternative directions of technological change in the energy system. The three A1 groups are distinguished by their technological emphasis: fossil intensive (A1FI), non-fossil energy sources (A1T), or a balance across all sources (A1B) (where balanced is defined as not relying too heavily on one particular energy source, on the assumption that similar improvement rates apply to all energy supply and end use technologies).

A2. The A2 storyline and scenario family describes a very heterogeneous world. The underlying theme is self reliance and preservation of local identities. Fertility patterns across regions converge very slowly, which results in continuously increasing population. Economic development is primarily regionally oriented and per capita economic growth and technological change more fragmented and slower than other storylines.

B1. The B1 storyline and scenario family describes a convergent world with the same global population, that peaks in mid-century and declines thereafter, as in the A1 storyline, but with rapid change in economic structures toward a service and information economy, with reductions in material intensity and the introduction of clean and resource efficient technologies. The emphasis is on global solutions to economic, social and environmental sustainability, including improved equity, but without additional climate initiatives.

B2. The B2 storyline and scenario family describes a world in which the emphasis is on local solutions to economic, social and environmental sustainability. It is a world with continuously increasing global population, at a rate lower than A2, intermediate levels of economic development, and less rapid and more diverse technological change than in the B1 and A1 storylines. While the scenario is also oriented towards environmental protection and social equity, it focuses on local and regional levels.

An illustrative scenario was chosen for each of the six scenario groups A1B, A1FI, A1T, A2, B1 and B2. All should be considered equally sound.

The SRES scenarios do not include additional climate initiatives, which means that no scenarios are included that explicitly assume implementation of the United Nations Framework Convention on Climate Change or the emissions targets of the Kyoto Protocol.

- Dynamical processes related to ice flow not included in current models but suggested by recent observations could increase the vulnerability of the ice sheets to warming, increasing future sea level rise. Understanding of these processes is limited and there is no consensus on their magnitude. {4.6, 10.7}
- Current global model studies project that the Antarctic ice sheet will remain too cold for widespread surface melting and is expected to gain in mass due to increased snowfall. However, net loss of ice mass could occur if dynamical ice discharge dominates the ice sheet mass balance. {10.7}
- Both past and future anthropogenic carbon dioxide emissions will continue to contribute to warming and sea level rise for more than a millennium, due to the timescales required for removal of this gas from the atmosphere. {7.3, 10.3}

Notes

1. *Climate change* in IPCC usage refers to any change in climate over time, whether due to natural variability or as a result of human activity. This usage differs from that in the Framework Convention on Climate Change, where climate change refers to a change of climate that is attributed directly or indirectly to human activity that alters the composition of the global atmosphere and that is in addition to natural climate variability observed over comparable time periods.
2. *Radiative forcing* is a measure of the influence that a factor has in altering the balance of incoming and outgoing energy in the Earth-atmosphere system and is an index of the importance of the factor as a potential climate change mechanism. Positive forcing tends to warm the surface while negative forcing tends to cool it. In this report radiative forcing values are for 2005 relative to pre-industrial conditions defined at 1750 and are expressed in watts per square metre ($W\ m^{-2}$). See Glossary and Section 2.2 for further details.
3. ppm (parts per million) or ppb (parts per billion, 1 billion = 1,000 million) is the ratio of the number of greenhouse gas molecules to the total number of molecules of dry air. For example: 300 ppm means 300 molecules of a greenhouse gas per million molecules of dry air.
4. Fossil carbon dioxide emissions include those from the production, distribution and consumption of fossil fuels and as a by-product from cement production. An emission of 1 GtC corresponds to 3.67 $GtCO_2$.
5. In general, uncertainty ranges for results given in this Summary for Policymakers are 90% uncertainty intervals unless stated otherwise, i.e., there is an estimated 5% likelihood that the

value could be above the range given in square brackets and 5% likelihood that the value could be below that range. Best estimates are given where available. Assessed uncertainty intervals are not always symmetric about the corresponding best estimate. Note that a number of uncertainty ranges in the Working Group I TAR corresponded to 2-sigma (95%), often using expert judgement.

6. In this Summary for Policymakers, the following terms have been used to indicate the assessed likelihood, using expert judgement, of an outcome or a result: *Virtually certain* > 99% probability of occurrence, *Extremely likely* > 95%, *Very likely* > 90%, *Likely* > 66%, *More likely than not* > 50%, *Unlikely* < 33%, *Very unlikely* < 10%, *Extremely unlikely* < 5%. (See Box TS.1.1 for more details).
7. In this Summary for Policymakers the following levels of confidence have been used to express expert judgments on the correctness of the underlying science: *very high confidence* at least a 9 out of 10 chance of being correct; *high confidence* about an 8 out of 10 chance of being correct. (See Box TS.1.1)
8. Halocarbon radiative forcing has been recently assessed in detail in IPCC's Special Report on Safeguarding the Ozone Layer and the Global Climate System (2005).
9. The average of near surface air temperature over land, and sea surface temperature.
10. Tropical cyclones include hurricanes and typhoons.
11. The assessed regions are those considered in the regional projections Chapter of the TAR and in Chapter 11 of this Report.
12. Consideration of remaining uncertainty is based on current methodologies.
13. In particular, the Southern and Northern Annular Modes and related changes in the North Atlantic Oscillation. {3.6, 9.5, Box TS.3.1
14. SRES refers to the IPCC Special Report on Emission Scenarios (2000). The SRES scenario families and illustrative cases, which did not include additional climate initiatives, are summarized in a box at the end of this Summary for Policymakers. Approximate CO_2 equivalent concentrations corresponding to the computed radiative forcing due to anthropogenic greenhouse gases and aerosols in 2100 (see p. 823 of the TAR) for the SRES B1, A1T, B2, A1B, A2 and A1FI illustrative marker scenarios are about 600, 700, 800, 850, 1250 and 1550 ppm respectively. Scenarios B1, A1B, and A2 have been the focus of model inter-comparison studies and many of those results are assessed in this report.
15. TAR projections were made for 2100, whereas projections in this Report are for 2090–2099. The TAR would have had similar ranges to those in Table SPM-2 if it had treated the uncertainties in the same way.
16. Decreases in pH correspond to increases in acidity of a solution. See Glossary for further details.
17. Emission scenarios are not assessed in this Working Group One report of the IPCC. This box summarizing the SRES scenarios is taken from the TAR and has been subject to prior line by line approval by the Panel.

Global Warming Battlefields: How Climate Change Threatens Security

MICHAEL T. KLARE

By any reckoning, global climate change poses a threat to world security writ large. Because it will imperil food production around the world and could render many heavily populated areas uninhabitable, it has the potential to endanger the lives and livelihoods of hundreds of millions of people. So far, most experts' warnings have naturally tended to focus on the large-scale, nontraditional security implications of global warming: mass starvation resulting from persistent drought, humanitarian disasters caused by severe hurricane and typhoon activity, the inundation of coastal cities, and so on. Just as likely, however, is an increase in more familiar security threats: war, insurgency, ethnic conflict, state collapse, and civil violence. The Nobel committee affirmed as much in October when it awarded the Peace Prize to former Vice President Al Gore and the Intergovernmental Panel on Climate Change for their efforts to raise awareness about global warming. The prize committee cited "increased danger of violent conflicts and wars, within and between states."

Climate change will increase the risk of conflict because it is almost certain to diminish the supply of vital resources—notably food, water, and arable land—in areas of the planet that already are suffering from resource scarcity, thus increasing the risk that desperate groups will fight among themselves for whatever remains of the means of survival. In wealthier societies, such conflicts can be mitigated by food and housing subsidies provided by the central governments and by robust schemes for relocation and reconstruction. In poorer countries, where little or no such capacity exists, the conflicts are more likely to be decided by ethnic or religious militias and the power of the gun.

Violent conflict over vital resources has, of course, been a characteristic of the human condition since very ancient times. Archaeological remains and the oldest written records attest to the fact that early human communities fought for control over prime growing areas, hunting zones, timber stands, and so on. A growing body of evidence also suggests that severe climate changes—for example, the "little Ice Age" of circa AD 1300–1700—have tended to increase the risk of resource-related conflict. Steven A. LeBlanc of the Peabody Museum of Archaeology and Ethnology at Harvard has noted, for example, that conflict among the Anasazi people of the American Southwest appears to have increased substantially with the cooling trend (and reduced food output) of the early 1300s, as indicated by the abandonment of exposed valley-floor settlements in favor of more defensible cliff dwellings.

Resource conflict has continued into more recent times, growing even more pronounced as European adventurers and settlers invaded Africa, Asia, and the Americas in search of gold, furs, spices, timber, land, human chattel, iron, copper, and oil—often encountering fierce resistance in the process. Today, indigenous peoples are still battling to preserve their lands and traditional means of livelihood in the few remaining unexploited tropical forests, mountain highlands, and other wilderness areas left on the planet.

As climate change kicks in, the risk of resource wars will grow many times over.

Elsewhere, many of those on the bottom rungs of the socioeconomic ladder—especially those who depend on agriculture or herding for their livelihoods—are also caught up in perennial conflict over access to land, water, energy, and other resources. Even as inter-state wars have diminished in number worldwide, conflicts between various groups in recent times have been exacerbated by rapid population growth; increased competition from agribusiness and cheap imported foodstuffs; the growing popularity of militant ethnic, religious, and political ideologies; and other exogenous factors. Even without global warming, these factors will continue to increase the likelihood of intergroup conflict. As climate change kicks in, the risk of resource wars will grow many times over.

The Hardest Hit

Accelerated by the greenhouse effect (the warming produced as greenhouse gases such as carbon dioxide trap heat in the atmosphere), climate change will affect the global resource equation in many ways. Essentially, these can be grouped into four key effects: (1) *diminished rainfall* in many tropical and temperate areas, leading to more frequent and prolonged droughts; (2) *diminished river flow* in many of these same areas as a result of reduced rainfall or the shrinking of mountain glaciers, producing greater water scarcity in food-producing regions; (3) *a rising sea level,* leading to the inundation of coastal cities and farmlands; and (4) *more frequent and severe storm events,* producing widespread damage to farms, factories, and villages. These effects will vary in their application to different parts of the globe, with some areas experiencing greater trauma than others, but the net result will be a substantial reduction in life-sustaining resources for a good part of the earth's population.

The particular impacts of these global warming effects on various communities have been studied in a piecemeal fashion for some time, but were given their most systematic examination in the Fourth Assessment Report of the UN-sponsored Intergovernmental Panel on Climate Change (IPCC), released in April 2007. As part of this study—the most comprehensive of global warming yet conducted—the IPCC convened a task force on "Impacts, Adaptation, and Vulnerability," called Working Group II. Ecosystem by ecosystem, region by region, this group's report provides an extraordinary overview of what can be expected from global warming's long-term impact on natural habitats and human communities around the planet. Although dry and dispassionate in tone, the report of Working Group II is devastating in its conclusions. Among its principal findings:

On water scarcity: "By mid-century, annual average river runoff and water availability are projected to . . . decrease by 10 to 30 percent over some dry regions at mid-latitudes and in the dry tropics, some of which are presently water-stressed areas. . . . In the course of the century, water supplies stored in glaciers and snow cover are projected to decline, reducing water availability in regions supplied by meltwater from major mountain ranges, where more than one-sixth of the world population currently lives."

On food availability: "At lower latitudes, especially seasonally dry and tropical regions, crop productivity is projected to decrease for even small local temperature increases (1 to 2 degrees centigrade), which would increase the risk of hunger. . . . Increases in the frequency of droughts and floods are projected to affect local crop production negatively, especially in subsistence sectors at low latitudes.

On coastal inundation: "Many millions more people are projected to be flooded every year due to sea-level rise by the 2080s. Those densely populated and lowlying areas where adaptive capacity is relatively low, and which already face other challenges such as tropical storms or local coastal subsidence, are especially at risk."

Most of these effects will be felt across the entire planet, but the degree to which they produce death, injury, and suffering will vary with the relative wealth and resiliency of the societies involved. In general, affluent and well-governed societies will be better able to cope with trauma and provide for the minimum needs of their affected citizens; poor and inadequately governed nations will be much less able to cope. And it is in the latter countries where conflict is most likely to arise over the allocation of relief supplies and relocation options.

The pivotal relationship between climate change and the coping capacity of affected states will be especially pronounced in Africa. That continent is expected to suffer disproportionately from the direst effects of global warming—especially from prolonged drought and water scarcity—and it possesses the least capacity to mitigate these impacts. According to Working Group II, as early as 2020, between 75 million and 250 million Africans are expected to face increased water scarcity as a result of climate change; by the 2050s, this number is projected to range between 350 and 600 million people. Because food production in Africa is already stretched to the limit, the decline in water availability will reduce crop yields and greatly increase the risk of hunger and malnutrition. According to the Working Group II report, yields from rain-fed agriculture in some African countries could be reduced by as much as half by 2020. Increased rural unrest and conflicts over land are a likely result.

Parts of South and Central Asia could also suffer from violence related to global warming. A rise in average temperatures and a decline in water supplies are expected to produce a sharp reduction in cereal production throughout this vast region, which contains some of the most heavily populated countries in the world. In Bangladesh, for example, wheat production could decline by 32 percent by 2050 and rice production by 8 percent. For all of South Asia, according to the IPCC report, "net cereal production . . . is projected to decline at least between 4 to 10 percent by the end of this century under the most conservative climate change scenario." With many millions of subsistence farmers in these countries already struggling to survive, production declines on this scale will prove catastrophic. Large numbers of rural residents forced into destitution will no doubt migrate to cities in search of jobs. But some may be attracted to radical sects or ethnic bands that promise salvation through the seizure of less-affected lands held by wealthy landowners or other ethnic groups.

In general, adverse effects from global warming likely will produce suffering on an unprecedented scale. Many will starve; many more will perish from disease, flooding, or fire. Others, however, will attempt to survive in the same manner as their predecessors: by fighting among themselves for whatever food and water remains; by invading more favorable locales; or by migrating to distant lands, even in the face of violent resistance.

Resource Wars

The onset of severe climate change will increase the frequency and intensity of certain familiar types of conflict and also introduce some new or largely unfamiliar forms. Two kinds of conflicts—resource wars and ethnic warfare attendant on state collapse—are among the more familiar of these. A third, less familiar type of violence likely to increase as a result of global warming might best be described as migratory conflict.

Resource wars arise when competing states or ethnic enclaves fight over the possession of key resources—particularly water supplies, oil reserves, diamond fields, timber stands, and mineral deposits. Such conflicts, as noted, have been a feature of human behavior since time immemorial. Conflicts over resources were less prevalent during the cold war era, when ideological antagonisms were the driving force in world affairs, but they have become more conspicuous since the demise of the Soviet Union and the outbreak of fresh disputes in the developing world. Though often characterized as ethnic and religious wars, many of these newer conflicts have been, at root, disputes over the allocation of land, water, timber, or other valuable commodities. Bitter fighting in Angola and Sierra Leone, for example, was principally driven by competition over the illicit trade in diamonds. Struggles over diamonds, timber, and coltan (a critical ingredient in the manufacture of cell phones) have fueled the ongoing violence in Congo. Wars in Somalia, Ethiopia, and the Darfur region of Sudan have largely been sparked by disputes over land and water rights.

Even without global warming, the incidence of intergroup wars like these is likely to increase because the demand for key resources is growing while supplies, in many cases, are shrinking. On the demand side of the ledger, many developing countries are expected to experience a sharp increase in population over the next several decades along with a steady increase in per capita consumption levels. On the supply side, many once-lucrative sources of oil, natural gas, uranium, copper, timber, fish, and underground water (aquifers) are expected to be depleted, producing significant scarcities of these materials. Virtually all states and societies are likely to experience some traumas and hardships as a result, but some groups will suffer far more than others. And because these disparities are likely to coincide with national, ethnic, and religious distinctions, they will provide ample fodder for those who seek justifications for waging war on "others" who can be portrayed as the cause of one's own hardships and misfortunes.

Add climate change to the equation, and the picture becomes much, much worse. While most of the world's regions are likely to experience a reduction in the supply of at least some critical resources, it is true that a few could see limited gains from global warming. Some countries in the far north, for example, could benefit from more rainfall and longer growing seasons, allowing for increased food output. Russia also hopes to benefit from the melting of the Arctic ice cap, which theoretically would allow oil and natural gas drilling in areas now covered year-round by thick ice. But even if these hypothetical advantages are not outweighed by other, less desirable consequences of global warming, any perception of a widening chasm between the "winners" and "losers" of climate change—when the overwhelming majority of the world's population is likely to fall in the latter category—could direct angry and potentially lethal attention toward the former.

Climate change will increase pressure on nearly every key resource used by humans, but land and fresh water will probably experience the greatest effects. Conflict over arable land has been one of the most persistent causes of warfare throughout history, and it is hard to imagine that global warming will not increase the likelihood of this type of conflict. If the projections by the IPCC's Working Group II prove accurate, vast inland areas of North and South America, Africa, and South and Central Asia are likely to suffer from diminished rainfall and recurring drought, turning once productive croplands into lifeless dustbowls. At the same time, many once reliable river systems will offer sharply reduced water flows, as the glaciers and snowpacks that feed them melt and recede or disappear. Again, not *every* area will suffer in this fashion: Some coastal highlands (for example, in the Horn of Africa) could experience increased precipitation and longer growing seasons, allowing greatly increased food production. Under these circumstances, those who feel cheated by the vagaries of climate change may feel impelled to invade and occupy the lands of those who, in their view, are unjustly blessed by the same fickle forces.

An area of particular concern among many climatologists is the Sahel region of Central Africa. The Sahel—the southern fringe of the Sahara Desert—stretches clear across Africa at its widest point from Senegal and Mauritania in the west through Mali, Niger, Chad, and Sudan (notably Darfur), to Eritrea and Ethiopia in the east. This is an area historically inhabited by Muslim pastoralists (mainly cattle herders) who are now being driven south by the steady advance of the Sahara into lands occupied by non-Muslim farmers—often provoking conflict in the process. The southward advance of the Sahara is believed to be one of the earliest observable effects of climate change and, as global temperatures rise, its rate of expansion is expected to increase. This, in turn, is likely to trigger intensified conflict from one end of the Sahel to the other.

Watching the River Flow

Global climate change is also likely to increase the risk of conflict over vital supplies of fresh water. Although states have rarely gone to war over disputed water supplies in recent times, they have often threatened to do so, and the risk factors appear to be growing. Water scarcity and stress are already a significant problem in many parts of the world, and are expected to become more so as a result of population growth, urbanization, and industrialization. Furthermore, many of the countries with the greatest exposure to water scarcity are highly dependent on river systems that arise outside their territory and pass through nations with which they have poor or unfriendly relations. Egypt, for example, is almost entirely dependent for its fresh water on the Nile, which arises in Central Africa (in the case of the White Nile) and Ethiopia (in the case of the Blue Nile). Iraq and Syria both depend for much of their water on the Tigris and Euphrates, which originate in Turkey. Israel relies on the Jordan River, which originates, in part, in Lebanon and Syria.

When an upstream country in any of these trans-boundary systems decides to dam a river and use it for domestic irrigation, the downstream country will inevitably experience a reduced flow, and this can be seen by that country as a significant threat to its well-being. It is not surprising, then, that downstream states like Egypt and Israel have threatened to go to war against any upstream state that endangers its water supply in this manner. "Water for Israel is not a luxury," former Prime Minister Moshe Sharett once proclaimed. "It is not just a desirable and helpful addition to our natural resources. Water is life itself."

How, exactly, global warming will affect any particular river system cannot be predicted with absolute assurance. However, it is clear from the IPCC's Fourth Assessment Report that many of the world's most important trans-boundary river systems in tropical and mid-temperate areas are likely to experience reduced flows as a result of climate change. This will significantly increase the competition among states for the ever-diminishing supply, thus increasing the risk of conflict.

Egypt is a source of particular concern in this regard, both because of its extreme reliance on the Nile—which originates far from its own territory—and because of its repeated threats to attack any upstream country that attempts to interfere with the river's flow. The Nile is the world's longest river and travels for more than a thousand miles through nearly cloudless desert before reaching Cairo, making it especially vulnerable to higher evaporation rates as global temperatures rise. Any reduction in upstream rainfall would also reduce its flow, increasing Egypt's vulnerability. Meanwhile, as water scarcity grows, so will Egypt's population, which is projected to increase from 80 million today to between 115 million and 179 million by 2050. Under these circumstances, any efforts by Ethiopia, Sudan, Uganda, or any of the other upstream states to divert the Nile's flow to meet the needs of their own soaring populations would almost assuredly trigger a panicky and quite possibly violent response from Egypt.

The Mogadishu Effect

A second type of violence that likely will increase as a result of global climate change is the sort of militia rule and gang warfare that prevail today in Mogadishu, the Somali capital. This is a condition in which the established government no longer exists or exercises effective authority; as a consequence, armed bands of one kind or another control access to critical resources, and these gangs or militias constantly fight among themselves over what little remains. Such violence has become familiar in the post–cold war era, as once vigorous states have collapsed and ethnic militias—often allied with or built around criminal associations—have arisen in their place. Multinational peacekeeping forces have confronted such bands in Somalia, Sierra Leone, Congo, Bosnia, Haiti, Sudan, Rwanda, and elsewhere, often with discouraging results.

Much research has been devoted to the causes of state collapse and the rise of ethnic militias in the post–cold war era, but this research has not managed to identify a clear, consistent set of precipitating factors. Corruption, authoritarianism, endemic poverty, ethnic favoritism, and poor social services are often common factors, but each case has its own distinctive features. What is true in all of these instances, however, is that the central government proves incapable of coping with an onslaught of powerful socioeconomic forces and either disintegrates entirely (as in Somalia) or loses control of large regions of the country—sometimes everything but the capital itself. Looking around the world, one can identify any number of countries that could, under existing circumstances, fall prey to these types of forces. Add global warming to the mix, and the pressures on these already vulnerable states grow much stronger.

Global warming will contribute to the propensity for weak states to collapse and give rise to militia rule and ethnic conflict.

Climate change will contribute to the propensity for weak states to collapse and give rise to militia rule and ethnic conflict for a variety of reasons. Consider any nation in the tropical or sub-temperate regions that depends for a significant share of its gross domestic product on farming, herding, forestry, and fishing, and that encompasses within its population more than one major ethnic, religious, or linguistic community. As indicated in the report of Working Group II, global warming is likely to harm some if not all of these livelihoods, though not to the same extent and not all at once. Also, some outlying parts of the country may become virtually uninhabitable, forcing people to migrate to the major city (or cities), often the capital or major port; or to areas more fortunate, which may be occupied by people of a different ethnicity (or religion, language group, and so forth). The decline in farming, fishing, and other livelihoods will contribute to a reduction in GDP, diminishing the revenues of the central government and thus its ability to shoulder additional burdens. Meanwhile, the movement of desperate refugees to the cities or other areas will produce an enormous need for relief services and exacerbate inter-group tensions.

All this would be a Herculean challenge for even the most affluent and capable governments, as the aftermath of Hurricane Katrina showed Americans. For poor, weak, and divided governments, the challenge could well prove insurmountable. As states collapse under the strain, what might be called the "Mogadishu effect" will kick in. Armed groups will coalesce around clans, tribes, village ties, and so on, as each group strives to ensure its own survival, at whatever price in bullets and blood. It is in precisely these circumstances, moreover, that extremist movements take root. With food and housing in short supply and city streets clogged with refugees, it is easy for a demagogue to blame another group or tribe for his own group's misfortunes and to call for violent action to redress grievances.

Because existing models cannot pinpoint future climate trends at the local level, it is not possible to predict where this combination of effects is most likely to result in state collapse, militia rule, and warfare among armed bands in the years ahead. Nonetheless, in its carefully worded manner, the IPCC in its Fourth Assessment Report provides some revealing hints. Speaking of the "negative effects" of climate change on human populations, it concludes: "The most vulnerable industries, settlements, and societies are generally those in coastal and river flood plains, those whose economies are closely linked with climate-sensitive resources, and those in areas prone to extreme weather events, especially where rapid urbanization is occurring." In most nations, moreover, vulnerable areas are not isolated but are linked in extensive and complex ways to other parts of their countries and to the surrounding regions.

As the impacts from climate change spread and state systems collapse in their wake, giving rise to gang and militia rule, violence might take many forms. Street-to-street combat for control over particular neighborhoods (and the distribution of relief supplies or vital commodities) is one common form. Fighting for control over desert oases, as in Darfur, is another. Impoverished farmers driven from their own territories by drought may invade neighboring lands. All of this will cause immense suffering and generate more refugees, creating ripple effects and sparking calls for international humanitarian intervention, as in the case of Darfur. This could lead as well to the deployment of international peacekeeping forces, and so result in clashes of yet another sort. The breakdown of state rule in affected areas will also facilitate the activities of criminals, mercenaries, drug traffickers, and others who flourish in an atmosphere of chaos—including international terrorists.

Migratory Conflicts

Yet a third type of conflict arising from global climate change might best be described as migratory warfare. This is armed violence provoked by efforts by large groups of people to migrate from environmentally devastated areas to less affected regions in the face of armed resistance by those inhabiting the more privileged locales. Of course, undocumented migrants from drought-prone areas of Mexico and Africa are already meeting significant resistance in their efforts to enter the United States and Europe, respectively. Many have perished in desperate attempts to circumvent the fences, border patrols, and other means employed to impede them.

But these struggles take place largely on an individual basis, or in groups of ten or a hundred at a time. Once global warming occurs on a massive scale, it is possible to conceive of migratory movements encompassing entire communities or regions, involving tens of thousands of people—some equipped with arms or formed into militias. In such scenarios, the struggle to cross national borders and settle in new lands could take on the form of ragged military campaigns—not unlike the travails experienced by the Israelites on entering the "Promised Land" (the Jordan River valley) as recounted in the Jewish scriptures.

The growing prevalence of migratory pressures like these could produce a number of worrisome phenomena. On one hand, destination countries like the United States, Spain, France, and Italy could choose to place even greater emphasis on the physical sealing-off of their borders and the use of military or paramilitary forces to stem the flood of immigrants. Even as experts debate the utility of such measures—especially in the face of vastly increased migratory pressures—more forcible means are certain to produce higher levels of death (whether through drowning, heat prostration, or shootouts with border guards), raising difficult moral questions. On the other hand, the growing intrusion of people with alien backgrounds could further inflame anti-immigrant sentiment in these countries, possibly leading to anti-immigrant violence or the ascendancy of ultranationalistic or even neo-Nazi political parties (with attendant social disorder and the possibility of increased militarism). However this plays out, migratory pressures on this scale are certain to prove highly destabilizing in many parts of the world.

Migratory pressures are likely to be among the most destabilizing impacts of global warming.

The global security implications of such phenomena were first accorded serious consideration in a 2007 study conducted by a group of retired US military officers convened by the CNA Corporation, a not-for-profit military contractor. In *National Security and the Threat of Climate Change,* the officers, led by former Army Chief of Staff General Gordon R. Sullivan, concluded that migratory pressures are likely to be among the most destabilizing impacts of global warming. For the United States, the greatest national security problem arising out of climate change "may be an increased flow of migrants northward into the U.S." Despite vigorous efforts to stem the flood of illegal immigrants, the rate of immigration is likely to rise "because the water situation in Mexico is already marginal and could worsen with less rainfall and more droughts." This, in turn, is likely to produce increased political outrage among long-term residents of border states, along with intensified efforts to exclude illegal migrants—with questionable results.

A similar dynamic is developing in the wealthier parts of Europe. "The greater threat to Europe lies in migration of people from across the Mediterranean, from the Maghreb, the Middle East, and sub-Saharan Africa," the CNA study observed. "As more people migrate from the Middle East because of water shortages and loss of their already marginal agricultural lands (as, for instance, if the Nile Delta disappears under the rising sea level), the social and economic stress on European nations will grow." Such stress can take many forms, but anti-immigrant violence is sure to be among the more likely outcomes.

A sharp increase in international migrations brought about by climate change is likely to have other baneful effects for world stability. As countries act to seal their borders and prevent the in-migration of environmental refugees, they are far less likely to participate in inclusionary political projects like the further expansion of the European Union—which, by its nature, would facilitate migration. Indeed, one might expect the EU to fall apart, with the original core of Western European countries breaking away from Greece and newer members in southeastern Europe, which are expected to suffer from drought and water scarcity as a result of global warming. This, in turn, could give rise to the emergence of new military blocs, with Russia keen to restore its influence in east-central Europe. Other efforts to build and expand regional partnerships—for example, the Association of Southeast Asian Nations—could suffer a similar fate, as politically insecure states seek to drive off migrants from less-fortunate neighboring nations. Isolationism and xenophobia historically have been harbingers of conflict.

Looking Ahead

As the effects of global warming become more pronounced, the world community will have to cope with a wide range of extreme environmental perils: prolonged droughts, intense storms, extensive coastal flooding, and so on. In practical terms, however, it may well be that the most costly and challenging consequence of climate change will be an increase in violent conflict and all the humanitarian trauma this brings with it. The war in Darfur provides a sobering indication of what this is likely to entail: Although this tragic conflict, affecting hundreds of thousands of still-vulnerable civilians, cannot be attributed to climate change alone, it comes as close as any contemporary contest to the global warming battlefields of the future.

It may well be that the most costly and challenging consequence of climate change will be an increase in violent conflict and all the humanitariantrauma this brings with it.

All this should lead us to think differently about both "national security" and global warming itself. In traditional national security discourse, the overarching challenge is how best to protect the nation against identifiable armed adversaries. But in a world of severe climate change, the greatest challenge is not likely to come from well-equipped hostile powers. More likely it will emerge from chaos and violence and civil conflicts arising from the breakdown of states and the ensuing struggles over scarce and diminishing resources.

Under these circumstances, national security will take on an entirely new meaning—requiring, for one thing, a corresponding transformation of the role and organization of a nation's armed forces. One would expect, for instance, a much greater emphasis on civil-defense functions: flood control, emergency response in times of disaster, anti-looting operations, refugee protection and resettlement, and so forth. The international community will also need to be much better prepared for humanitarian interventions and peacekeeping in the event of environmental-related conflicts.

Finally, it should be apparent that a world of unceasing resource wars, state disintegration, militia rule, and migratory conflicts will not be a safe or desirable world for anyone. These may not be the images one most associates with global warming, but they may, in fact, prove to be the most immediate and concrete consequences of climate change. As if there were not already enough good reasons to take swift action to curb emissions of greenhouse gases, the prospect of increased armed conflict should, one would hope, convince those still unaware of the magnitude of the danger.

Michael T. Klare, a *Current History* contributing editor, is a professor at Hampshire College and author of the forthcoming *Rising Powers, Shrinking Planet: The New Geopolitics of Energy* (Metropolitan Books, 2008).

Article 3

China Needs Help with Climate Change

KELLY SIMS GALLAGHER

This year, China will have become the largest aggregate emitter of greenhouse gases in the world, surpassing the United States for the first time since records have been kept. America will be the largest per-capita emitter and the second-largest aggregate emitter. These two countries have the unique ability to make or break the global climate change problem.

While it has long been recognized that China would be a pivotal nation in terms of dealing with climate change, the rate of growth of greenhouse gas emissions in China has been breathtaking, even to the Chinese themselves. At a minimum, it is now imperative to find incentives and mechanisms to induce China to reduce the growth of its emissions in the near term and, ultimately, to significantly reduce emissions below current levels.

The rate of growth of greenhouse gas emissions in China has been breathtaking, even to the Chinese themselves.

It is also imperative to assist China in this endeavor. Indeed, increased cooperation between Washington and Beijing is probably necessary if the climate change threat is to be effectively addressed. This will require that the two countries stop using each other as an excuse for inaction and instead form a partnership to ameliorate global warming.

Based on China and America's shared challenges of reducing greenhouse gas emissions and their economies' reliance on coal, a climate partnership should include high-level policy coordination and the establishment of a fund to provide low-cost financing for low-carbon projects in both countries. It should include capacity-building measures to help enhance the effectiveness of China's institutions, policies, and enforcement measures to reduce emissions. And it should include a joint innovation initiative to promote pre-commercial research, development, and demonstration of low-carbon technologies, particularly focused on carbon capture and storage, renewable energy, and energy efficiency technologies.

A Big-Time Emitter

China's contribution to the gases that are warming the world by trapping heat in the atmosphere is a direct result of the country's astonishingly rapid economic growth and rising demand for energy. Along with the United States, China is now one of the world's two largest energy producers and consumers. In terms of oil and electricity consumption, the People's Republic remains somewhat behind the United States. It consumes two-thirds as much commercial energy as America does, consumes one-third as much oil, imports one-third as much petroleum (although China's oil import growth rate has been much faster in recent years), and uses two-thirds as much electricity.

But when it comes to coal, the picture is different. China consumes twice as much coal as does the United States, though it has only 13 percent of the world's coal reserves, compared with 27 percent for the United States. Coal absolutely dominates the energy picture in China, accounting for 70 percent of its commercial energy supply. In 2006, China reportedly consumed 2.8 billion metric tons of coal, mostly for power plants and industry. By comparison, the United States consumed 1.3 billion metric tons, nearly all of which was used by power plants. In the United States, coal accounts for one-third of the total energy supply and half the country's electricity generation.

The growth in China's power sector has been almost unbelievably fast. Between 2005 and 2006, for example, electricity capacity increased by about 20 percent, from 517 gigawatts (gw) to 622 gw, nearly all of which was coal-fired. At this growth rate, China's total power sector capacity increased by double every three and a half years.

The transportation sector at this point is a relatively small consumer of energy in China, accounting for less than 10 percent of overall consumption. China has adopted new fuel-efficiency standards for passenger cars and tax policies for fuel consumption that should help to avoid a big increase in oil consumption. But the potential market for automobiles in China is huge. Fuel efficiency standards will need to be further strengthened and complementary measures introduced to reduce demand for cars if China is to avoid becoming the biggest oil-consuming country in the world.

As a result of all this industrial development, the growth in China's greenhouse emissions has been considerable. According to the latest official data from America's Oak Ridge National Laboratory, China in 2004 accounted for approximately 18 percent of the world's total carbon dioxide emissions from fossil fuel–burning and cement production, as compared with 22 percent for the United States. As of 2007, we now know that China has caught up to the United States. (China's per-capita emissions, on the other hand, are only one-fifth those of the United States.)

Both countries have signed and ratified the United Nations Framework Convention on Climate Change. Both have also signed the Kyoto Protocol, but only China has ratified it. At the central government level, neither country has binding policies aimed specifically at reducing greenhouse gas emissions, although both have efficiency-oriented policies that have the benefit of reducing carbon emissions.

Many local governmental bodies ignore or flout even the most basic energy-efficiency policies.

In June, the Chinese government for the first time issued a specific package of voluntary measures aimed at cutting greenhouse gas emissions; several sets of voluntary policies have been promoted as well by the past three presidential administrations in the United States. At the state and local levels, many US governmental entities have passed regulations to reduce carbon dioxide emissions. This is not true of China's provinces and localities, where many local governmental bodies ignore or flout even the most basic energy-efficiency policies.

Energy Challenges

China's energy-related challenges are many. They include the country's need for energy to sustain economic growth, its increasing dependency on foreign oil and gas, its aspiration to provide modern forms of energy to the poor, its increasingly severe urban air pollution, and its already massive acid deposition (dispersed in rain or deposited on surfaces). This is not to dismiss growing domestic and international concerns about global climate change or the need for affordable, advanced energy technologies to address all of these challenges. However, as China begins to consider how to address global warming, it will be simultaneously weighing the competing energy-related challenges, all of which are seen as more pressing by the Chinese government today.

Economically, China's growing energy consumption presents both challenges and opportunities. One concern is that as China imports greater amounts of energy, prices of these commodities could rise until supply catches up, and price spikes will be especially likely during supply disruptions. At the same time, there is a pressing need simply to supply enough energy, especially in the form of electricity, to meet the very high demand created by Chinese industry. The power sector has been through several boom-and-bust cycles because, when electricity shortages emerge, the power industry responds by adding huge quantities of new capacity as fast as it can. This causes oversupply for a time until the economy catches up and a new shortage emerges. The shortages have been harmful to the Chinese economy intermittently, whenever electricity has been rationed and factories have been forced to shut down.

On the opportunity side, the Chinese energy sector is already large and is growing rapidly, so it represents a remarkable market opportunity for both Chinese and foreign energy services companies. In 2006 alone, China installed 101 gw of new power capacity, 90 gw of which was coal-fired. To put this astounding number in perspective, India's entire electricity system, as of 2004, was 131 gw.

Despite the perception that China has become an industrial powerhouse, 135 million Chinese still live in absolute poverty (on less than $1 a day) and millions more remain just above that arbitrary poverty divide, so there is a tremendous imperative to foster economic development and high growth rates. In addition, the need to provide better energy services to the poor—to improve the quality of life for those still reliant on traditional forms of energy such as charcoal, crop wastes, and dung—remains very much a preoccupation of the Chinese government. Because of the country's gigantic population, China's total energy consumption and greenhouse gas emissions would still be large even if everyone consumed a very small amount of energy.

Since the beginning of the twenty-first century, China has emerged as a major consumer of oil, and there is strong potential for China to become a major natural gas consumer as well, especially when it gets serious about reducing its greenhouse gas emissions. China became a net importer of oil in the mid-1990s. It is now the world's second-largest consumer of oil, and the third-largest oil importer.

About half of China's oil imports come from the Middle East. However, Angola became its largest supplier in 2006, and China has invested heavily in energy resources in Africa. Although there have been several new oil discoveries in China recently, reserves there are on the decline. China has relatively few natural gas reserves, and therefore uses virtually no natural gas in its power sector. It is trying to increase production of coal-bed methane. If China decides to increase its use of natural gas, it will likely import it through liquid natural gas import terminals on the coast or by overland pipeline from Central Asia or Russia. In any event, China's long-term energy security depends not only on its having sufficient supplies of energy to sustain its rapid economic growth, but also on its ability to manage the growth in energy demand. Unmanaged demand, it is becoming clear, will cause intolerable environmental damage.

Coal is at the heart of many of China's environmental woes. Particulate matter from coal is a major air pollutant. Sulfur dioxide emissions from coal combustion, the source of most acid deposition, rose 27 percent between 2001 and 2005. Coal is also the most carbon-intensive of the fossil fuels, and it is China's main source of energy. It accounts for four-fifths of China's CO_2 emissions, most of which come from the industrial and electricity sectors. As of 2000, electricity accounted for 52 percent of China's CO_2 emissions (and 75 percent of China's electricity is consumed by industry), while cement production accounted for 28 percent, iron and steel production for 9 percent, and transportation for 8 percent. Already today, annual emissions from Chinese coal are three times as great as US emissions from transportation, although US transportation emissions are 17 times higher than Chinese transportation emissions.

The possible impact of climate change on China itself has not been studied as well as the possible impact on the United States, but it is clear that we could see very adverse effects on China's water supply, agriculture, and sea levels. Between 1956 and 2000, precipitation decreased 50–120 millimeters per year along the northern Yellow River, an already arid region. During the same period, precipitation increased 60–130 millimeters per year along the southern Yangtze River, an area that has long been plagued by heavy flooding. The mountain glaciers on the Tibetan plateau are receding rapidly, which carries major implications for fresh water supply in already water-stressed northern China. The glacier in the Tianshan Mountains that is the source of the Urumqi River, for example, shrank 11.3 percent between 1962 and 2001. Meanwhile, a sea level rise of 30 centimeters would cause massive coastal inundation. Chinese analysts have estimated this would cause the equivalent of $7.5 billion in economic losses to the Pearl River Delta area, $1.3 billion for the Yangtze Delta area, and $6.9 billion for the Yellow River Delta (including the Bohai Sea).

The Chinese Approach

China has already taken important steps toward moderating future growth in greenhouse gas emissions, largely through energy efficiency and renewable energy measures. Energy intensity (the amount of energy used to generate economic activity, usually calculated as total energy consumption divided by GDP) dramatically declined in China from 1980 to 2004. This means that China's overall energy efficiency improved and that significant growth in greenhouse gas emissions was avoided. Despite this improvement, however, China's overall energy efficiency remains considerably lower than most industrialized countries' and, unfortunately, it appears to have worsened in the past two years.

The central government has set forth some aggressive policies and targets for energy efficiency for the coming years. Because so much of China's energy is derived from coal, efficiency measures that reduce coal combustion will greatly help to reduce greenhouse gas emissions. China's 11th Five-Year Plan (2006–2010) called for a 20 percent reduction in energy intensity by 2010. This goal is already proving hard to achieve—last year's efficiency improvements fell short of the plan's objective, and in 2005 China's energy intensity actually *increased* slightly. Even so, the energy intensity target was at the heart of the climate change plan that the Chinese government announced in June 2007 in advance of the Group of Eight summit. By improving thermal efficiency, Beijing estimated that it could reduce China's carbon dioxide emissions by a total of 110 million tons by 2010.

The Chinese government issued its first fuel efficiency standards for passenger cars in 2005, and they will be strengthened in 2008. China also has implemented vehicle excise taxes so that the purchase of a car or sport utility vehicle with a big engine requires a much higher tax payment than does the purchase of a car with a small, energy-efficient engine. In the case of both fuel efficiency standards and excise taxes, China's policies are more stringent than comparable ones in the United States. And Beijing has adopted strong efficiency standards for appliances as well. The China Energy Group at Lawrence Berkeley National Laboratory estimates that, by 2010, those standards will have reduced carbon dioxide emissions in China by 40 million tons. By comparison, the US appliance standards will have saved 50 million tons of CO_2 by 2010.

The Chinese government has also aggressively promoted low-carbon energy supply options, especially renewable energy, hydropower, and nuclear energy. If you exclude large hydropower but include small hydropower, China has twice as much installed renewable power capacity as the United States. In fact, as of 2005, China led the world in total installed renewable energy capacity at 42 gw, compared to 23 gw in the United States. China accounts for 63 percent of the solar hot water capacity in the world, and as of 2005 it had installed 1.3 gw of wind capacity.

China in 2005 enacted a Renewable Energy Law that requires grid operators to purchase electricity from renewable generators. It sets a target of 10 percent of electric power generation capacity coming from renewable energy sources by 2010 (not including large hydro). By expanding bioenergy, solar, wind, geothermal, and tidal energy sources, the government estimates it can reduce CO_2 emissions another 90 million tons by 2010. The government has exploited its large hydropower resources at some social and ecological cost, such as forced relocations of communities, loss of ecosystems, and decreased river flow, but it believes it has substantial scope for increasing hydropower further still. In fact, it estimated this year that it could achieve a reduction of 500 million tons of carbon dioxide by 2010 with increased hydropower.

Compared with coal and hydro, China has scarcely begun its expansion of nuclear power. By 2020, the government plans to have built 40 gw of new nuclear power plants. But even if Beijing meets that goal, the 40 gw would only account for about 4 percent of the total electric capacity anticipated to exist by then.

The Chinese government is also devoting a substantial portion of its R&D dollars to the research, development, and demonstration of advanced energy technologies. During the period covered by the 11th Five-Year Plan, the Ministry of Science and Technology's budget for energy research, development, and demonstration is about 3.5 billion yuan (about $466 million). The budget for advanced coal technology is about 700 million yuan (about $93 million). Five coal co-production and gasification demonstration projects are planned for the next five years, in collaboration with Chinese industry. If all are actually built, there will be more coal gasification and co-production plants in China than in the United States.

Powered by Coal

Despite all of the Chinese government's laudable efforts to improve energy efficiency and expand the use of low-carbon energy sources such as renewable energy, nuclear power, and hydro-electric power, China's carbon dioxide emissions grew at the worrying rate of 9 percent per year from 1999 to 2004. At this rate, the Chinese will double emissions by 2009. The main drivers of this growth are heavy reliance on coal using conventional technologies, the still relatively poor efficiency of most power-plant technologies, and the weakness of government institutions when it comes to implementing and enforcing policies.

China's heavy reliance on the most greenhouse gas–intensive fuel of all—namely, coal—compounds the challenge of addressing global warming. Even with all of the measures taken so far to diversify energy supply, coal still accounts for 80 percent of China's greenhouse gas emissions. Even if a large increase in consumption of natural gas were achieved, it would not make much of an impact on emissions because it currently represents such a small portion of China's energy supply (less than 5 percent).

The other way to reduce emissions of heat-trapping gases, other than switching to cleaner fuels, is to improve the energy efficiency of China's power plants, as the government has recognized. More than half of China's power plants are smaller than 300 megawatts (mw)—in fact, more than 5,000 plants are smaller than 100 mw. This results in very poor energy efficiency. There are a handful of supercritical (high-efficiency) plants, and the first ultra-supercritical pulverized coal plant (the Huaneng Group's Yu-Huan plant) came on line in November 2006. Thirty-four additional ultra-supercritical plants are under construction. Yet, as impressive as this seems, China built three times that number of traditional coal plants last year alone. A more comprehensive policy is needed, one that will result in all new plants being high-efficiency or capable of capturing carbon.

Each new coal-fired power plant represents a 50- to 75-year commitment (and source of emissions) because these plants are unlikely to be prematurely retired. The International Energy Agency estimates that 55 percent of the new coal-fired power plants that will be constructed in the world between now and 2030 will be built in China. By using the cheapest technologies currently available for its power plants and industrial facilities (which is perfectly rational in strict economic terms), China is effectively locking itself and the world into high greenhouse gas emissions in the coming decades. This is so because technologies to economically capture carbon from conventional power plants do not currently exist. (In fact, this reflects an urgent need for additional R&D.)

Therefore, since the rapid growth in power plants and related infrastructure in China is expected to continue, "leapfrogging" to lower-carbon technologies in the near term is critical. Will all of the new plants that China is building use conventional high-carbon technologies or best-available low-carbon technologies? Much depends on the answer. Within five to ten years, if it maintains its recent rate of growth, China's energy sector will have installed the same amount of electricity capacity as the United States currently has (992 gw), virtually all of it in conventional coal-fired power.

We have learned that technological leapfrogging is not an automatic process. Developing countries either lack technical capabilities or cannot afford the costs of more advanced technologies. To achieve leapfrogging in coal-fired power plant technologies, the Chinese will need policies and incentives that mandate or promote the use of lower-carbon technologies.

What Now?

China may be more likely to develop these policies and incentives, however, if they are part of an international initiative that features a new climate partnership between the United States and China. Given the stakes in global warming, there is little time to lose in starting to build such a partnership. First, as soon as the United States has established a domestic mandatory program to reduce greenhouse gases, Washington should ask Beijing to adopt one as well, unique to its own circumstances.

In addition, the United States should consider forming a bilateral or multilateral investment fund to accelerate the deployment of low-carbon technologies in China. This fund could provide low- or no-interest loans or direct grants for major new industrial facilities or power plants that use low-carbon technologies. Without policies in place that effectively require the use of low-carbon technologies (for example, carbon dioxide performance standards or carbon taxes), or incentive programs that make the use of low-carbon technologies financially attractive (for example, special loans for coal gasification with carbon capture), the private sector will have little or no incentive to develop, transfer, and deploy low-carbon energy technologies in China. The establishment of concrete greenhouse gas policies and financial incentive programs in both the United States and China is of the utmost urgency.

Meanwhile, there is considerable scope for enhanced energy-technology cooperation between the two countries. Joint research, development, and demonstration projects can be valuable for both China and the United States. They can also provide a mechanism for bringing the US private sector in contact with Chinese partners. While there has been ongoing technology cooperation between the US Department of Energy and the Chinese Ministry of Science and Technology through protocols on fossil energy, energy efficiency, and renewable energy, this cooperation has been inadequate, underfunded, and a low priority for the US government. The areas that should be given high priority include research, development, and demonstration of carbon capture and storage, renewable energy, energy storage, and energy efficiency technologies.

As a developing country, China still lacks many of the institutions, policies, and enforcement mechanisms that are needed to foster technology transfer and environmental protection.

Finally, the United States needs to significantly bolster its cooperative activities related to increasing China's capacity for energy and environmental data collection and reporting, policy making, institution building, and regulatory enforcement. As a developing country, China still lacks many of the institutions, policies, and enforcement mechanisms that are needed to foster technology transfer and environmental protection. This is particularly the case at the provincial and county levels, and it has become one of the biggest obstacles to the achievement of real gains in energy efficiency and reduced greenhouse gas emissions.

Moral and Practical

Chinese leaders increasingly are expressing concern about the effects of global warming on China itself, while also worrying about the general deterioration of China's air and water quality. As international pressure builds because of China's new status as the largest overall emitter, and as scientific evidence accumulates regarding China's own vulnerability to climate change, the government in Beijing will likely be looking for help and ideas as to how to reduce emissions. As a matter of morality, the United States would do well to acknowledge that it put the largest portion of greenhouse gases into the atmosphere during the twentieth century, just as China will be the dominant emitter during the twenty-first century, and so it has an obligation to help China, still very much a developing country, confront this challenge.

Similarly, the pragmatic response would be to acknowledge that, since the two countries are the world's biggest emitters, the United States might as well form a partnership with China to develop creative ideas, technologies, and policies for preventing dangerous climate change in ways that are designed to produce mutual benefits. Such a partnership could help produce innovative low-carbon technologies for public and private benefit, wider and more open markets for advanced energy technologies, investment opportunities for Wall Street, and a more effective governance system in China. Catastrophic climate change might still be avoided if, but only if, the United States and China both act in time to reduce their emissions.

Kelly Sims Gallagher is director of the Energy Technology Innovation Policy program at the Belfer Center for Science and International Affairs at Harvard University's John F. Kennedy School of Government.

Where Oil and Water Do Mix

Environmental Scarcity and Future Conflict in the Middle East and North Africa

Jason J. Morrissette and Douglas A. Borer

"Many of the wars of the 20th century were about oil, but wars of the 21st century will be over water."

—Isamil Serageldin
World Bank Vice President

In the eyes of a future observer, what will characterize the political landscape of the Middle East and North Africa? Will the future mirror the past or, as suggested by the quote above, are significant changes on the horizon? In the past, struggles over territory, ideology, colonialism, nationalism, religion, and oil have defined the region. While it is clear that many of those sources of conflict remain salient today, future war in the Middle East and North Africa also will be increasingly influenced by economic and demographic trends that do not bode well for the region. By 2025, world population is projected to reach eight billion.[1] As a global figure, this number is troubling enough; however, over 90 percent of the projected growth will take place in developing countries in which the vast majority of the population is dependent on local renewable resources. For instance, World Bank estimates place the present annual growth rate in the Middle East and North Africa at 1.9 percent versus a worldwide average of 1.4 percent.[2] In most of these countries, these precious renewable resources are controlled by small segments of the domestic political elite, leaving less and less to the majority of the population. As a result, if present population and economic trends continue, we project that many future conflicts throughout the region will be directly linked to what academic researchers term "environmental scarcity"[3]—the scarcity of renewable resources such as arable land, forests, and fresh water.

The purpose of this article is twofold. In the first section, we conceptualize how environmental scarcity is linked to domestic political unrest and the subsequent crisis of domestic political legitimacy that may ultimately result in conflict. We review the academic literature which suggests that competition over water is the key environmental variable that will play an increasing role in future domestic challenges to governments throughout the region. We then describe how these crises of domestic political legitimacy may result in both intrastate and interstate conflict. Even though the Middle East can generally be characterized as an arid climate, two great river systems, the Nile and the Tigris/Euphrates, serve to anchor the major population centers in the region. Conflict over the water of the Nile may someday come to pass between Egypt, Sudan, and Ethiopia; while Turkey, Syria, and Iraq all are located along the Tigris/Euphrates watershed and compete for its resources. Further conflict over water may embroil Israel, Syria, and the Palestinians.

Despite many existing predictions of war over water, we investigate the intriguing question: How have governments in the Middle East thus far avoided conflict over dwindling water supplies? In the second section of the article, we discuss the concept of "virtual water" and use this concept to illustrate the important linkages between water usage and the global economy, showing how existing tangible water shortages have been ameliorated by a combination of economic factors, which may or may not be sustainable into the future.

Environmental Scarcity and Conflict: An Overview

Mostafa Dolatyar and Tim Gray identify water resources as "the principal challenge for humanity from the early days of civilization."[4] The 1998 United Nations Development Report estimates that almost a third of the 4.4 billion people currently living in the developing world have no access to clean water. The report goes on to note that the world's population tripled in the 20th century, resulting in a corresponding sixfold increase in the use of water resources. Moreover, infrastructure problems related to water supply abound in much of the developing world; the United Nations estimates that between 30 and 50 percent of the water presently diverted for irrigation purposes is lost through leaking pipes alone. In turn, roughly 20 countries in the developing world presently suffer from water stress (defined as having less than 1,000 cubic meters of available freshwater per capita), and 25 more are expected to join that list by 2050.[5] In response to these trends, the United Nations resolved in 2002 to reduce by half the proportion of people in the developing world who are unable to reach—or afford—safe drinking water.

In turn, numerous scholars in recent years have conceptualized water in security terms as a key strategic resource in many regions of the world. Thomas Naff maintains that water scarcity holds significant potential for conflict in large part because it is fundamentally essential to life. Naff identifies six basic characteristics that distinguish water as a vital and potentially contentious resource. (1) Water is necessary for sustaining life and has no substitute for human or animal use. (2) Both in terms of domestic and international policy, water issues are typically addressed by policymakers in a piecemeal fashion rather than comprehensively. (3) Since countries typically feel compelled by security concerns to control the ground on or under which water flows, by its nature, water is also a terrain security issue. (4) Water issues are frequently perceived as zero-sum, as actors compete for the same limited water resources. (5) As a result of the competition for these limited resources, water presents a constant potential for conflict. (6) International law concerning water resources remains relatively "rudimentary" and "ineffectual."[6] As these factors suggest, water is a particularly volatile strategic issue, especially when it is in severe shortage.

Arguing that environmental concerns have gained prominence in the post-Cold War era, Alwyn R. Rouyer establishes a basic paradigm of contemporary environmental conflict. Rouyer argues that "rapid population growth, particularly in the developing world, is putting severe stress on the earth's physical environment and thus creating a growing scarcity of renewable resources, including water, which in turn is precipitating violent civil and international conflict that will escalate in severity as scarcity increases."[7] Rouyer goes on to assert that this potential conflict over scarce resources will likely be most disruptive in states with rapidly expanding populations in which policymakers lack the political and economic capability to minimize environmental damage.

Almost a third of the 4.4 billion people currently living in the developing world have no access to clean water.

Security concerns linked fundamentally to environmental scarcity are far from a contrivance of the post-Cold War era, however. Ulrich Küffner asserts that conflicts over water "have occurred between many countries in all climatic regions, but between countries in arid regions they appear to be unavoidable. Claims over water have led to serious tensions, to threats and counter threats, to hostilities, border clashes, and invasions."[8] Moreover, as Miriam Lowi notes, "Well before the emergence of the nation-state, the arbitrary political division of a unitary river basin . . . led to problems regarding the interests of the states and/or communities located within the basin and the manner in which conflicting interests should be resolved."[9] Lowi fundamentally frames the issue of water scarcity in terms of a dilemma of collective action and failed cooperation—the archetypal "Tragedy of the Common"—in which communal resources are abused by the greediness of individuals. In many regions of the world, the international agreements and coordinating institutions necessary to lower the likelihood of conflict over water are either inadequate or altogether nonexistent.[10]

Thomas Homer-Dixon argues that the environmental resource scarcity that potentially results in conflict, including water scarcity, fundamentally derives from one of three sources. The first, *supply-induced scarcity,* is caused when a resource is either degraded (for example, when cropland becomes unproductive due to overuse) or depleted (for example, when cropland is converted into suburban housing). Throughout most of the Middle East and North Africa countries, both environmental and resource degradation and depletion are of relevant concern. For instance, many of these countries face significant decreases in the agricultural productivity of their arable soil as a result of ongoing trends of desertification, soil erosion, and pollution. This problem is coupled with the continued loss of croplands to urbanization, as rural dwellers move to cities in search of employment and opportunity. The second source of environmental scarcity, *demand-induced scarcity,* is caused by either an increase in per-capita consumption or by simple population growth. If the supply remains constant, and demand increases by existing users consuming more, or more users each consuming the same amount, eventually scarcity will result as demand overtakes supply. The third type of environmental scarcity is known as *structural scarcity,* a phenomenon that results when resource supplies are unequally distributed. In this case the "haves" in any given society generally control and consume an inordinate amount of the existing supply, which results in the more numerous "have-nots" experiencing the scarcity.[11]

These three sources of scarcity routinely overlap and interact in two common patterns: "resource capture" and ecological marginalization. Resource capture occurs when both demand-induced and supply-induced scarcities interact to produce structural scarcity. In this pattern, powerful groups within society foresee future shortages and act to ensure the protection of their vested interests by using their control of state structures to capture control of a valuable resource. An example of this pattern occurred in Mauritania (one of Algeria's neighbors) in the 1970s and 1980s when the countries bordering the Senegal River built a series of dams to boost agricultural production. As a result of the new dams, the value of land adjacent to the river rapidly increased—an economic development that motivated Mauritanian Moors to abandon their traditional vocation as cattle grazers located in the arid land in the north and, instead, to migrate south onto lands next to the river. However, black Mauritanians already occupied the land on the river's edge. As a result, the Moorish political elite that controlled the Mauritanian government rewrote the legislation on citizenship and land rights to effectively block black Mauritanians from land ownership. By declaring blacks as non-citizens, the Islamic Moors managed to capture the land through nominally legal (structural) means. As a result, high levels of violence later arose between Mauritania and Senegal, where hundreds of thousands of the black Mauritanians had become refugees after being driven from their land.[12]

The second pattern, ecological marginalization, occurs when demand-induced and structural scarcities interact in a way that results in supply-induced scarcity. An example of this pattern comes from the Philippines, a country whose agricultural lands traditionally have been controlled by a small group of dominant landowners who, prior to the election of former President Estrada, have controlled Filipino politics since colonial times. Population growth in the 1960s and 1970s forced many poor peasants to settle in the marginal soils of the upland interior. This more mountainous land could not sustain the lowland slash-and-burn farming practices that they brought with them. As a result, the Philippines suffered serious ecological damage in the form of water pollution, soil erosion, landslides, and changes in the hydrological cycle that led to further hardship for the peasantry as the land's capacity shrunk. As a result of their economic marginalization, many upland peasants became increasingly susceptible to the revolutionary rhetoric promoted by the communist-led New People's Army, or they supported the "People Power" movement that ousted US-backed Ferdinand Marcos from power in 1986.[13]

Thus, as shown in the Philippines, social pressures created by environmental scarcity can have a direct influence on the ruling legitimacy of the state, and may cause state power to crumble. Indeed, reductions in agricultural and economic production can produce objective socio-economic hardship; however, deprivation does not necessarily produce grievances against the government that result in serious domestic unrest or rebellion. One can look at the relative stability in famine-stricken North Korea as a poignant example of a polity whose citizens have suffered widespread physical deprivation under policies of the existing regime, but who are unwilling or unable to risk their lives to challenge the state.

This phenomenon is partly explained by conflict theorists who argue that individuals and groups have feelings of "relative deprivation" when they perceive a gap between what they believe they deserve and what in reality they actually have achieved.[14] In other words, can a government meet the expectations of the masses enough to avoid conflict? For example, in North Korea—a regime that tightly controls the information that its people receive—many people understand that they are suffering, but they may not know precisely how much they are suffering relative to others, such as their brethren in the South. The North Korean government indoctrinates its people to expect little other than hardship, which in turn it blames on outside enemies of the state. Thus, the people of North Korea have very low expectations, which their government has been able to meet. More important, then, is the question of whom do the people perceive as being responsible for their plight? If the answer is the people's own government—whether as a result of supply-induced, demand-induced, or structural resource scarcity—then social discord and rebellion are more likely to result in intrastate conflict, as citizens challenge the ruling legitimacy of the state itself. If the answer is someone else's government, then interstate conflict may result.

On numerous occasions, history has shown that governments whose people are suffering can remain in power for long periods of time by pointing to external sources for the people's hardship.[15] As noted above regarding political legitimacy, perception is politically more important than any standard of objective truth.[16] When faced with a crisis of legitimacy derived from environmental resource scarcity, any political regime essentially has a choice of two options in dealing with the situation. The regime may choose temporarily not to respond to looming challenges to its authority because water-induced stress may in fact pass when sufficient heavy rainfall occurs. However, most regimes in the Middle East and North Africa have sought more proactive ways to ensure their survival. Indeed, a people might forgive its government for one drought, but if governmental action is not taken, a subsequent drought-induced crisis of legitimacy could result in significant social upheaval by an unforgiving public. Furthermore, if the government itself is perceived to be the direct source of the scarcity—through structural arrangements, resource capture, or other means—these trends of social unrest are likely to be exacerbated. Thus, in order to survive, most states have developed policies to increase their water supplies and to address issues of environmental scarcity. The problem with doing so throughout most of the Middle East and North Africa, however, is that increasing supply in one state often creates environmental scarcity problems in another. If Turkey builds dams, Iraq and Syria are vulnerable; if Ethiopia or the Sudan builds dams, Egypt feels threatened. Thus far, interstate water problems leading to war have been avoided due to the economic interplay between oil wealth and the importation of "virtual water," which will be discussed at greater length below.

As noted above, resource scarcity issues centered on water are particularly prominent in the Middle East and North Africa. Ewan Anderson notes that resource geopolitics in the Middle East "has long been dominated by one liquid—oil. However, another liquid, water, is now recognized as the fundamental political weapon in the region."[17] Ecologically speaking, water scarcity in the Middle East and North Africa results from four primary causes: fundamentally dry climatic conditions, drought, desiccation (the degradation of land due to the drying up of the soil), and water stress (the low availability of water resulting from a growing population).[18] These resource scarcity problems are exacerbated in the Middle East by such factors as poor water quality and inadequate—and, at times, purposefully discriminatory—resource planning. As a result of these ecological and political trends, Nurit Kliot states, "water, not oil, threatens the renewal of military conflicts and social and economic disruptions" in the Middle East.[19] In the case of the Arab-Israeli conflict, Alwyn Rouyer suggests that "water has become inseparable from land, ideology, and religious prophecy."[20] Martin Sherman echoes these sentiments in the following passage, describing specifically the Arab-Israeli conflict:

> In recent years, particularly since the late 1980s, water has become increasingly dominant as a bone of contention between the two sides. More than one Arab leader, including those considered to be among the most moderate,

such as King Hussein of Jordan and former UN Secretary General, Boutros Boutros-Ghali of Egypt, have warned explicitly that water is the issue most likely to become the cause of a future Israeli-Arab war.[21]

Water is a particularly volatile strategic issue, especially when it is in severe shortage.

While Jochen Renger contends that a conflict waged explicitly over water may not lie on the immediate horizon, he notes that "it is likely that water might be used as leverage during a conflict."[22] As a result of such geopolitical trends, managing these water resources in the Middle East and North Africa—and, in turn, managing the conflict over these resources—should be considered a primary concern of both scholars and policymakers.

Keeping the Peace: The Importance of Virtual Water

The warning signals that war over water may replace war over oil and other traditional sources of conflict are very real in recent history. Yet, for more than 25 years, despite increasing demand, water has not been the primary cause of war in the Middle East and North Africa. The scenarios outlined in the preceding section have yet to fully address the fundamental questions of why and how governments in the region have thus far avoided major interstate conflict over water. In order to understand the likelihood of war, we must address the foundation of the past peace, testing whether or not this foundation remains strong for the foreseeable future. How have the governments of the region been able to avoid the apparently inevitable consequences of conflict that derive from the interlinked problems of water deficits, population growth, and weak economic performance? In this section of the article, we turn our attention to the important linkages between water usage and the global economy, showing how existing water shortages have been ameliorated by a combination of economic factors.

To understand the politics of water in the Middle East and North Africa, one must first look at the region's most fungible resource: oil. For much of the post-World War II era, the growing need for oil to fuel economic growth has served as the dominant motivating factor in US security policy in the Middle East. Conventional wisdom in the United States holds that US dependency on Middle Eastern oil is a strategic weakness. Indeed, the specter of a regional hegemonic power that controls the oil and that is also hostile to the United States strikes fear into the hearts of policymakers in Washington. Thus, for roughly the past 50 years, the United States has sought to prop up "moderate" (meaning pro-US) regimes while denying hegemony to "radical" (meaning anti-US) regimes.[23] However, we contend that both policymakers and the public at large in the United States generally misunderstand the politics of oil as they relate to water in the Middle East.

Country	Total Water Resources per Capita (cubic meters) in 2000	Percent of Population with Access to Adequate, Improved Water Source, 2000	GDP per Capita, 2000 Estimate
Algeria	477	94%	$5,500
Egypt	930	95%	$3,600
Iran	2,040	95%	$6,300
Iraq	1,544	85%	$2,500
Israel	180	99%	$18,900
Jordan	148	96%	$3,500
Kuwait	—	100%	$15,000
Lebanon	1,124	100%	$5,000
Libya	148	72%	$8,900
Morocco	1,062	82%	$3,500
Oman	426	39%	$7,700
Qatar	—	100%	$20,300
Saudi Arabia	119	95%	$10,500
Sudan	5,312	—	$1,000
Syria	2,845	80%	$3,100
Tunisia	434	—	$6,500
Turkey	3,162	83%	$6,800
Yemen	241	69%	$820

Figure 1 Water Resources and Economics in the Middle East and North Africa.

Sources: World Bank Development Indicators, Country-at-a-Glance Tables, Freshwater Resources, and *CIA World Factbook,* at http://www.worldbank.org and http://www.cia.gov

In absolute terms, problems arising from US vulnerability to foreign oil are basically true—it would be better to be free of dependency on oil from any foreign source than to be dependent. However, the other side of the equation is often forgotten: oil-producing states are dependent on the United States and other major oil importers for their economic livelihood. More bluntly put, oil-exporting states are dependent on the influx of dollars, euros, and yen to purchase goods, services, and commodities that they lack. Thus, oil-producing countries in the Middle East and North Africa, few of whom have managed to successfully diversify their economies beyond the petroleum sector, exist in an interdependent world economy. The world depends on their oil, and they depend on the world's goods and services—including that most valuable life-sustaining resource, water.

On the surface, this perhaps seems to be a contentious claim. Outgoing oil tankers do not return with freshwater used to grow crops, and Middle East countries do not rely on the importation of bottled water for their daily consumption needs. However, according to hydrologists, each individual needs approximately 100 cubic meters of water each year for personal needs, and an additional 1,000 cubic meters are required to grow the food that person consumes. Thus, every person alive requires approximately 1,100 cubic meters of water every year. In 1970, the water needs of most Middle Eastern and North African countries could be met from sources within the region. During the colonial and early post-colonial eras, regional governments and their engineers had effectively managed supply to deliver new water to meet the requirements of the growing urban populations, industrial requirements, agricultural needs, and other demand-induced factors. What is clear is that in the past 30 years, the status of the region's water resources has significantly worsened as populations have increased (an example of demand-induced scarcity). Since the mid-1970s, most countries have been able to supply daily consumption and industrial needs; however, as indicated in Figure 1, the approximate 1,000 cubic meters of water per capita that is required for self-sufficient agricultural production represents a seemingly impossible challenge for some Middle Eastern and North African economies.

Simply put, many countries of the region cannot presently meet the irrigation requirements needed to feed their own growing populations.[24] Furthermore, for those countries that have sufficient resources to meet this need in aggregate (such as Syria), resource capture and structural distribution problems keep water out of the hands of many citizens. If this situation has been deteriorating for nearly three decades, the question remains: Why has there been no war over water? The answer, according to Tony Allen, lies in an extremely important hidden source of water, which he describes as "virtual water."[25] Virtual water is the water contained in the food that the region imports—from the United States, Australia, Argentina, New Zealand, the countries of the European Union, and other major food-exporting countries. If each person of the world consumes food that requires 1,000 cubic meters of water to grow, plus 100 additional cubic meters for drinking, hygiene, and industrial production, it is still possible that any country that cannot supply the water to produce food may have sufficient water to meet its needs—if it has the economic capacity to buy, or the political capacity to beg, the remaining virtual water in the form of imported food.

According to Allen, more water flows into the countries of the Middle East and North Africa as virtual water each year than flows down the Nile for Egypt's agriculture. Virtual water obtained in the food available on the global market has enabled the governments of the region's countries to augment their inadequate and declining water resources. For instance, despite its meager freshwater resources of 180 cubic meters per capita, Israel—otherwise self-sufficient in terms of food production—manages its problems of water scarcity in part by importing large supplies of grain each year. As noted in Figure 1, this pattern is replicated by eight other countries in the region that have less than 1,100 cubic meters of water per person. Thus, the global cereal grain commodity markets have proven to be a very accessible and effective system for importing virtual water needs. In the Middle East and North Africa, politicians and resource managers have thus far found this option a better choice than resorting to war over water with their neighbors. As a result, the strategic imperative for maintaining peace has been met through access to virtual water in the form of food imports from the global market.[26]

The global trade in food commodities has been increasingly accessible, even to poor economies, for the past 50 years. During the Cold War, food that could not be purchased was often provided in the form of grants by either the United States or the Soviet Union, and in times of famine, international relief efforts in various parts of the globe have fed the starving. Over time, competition by the generators of the global grain surplus—the United States, Australia, Argentina, and the European Community—brought down the global price of grain. As a result, the past quarter-century, the period during which water conflicts in the Middle East and North Africa have been most insistently predicted, was also a period of global commodity markets awash with surplus grain. This situation allowed the region's states to replace domestic water supply shortages with subsidized virtual water in the form of purchases from the global commodities market. For example, during the 1980s, grain was being traded at about $100 (US) a ton, despite costing about $200 a ton to produce.[27] Thus, US and European taxpayers were largely responsible for funding the cost of virtual water (in the form of significant agricultural subsidies they paid their own farmers) which significantly benefited the countries of the Middle East and North Africa.

For the most part we concur with Allen's evaluation that countries have not gone to war primarily over water, and that they have not done so because they have been able to purchase virtual water on the international market. However, the key question for the future is, Will this situation continue? If the answer is yes, and grain will remain affordable to the countries of the region, then it is relatively safe to conclude that conflict derived from environmental scarcity (in the form of water deficits) will not be a significant problem in the foreseeable future. However, if the answer is no, and grain will not be as affordable as it has been in the past, then future conflict scenarios based on environmental scarcity must be seriously considered.

Global Economic Restructuring: The World Trade Organization's Impact on Subsidies

Regrettably, a trend toward the answer "no" appears to be gaining some momentum due to ongoing structural changes in the global economy. The year 1995 witnessed a dramatic change in the world grain market, when wheat prices rose rapidly, eventually reaching $250 a ton by the spring of 1996. With the laws of supply and demand kicking in, this increased price resulted in greater production; by 1998, world wheat prices had fallen back to $140 a ton, but had risen again to over $270 by June 2001.[28] These rapid wheat price fluctuations reemphasize the strategic importance and volatility of virtual water. If the global price of food staples remains affordable, many countries in the Middle East and North Africa may struggle to meet the demand-induced scarcity resulting from their growing populations, but they most likely will succeed. However, if basic food staple prices rise significantly in the coming decades and the existing economic growth patterns that have characterized the region's economies over the past 30 years remain constant, an outbreak of war is more likely.

It is clear that recent structural changes in the world economy do not favor the continuation of affordable food prices for the region's countries in the future. As noted above, wheat that costs $200 a ton to produce has often been sold for $100 a ton on world markets. This situation is possible only when the supplier is compensated for the lost $100 per ton in the form of a subsidy. Historically, these subsidies have been paid by the governments of major cereal grain-producing countries, primarily the United States and members of the European Union. Indeed, for the last 100 years, farm subsidies have been a bedrock public policy throughout the food-exporting countries of the first world. However, with the steady embrace of global free-trade economics and the establishment of the World Trade Organization (WTO), agricultural subsidies have come under pressure in most major grain-producing countries. According to a recent US Department of Agriculture (USDA) study, "The elimination of agriculture trade and domestic policy distortions could raise world agriculture prices about 12 percent.[29]

Many countries of the region cannot presently meet the irrigation requirements needed to feed their own growing populations.

Thus, as the WTO gains systematic credibility over the coming decades, its free-trade policies will further erode the practice of farm price supports, and it is highly unlikely that the aggregate farm subsidies of the past will continue at historic levels in the future. Under the new WTO regime, global food production will be increasingly based on the real cost of production plus whatever profit is required to keep farmers in business. Therefore, as global food prices rise in the future, and American and European governments are restricted by the new global trading regime from subsidizing their farmers, the price of virtual water in the Middle East and North Africa and throughout the food-importing world will also rise. According to the USDA report mentioned above, both developed and developing countries will gain from WTO liberalization. Developed countries that are major food exporters will gain immediately from the projected $31 billion in increased global food prices, of which they will share $28.5 billion ($13.3 billion to the United States), with $2.6 billion going to food exporters in the developing world. However, the report also claims food-importing countries will gain because global food price increases will spur more efficient production in their own economies, thus enabling them a "potential benefit" of $21 billion.[30] Even if accepted at face value, it is clear that such benefits will occur mostly in those developing countries with an abundance of water resources. Indeed, developing countries that produce fruits, vegetables, and other high-value crops for export to first world markets may indeed benefit from the reduction of farm subsidies, which today undercut their competitive advantage. But when it comes to basic foodstuffs—wheat, corn, and rice—the cereal grains that sustain life for most people, the developing world cannot compete with the highly efficient mechanized corporate farms of the first world.

In future research, basic intelligence is needed on two fronts. First, we must obtain a clearer understanding of the capacity of global commodity markets to meet future virtual water needs in the form of food. Second, we must identify which Middle East and North Africa governments will most likely have the economic capacity to meet their virtual water needs though food purchases—or, perhaps more important, which ones will not. In short, is there food available in the global market, and can countries afford to buy it? Countries that cannot afford virtual water may choose instead to pursue war as a means of achieving their national interest goals. Clearly the strongest countries, or those least susceptible to intrastate or interstate conflict arising from environmental scarcity, are those that have significant water resources or the economic capacity to purchase virtual water. However, it is also clear that the relative condition of peace that has existed in the Middle East and North Africa has been maintained historically through deeply buried linkages between American and European taxpayers, their massive farm subsidies programs, and world food prices. In the future, it appears that these hidden links may be radically altered if not broken by the World Trade Organization, and, as a result, the likelihood of conflict will increase.

Conclusion: Why War Will Come

Having moved away from the conventional understanding of water strictly as a zero-sum environmental resource by reconceptualizing it in more fungible economic terms, we nevertheless believe two incompatible social trends will collide to make war in the Middle East and North Africa virtually inevitable in the future. The first trend is economic globalization. As capitalism becomes ever more embraced as the global economic philosophy, and the world increasingly embraces free-trade economics, economic growth is both required and is inevitable. The WTO will facilitate this aggregate global growth, which, on the plus side, will undoubtedly increase the basic standard of living for the average world citizen. However, the global economy will be required to meet the needs of an estimated eight billion citizens in the year 2025. Achieving growth will demand an ever-greater share of the world's existing natural resources, including water. Thus, if present regional economic and demographic trends continue, resource shortfalls will occur, with water being the most highly stressed resource in the Middle East and North Africa.

Globalization is both a cause and a consequence of the rapid spread of information technology. Thus, in the globalized world, the figurative distance between cultures, philosophies of rule, and, perhaps more important, a basic understanding of what is possible in life, becomes much shorter. Personal computers, the internet, cellular phones, fax machines, and satellite television are all working in partnership to rewire the psychological infrastructure of the citizens of the Middle East and North Africa, and the world at large. As a result, by making visible what is possible in the outside world, this cognitive liberation will bring heightened material expectations of a better life, both economically and politically. Consequently, citizens will demand more from their governments. This emerging reality will collide head-on with the second trend—political authoritarianism—that characterizes most Middle East and North Africa governments.

Throughout the region there are few governments that allow for public expression of dissent. Although Turkey, Algeria, Tunisia, and Egypt are democracies in name, these states have exhibited a propensity to revert to authoritarian tactics when deemed necessary to limit political activity among their respective populaces.[31] Likewise, while Israel is institutionally a democracy, ethnic minorities are all but excluded from the democratic process. The remainder of the Middle East and North Africa states can be described only as authoritarian regimes. In retrospect, the most fundamental common denominator of all authoritarian regimes throughout history is their fierce resistance to change. Change is seen as a threat to the regime because most authoritarian regimes base their right to rule in some form of infallibility: the infallibility of the sultan, the king, or the ruling party and its ideology. Any admission that change is needed strikes at the foundation of this inflexible infallibility. Historically, most change has occurred in the Middle East and North Africa during times of intrastate unrest and interstate war. In the coming decades, globalization will bring change that will be resisted by governments of the region. As a result, to the distant observer the future will resemble the past: periods of wholesale peace will be a rare occurrence, intense competition and low-intensity conflict will be the norm, and major wars will occur at sporadic intervals.

The wild card in this equation may be post-2004 Iraq. Operation Iraqi Freedom and the ouster of Saddam Hussein have altered the strategic political landscape. If a sustainable democracy indeed emerges in Iraq, the country may turn away from future conflicts with its neighbors. Potential conflict between Turkey and Iraq over water may now be averted due to the fact that both countries may choose nonviolent solutions to their disputes. If President Bush's vision of a democratic Middle East comes to fruition, war may be averted. After all, there is a rich body of scholarly research regarding the "democratic peace" that suggests liberal democracies are significantly less likely to resort to war to resolve interstate disputes, and post-Saddam Iraq could serve as a key litmus test for the future of democratic reform in the region. However, it is also highly unlikely that regime change will come quickly to the moderate authoritarian states of the region that are also US allies. Decisionmakers in Washington may be able to dictate the political future of Iraq, but even America's mighty arsenal of political, economic, and military power cannot alter the basic demographic and environmental trends in the region.

Notes

1. Alex Marshall, ed., *The State of World Population 1997* (New York: United Nations Population Fund, 1997), p. 70.
2. "The World Bank: Middle East and North Africa Data Profile," *The World Bank Group Country Data* (2000), http://www.worldbank.org/data/countrydata/countrydata.html
3. The leading scholar in this area is Thomas Homer-Dixon. For example, see his recent book (coedited with Jessica Blitt), *Ecoviolence: Links Among Environment, Population, and Security* (New York: Rowman & Littlefield, 1998), which focuses on Chiapas, Gaza, South Africa, Pakistan, and Rwanda.
4. Mostafa Dolatyar and Tim S. Gray, *Water Politics in the Middle East: A Context for Conflict or Co-operation?* (New York: St. Martin's Press, 2000), p. 6.
5. *Human Development Report: Consumption for Human Development* (New York: United Nations Development Programme, Oxford Univ. Press, 1998), p. 55; "Water Woes Around the World," MSNBC, 9 September 2002, http://www.msnbc.com/news/802693.asp
6. Thomas Naff, "Conflict and Water Use in the Middle East," in *Water in the Arab World: Perspectives and Prognoses,* ed. Peter Rogers and Peter Lydon (Cambridge, Mass.: Harvard Univ. Press, 1994), p. 273.
7. Alwyn R. Rouyer, *Turning Water into Politics: The Water Issue in the Palestinian-Israeli Conflict* (New York: St. Martin's Press, 2000), p. 7.
8. Ulrich Küffner, "Contested Waters: Dividing or Sharing?" in *Water in the Middle East: Potential for Conflicts and Prospects for Cooperation,* ed. Waltina Scheumann and Manuel Schiffler (New York: Springer, 1998), p. 71.
9. Miriam R. Lowi, *Water and Power: The Politics of a Scarce Resource in the Jordan River Basin* (Cambridge, Eng.: Cambridge Univ. Press, 1993), p. 1.
10. Ibid., pp. 2ff.
11. Thomas Homer-Dixon and Jessica Blitt, "Introduction: A Theoretical Overview," in *Ecoviolence: Links Among Environment, Population,and Scarcity,* ed. Thomas Homer-Dixon and Jessica Blitt (New York: Rowman & Littlefield, 1998), p. 6.
12. Thomas Homer-Dixon and Valerie Percival, "The Case of Senegal-Mauritania," in *Environmental Scarcity and Violent*

Conflict: Briefing Book (Washington: American Association for the Advancement of Science and the University of Toronto, 1996), pp. 35–38.

13. Douglas Borer witnessed this agricultural problem while visiting rural areas on the Bataan peninsula in late 1985 and early 1986. The members of the New People's Army which he met were uninterested in Marxism, but they were very interested in ridding themselves of the Marcos regime. See Thomas Homer-Dixon and Valerie Percival, "The Case of the Philippines," ibid., p. 49.
14. Ted Gurr, *Why Men Rebel* (Princeton, N.J.: Princeton Univ. Press, 1970).
15. One need only look 90 miles southward from the Florida coast to find proof of this reality in Castro's Cuba.
16. Thus, Saddam Hussein was able to remain in power in Iraq until 2003 due to two essential factors. First, as noted in a recent article by James Quinlivan, Saddam had created "groups with special loyalties to the regime and the creation of parallel military organizations and multiple internal security agencies," that made Iraq essentially a "coup-proof" regime. (See James T. Quinlivan, "Coup-Proofing: Its Practice and Consequences in the Middle East," *International Security,* 24 [Fall 1999], 131–65.) Second, Saddam had convinced a significant portion of his people that the United States (and Britain) were responsible for their suffering. Thus, as long as these perceptions held and Saddam was able to command loyalty of the inner regime, his ouster from power by domestic sources remained unlikely.
17. Ewan W. Anderson, "Water: The Next Strategic Resource," in *The Politics of Scarcity: Water in the Middle East,"* ed. Joyce R. Starr and Daniel C. Stoll (Boulder, Colo.: Westview Press, 1988), p. 1.
18. Hussein A. Amery and Aaron T. Wolf, "Water, Geography, and Peace in the Middle East," in *Water in the Middle East: A Geography of Peace,* ed. Hussein A. Amery and Aaron T. Wolf (Austin: Univ. of Texas Press, 2000).
19. Nubit Kliot, *Water Resources and Conflict in the Middle East* (London and New York: Routledge, 1994), p. v, as quoted in Dolatyar and Gray, p. 9.
20. Rouyer, p. 9.
21. Martin Sherman, *The Politics of Water in the Middle East: An Israeli Perspective on the Hydro-Political Aspects of the Conflict* (New York: St. Martin's Press, 1999), p. xi.
22. Jochen Renger, "The Middle East Peace Process: Obstacles to Cooperation Over Shared Waters," in *Water in the Middle East: Potential for Conflict and Prospects for Cooperation,* ed. Waltina Scheumann and Manuel Schiffler (New York: Springer, 1998), p. 50.
23. Thus, even though the Saudi government is much more Islamized in religious terms than that of the Iraqis or Syrians, as long as the Saudi government is pro-US and serves US interests in supplying cheap oil, it receives the benevolent "moderate" label, while the more secularized Iraqis and Syrians have been labeled with the prerogative labels "radical" or "rogue-states."
24. Tony Allan, "Watersheds and Problemsheds: Explaining the Absence of Armed Conflict over Water in the Middle East," in *MERNIA: Middle East Review of International Affairs Journal,* 2 (March 1998), http://biu.ac.il/SOC/besa/meria/journal/1998/issue1/jv2n1a7.html
25. Ibid.
26. Ibid.
27. Ibid.
28. Prices from 26 June 2001 quoted at http://www.usafutures.com/commodityprices.htm
29. "Agricultural Policy Reform in the WTO The Road Ahead," in *ERS Agricultural Economics Report,* No. 802, ed. Mary E. Burfisher (Washington: US Department of Agriculture, May 2001), p. iii.
30. Ibid., p. 6.
31. For instance, as of 2004, Freedom House (http://www.freedomhouse.org) classifies Algeria, Egypt, and Tunisia as "not free," and Turkey as only "partly free."

Jason J. Morrissette is a doctoral candidate and instructor of record in the School of Public and International Affairs at the University of Georgia. He is currently writing his dissertation on the political economy of water scarcity and conflict. **Douglas A. Borer** (PhD, Boston University, 1993) is an Associate Professor in the Department of Defense Analysis at the Naval Postgraduate School. He recently served as Visiting Professor of Political Science at the US Army War College. Previously he was Director of International Studies at Virginia Tech, and he has taught overseas in Fiji and Australia. Dr. Borer is a former Fulbright Scholar at the University of Kebangsaan Malaysia, and has published widely in the areas of security, strategy, and foreign policy.

From *Parameters,* by Jason J. Morrissette and Douglas A. Borer (Winter 2004–2005, pp. 86–101). Published in 2005 by U.S. Army War College. Reprinted by permission.

UNIT 2

Population, Policy, and Economy

Unit Selections

5. **Do Global Attitudes and Behaviors Support Sustainable Development?,** Anthony A. Leiserowitz, Robert W. Kates, and Thomas M. Parris
6. **Paying for Climate Change,** Benjamin Jones, Michael Keen, and Jon Strand
7. **High-Tech Trash: Will Your Discarded TV or Computer End up in a Ditch in Ghana?,** Chris Carroll
8. **Down with Carbon: Scientists Work to Put the Greenhouse Gas in Its Place,** Sid Perkins

Key Points to Consider

- What policies should be enacted to combat environmental degradation? What changes in our understanding and concerns of these problems are necessary before we are willing to support climate change legislation?
- What is sustainable development? How can we achieve sustainable development at local and global levels with increasing population levels?
- How does an economist view climate change? Consider the benefits and negative consequences of effective climate change legislation. Who gains, who loses, who pays? Over what timescales should costs and benefits be calculated?
- How important is recycling? How can it be expanded, and how do we dispose of toxic byproducts?
- Can market forces solve the problem of global warming? A number of ideas have been developed regarding economic solutions to greenhouse gas emissions. Are voluntary emission cuts sufficient? What are the benefits of a carbon tax? Can buying carbon offsets contribute to reducing emissions?
- Are there other viable ways to reduce global warming without reducing our fossil fuel consumption? Is carbon sequestration possible? Clean coal? Are engineering solutions, such as placing light-reflecting aerosols in the stratosphere possible, and are they safe?

Student Website

www.mhhe.com/cls

Internet References

The Hunger Project
http://www.thp.org

Poverty Mapping
http://www.povertymap.net

World Health Organization
http://www.who.int

World Population and Demographic Data
http://geography.about.com/cs/worldpopulation

WWW Virtual Library: Demography & Population Studies
http://demography.anu.edu.au/VirtualLibrary

In Article 1, the severity of climate change is explored. The IPPC report is a "conservative" assessment of the ramifications of climate change. Other reports paint an even more dire picture of the consequences of inaction. In this section, the realities are explored. It would be naïve to simply say that we need to direct all of our resources—all of our assets—toward ameliorating the deleterious consequences of climate change. People are out of work, they feel overtaxed and there is a general uncertainty about what "needs to be done" versus "what should be done" when addressing our environmental issues.

In "Do Global Attitudes and Behaviors Support Sustainable Development," authors Leiserowitz *et al.* explore the issue of sustainability. In the time it takes to read this introduction, another 10,000 people will inhabit the planet. Thousands or even millions of people cannot do much environmental damage to our Earth, but with a population that will top 7 billion in May 2010, resources will be consumed without being replenished, and long-term environmental damage is almost unavoidable. The world's population is expected to approach 10 billion by the year 2050 (U.S. Census Bureau). One thing is clear: Population growth is undoubtedly limited by the availability of resources—and resources are finite. Sustainable development is necessary for our population to continue with a lifestyle that is not seriously compromised. As a global community, it is imperative that we address the question of sustainability. Our planet can only provide so much. Recognizing the limitations of our planet is a necessary first step for our long-term survival.

"Paying for Climate Change" looks at the position of climate remediation from an economic standpoint. Climate scientists deal with the facts of climate change. Inaction will lead to changes in climate, and change is not a good thing. Midwest farmers are used to growing certain crops. New York is used to being above water. If sea level had always been 20 meters higher than today, our cities would be at the water's edge. But if sea level rises, cities will have to be relocated to higher levels, and that is an insupportable change. Economists must step in to address these issues. What sacrifices are needed now to prevent future catastrophes? How much is this generation obligated to pay for future generations well being? And who should shoulder the burden to correct our impending economic disasters?

© Creatas/PunchStock

"High-Tech Trash" makes us all think about the everyday realities of garbage disposal. Anyone who reads this article (and follows up by checking out the accompanying photos at http://ngm.nationalgeographic.com/2008/01/high-tech-trash/essick-photography) will be shocked to see how our world deals with something as simple as computer disposal. This is just one example of the difficulties encountered when environmental stewardship clashes with economic opportunity.

"Down with Carbon" is far more upbeat. It proposes solutions to the deleterious effects of global warming through sequestration of carbon. These are the "high-tech" solutions. Rather than curtailing our burning of fossil fuels, Sid Perkins explores creative ideas revolving around actual removal of carbon from the atmosphere. One example is ocean fertilization, whereby nutrients are sown over the open ocean to promote algal blooms. The algae, through photosynthesis, draw CO_2 out of the atmosphere. Upon their death, the algae sink down to the depths of the ocean and successfully remove the carbon from the atmospheric system. Will ocean fertilization work? The jury is still out, but the ideas explored in this article are promising.

Do Global Attitudes and Behaviors Support Sustainable Development?

ANTHONY A. LEISEROWITZ, ROBERT W. KATES, AND THOMAS M. PARRIS

Many advocates of sustainable development recognize that a transition to global sustainability—meeting human needs and reducing hunger and poverty while maintaining the life-support systems of the planet—will require changes in human values, attitudes, and behaviors.[1] A previous article in *Environment* described some of the values used to define or support sustainable development as well as key goals, indicators, and practices.[2] Drawing on the few multinational and quasi-global-scale surveys that have been conducted,[3] this article synthesizes and reviews what is currently known about global attitudes and behavior that will either support or discourage a global sustainability transition.[4] (Table 1 provides details about these surveys.)

None of these surveys measured public attitudes toward "sustainable development" as a holistic concept. There is, however, a diverse range of empirical data related to many of the subcomponents of sustainable development: development and environment; the driving forces of population, affluence/poverty/consumerism, technology, and entitlement programs; and the gap between attitudes and behavior.

Development

Concerns for environment and development merged in the early concept of sustainable development, but the meaning of these terms has evolved over time. For example, global economic development is widely viewed as a central priority of sustainable development, but development has come to mean human and social development as well.

Economic Development

The desire for economic development is often assumed to be universal, transcending all cultural and national contexts. Although the surveys in Table 1 have no global-scale data on public attitudes toward economic development per se, this assumption appears to be supported by 91 percent of respondents from 35 developing countries, the United States, and Germany, who said that it is very important (75 percent) or somewhat important (16 percent) to live in a country where there is economic prosperity[5] What level of affluence is desired, how that economic prosperity is to be achieved, and how economic wealth should ideally be distributed within and between nations, however, are much more contentious questions. Unfortunately, there does not appear to be any global-scale survey research that has tried to identify public attitudes or preferences for particular levels or end-states of economic development (for example, infinite growth versus steady-state economies) and only limited or tangential data on the ideal distribution of wealth (see the section on affluence below).

Data from the World Values Survey suggest that economic development leads to greater perceived happiness as countries make the transition from subsistence to advanced industrial

Table 1 Multinational Surveys

Name	Year(s)	Number of Countries
One-time Surveys		
Pew Global Attitudes Project	2002	43
Eurobarometer	2002	15
International Social Science Program	2000	25
Health of the Planet	1992	24
Repeated Surveys		
GlobeScan International Environmental Monitor	1997–2003	34
World Values Survey	1981–2002	79
Demographic and Health Surveys	1986–2002	17
Organisation for Economic Co-operation and Development	1990–2002	22

Note. Before November 2003, GlobeScan, Inc. was known as Environics International. Surveys before this time bear the older name.

Source: For more detail about these surveys and the countries sampled, see Appendix A in A. Leiserowitz, R. W. Kates, and T. M. Parris, *Sustainability Values, Attitudes and Behaviors: A Review of Multi-national and Global Trends,* CID Working Paper No. 113 (Cambridge, MA: Science, Environment and Development Group, Center for International Development, Harvard University, 2004), http://www.cid.harvard.edu/cidwp/113.htm.

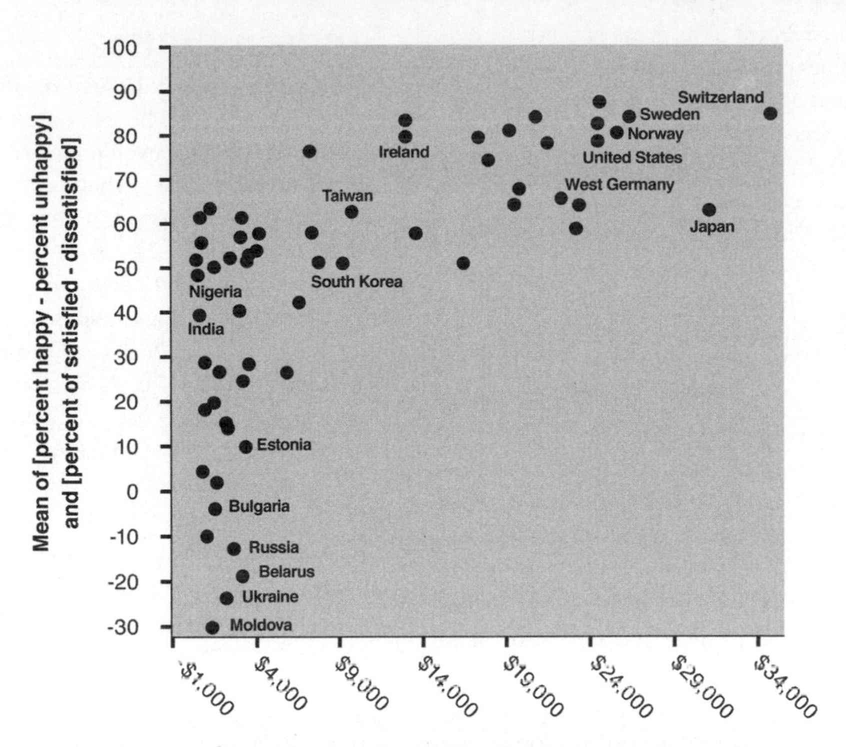

Figure 1 Subjective Well-being by Level of Economic Development.

Note. The subjective well-being index reflects the average of the percentage in each country who describe themselves as "very happy" or "happy" minus the percentage who describe themselves as "not very happy" or "unhappy"; and the percentage placing themselves in the 7–10 range, minus the percentage placing themselves in the 1–4 range, on a 10-point scale on which 1 indicates that one is strongly dissatisfied with one's life as a whole, and 10 indicates that one is highly satisfied with one's life as a whole.

Source: R. Inglehart, "Globalization and Postmodern Values," *Washington Quarterly* 23, no. 1 (1999): 215–228. Subjective well-being data from the 1990 and 1996 World Values Surveys. GNP per capita for 1993 data from *World Bank, World Development Report, 1995* (New York: Oxford University Press, 1995).

economies. But above a certain level of gross national product (GNP) per capita—approximately $14,000—the relationship between income level and subjective well-being disappears (see Figure 1). This implies that infinite economic growth does not lead to greater human happiness. Additionally, many of the unhappiest countries had, at the time of these surveys, recently experienced significant declines in living standards with the collapse of the Soviet Union. Yet GNP per capita remained higher in these ex-Soviet countries than in developing countries like India and Nigeria.[6] This suggests that relative trends in living standards influence happiness more than absolute levels of affluence, but the relationship between economic development and subjective well-being deserves more research attention.

Human Development

Very limited data is available on public attitudes toward issues of human development, although it can be assumed that there is near-universal support for increased child survival rates, adult life expectancies, and educational opportunities. However, despite the remarkable increases in these indicators of human well-being since World War II,[7] there appears to be a globally pervasive sense that human well-being has been deteriorating in recent years. In 2002, large majorities worldwide said that a variety of conditions had worsened over the previous five years, including the availability of well-paying jobs (58 percent); working conditions (59 percent); the spread of diseases (66 percent);the affordability of health care (60 percent); and the ability of old people to care for themselves in old age (59 percent). Likewise, thinking of their own countries, large majorities worldwide were concerned about the living conditions of the elderly (61 percent) and the sick and disabled (56 percent), while a plurality was concerned about the living conditions of the unemployed (42 percent).[8]

Development Assistance

One important way to promote development is to extend help to poorer countries and people, either through national governments or nongovernmental organizations and charities. There is strong popular support but less official support for development assistance to poor countries. In 1970, the United Nations General Assembly resolved that each economically advanced country would dedicate 0.7 percent of its gross national income (GNI) to official development assistance (ODA) by the middle of the 1970s—a target that has been reaffirmed in many subsequent international agreements.[9] As of 2004, only five countries had achieved this goal (Denmark, Norway, the Netherlands, Luxembourg, and Sweden). Portugal was close to the target at 0.63, yet all other countries ranged from a high of 0.42 percent (France) to lows of 0.16 and 0.15 percent (the United States and Italy respectively). Overall, the average ODA/GNI among the industrialized countries was only 0.25 percent—far below the UN target.[10]

By contrast, in 2002, more than 70 percent of respondents from 21 developed and developing countries said they would support paying 1 percent more in taxes to help the world's poor.[11] Likewise, surveys in the 13 countries of the Organisation for Economic Co-operation and Development's Development Assistance Committee (OECD-DAC) have found that public support for the principle of giving aid to developing countries (81 percent in 2003) has remained high and stable for more than 20 years.[12] Further, 45 percent said that their government's current (1999–2001) level of expenditure on foreign aid was too low, while only 10 percent said foreign aid was too high.[13] There is also little evidence that the public in OECD countries has developed "donor fatigue." Although surveys have found increasing public concerns about corruption, aid diversion, and inefficiency, these surveys also continue to show very high levels of public support for aid.

Public support for development aid is belied, however, by several factors. First, large majorities demonstrate little understanding of development aid, with most unable to identify their national aid agencies and greatly overestimating the percentage of their national budget devoted to development aid. For example, recent polls have found that Americans believed their government spent 24 percent (mean estimate) of the national budget on foreign assistance, while Europeans often estimated their governments spent 5 to 10 percent.[14] In reality, in 2004 the United States spent approximately 0.81 percent and the European Union member countries an average of approximately 0.75 percent of their national budgets on official development assistance, ranging from a low of 0.30 percent (Italy) to a high of 1.66 percent (Luxembourg).[15] Second, development aid is almost always ranked low on lists of national priorities, well below more salient concerns about (for example) unemployment, education, and health care. Third, "the overwhelming support for foreign aid is based upon the perception that it will be spent on remedying humanitarian crises," not used for other development-related issues like Third World debt, trade barriers, or increasing inequality between rich and poor countries—or for geopolitical reasons (for example, U.S. aid to Israel and Egypt).[16] Support for development assistance has thus been characterized as "a mile wide, but an inch deep" with large majorities supporting aid (in principle) and increasing budget allocations but few understanding what development aid encompasses or giving it a high priority.[17]

Environment

Compared to the very limited or nonexistent data on attitudes toward economic and human development and the overall concept of sustainable development, research on global environmental attitudes is somewhat more substantial. Several surveys have measured attitudes regarding the intrinsic value of nature, global environmental concerns, the trade-offs between environmental protection and economic growth, government policies, and individual and household behaviors.

Human-Nature Relationship

Most research has focused on anthropocentric concerns about environmental quality and natural resource use, with less attention to ecocentric concerns about the intrinsic value of nature. In 1967, the historian Lynn White Jr. published a now-famous and controversial article arguing that a Judeo-Christian ethic and attitude of domination, derived from Genesis, was an underlying historical and cultural cause of the modern environmental crisis.[18] Subsequent ecocentric, ecofeminist, and social ecology theorists have also argued that a domination ethic toward people, women, and nature runs deep in Western, patriarchal, and capitalist culture.[19] The 2000 World Values Survey, however, found that 76 percent of respondents across 27 countries said that human beings should "coexist with nature," while only 19 percent said they should "master nature" (see Figure 2). Overwhelming majorities of Europeans, Japanese, and North Americans said that human beings should coexist with nature, ranging from 85 percent in the United States to 96 percent in Japan. By contrast, only in Jordan, Vietnam, Tanzania, and the Philippines did more than 40 percent say that human beings should master nature.[20] In 2002, a national survey of the United States explored environmental values in more depth and found that Americans strongly agreed that nature has intrinsic value and that humans have moral duties and obligations to animals, plants, and non-living nature (such as rocks, water, and air). The survey found that Americans strongly disagreed that "humans have the right to alter nature to satisfy wants and desires" and that "humans are not part of nature" (see Figure 3).[21] This very limited data suggests that large majorities in the United States and worldwide now reject a domination ethic as the basis of the human-nature relationship, at least at an abstract level. This question, however, deserves much more cross-cultural empirical research.

Environmental Concern

In 2000, a survey of 11 developed and 23 developing countries found that 83 percent of all respondents were concerned a fair amount (41 percent) to a great deal (42 percent) about environmental problems. Interestingly, more respondents from developing countries (47 percent) were "a great deal concerned" about the environment than from developed countries (33 percent), ranging from more than 60 percent in Peru, the Philippines, Nigeria, and India to less than 30 percent in the Netherlands, Germany, Japan, and Spain.[22] This survey also

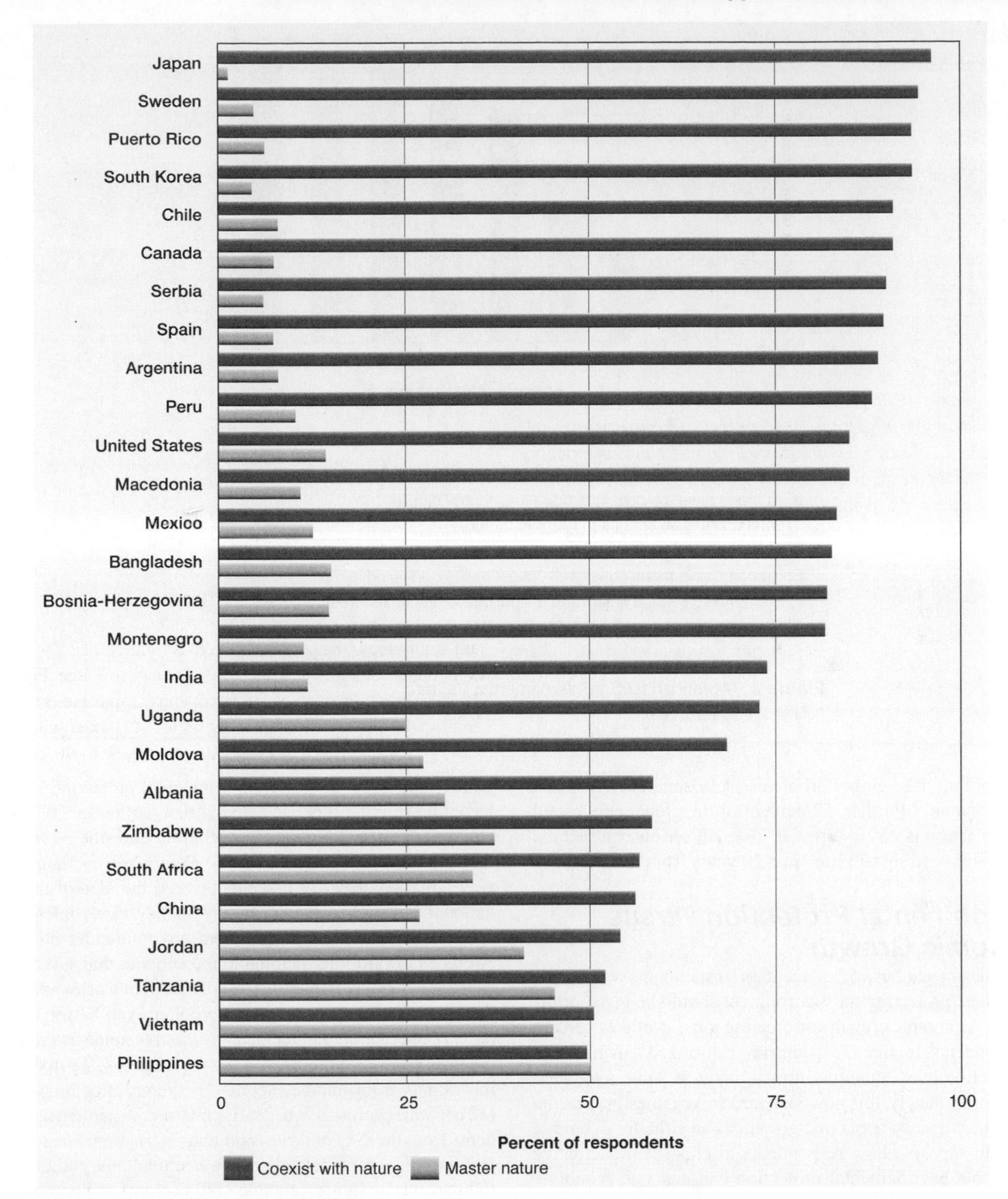

Figure 2 Human-Nature Relationship.

Note. The question asked, "Which statement comes closest to your own views: Human beings should master nature or humans should coexist with nature?"

Source: A. Leiserowitz, 2005. Data from world Values Survey, *The 1999–2002 Values Surveys Integrated Data File 1.0, CD-ROM in R. Inglehart, M. Basanez, J. Diez-Medrano, L. Halman, and R. Luijkx, eds., Human Beliefs and Values: A Cross-Cultural Sourcebook Based on the 1999–2002 Values Surveys, first edition* (Mexico City: Siglo XXI, 2004).

asked respondents to rate the seriousness of several environmental problems (see Figure 4). Large majorities worldwide selected the strongest response possible ("very serious") for seven of the eight problems measured. Overall, these results demonstrate very high levels of public concern about a wide range of environmental issues, from local problems like water

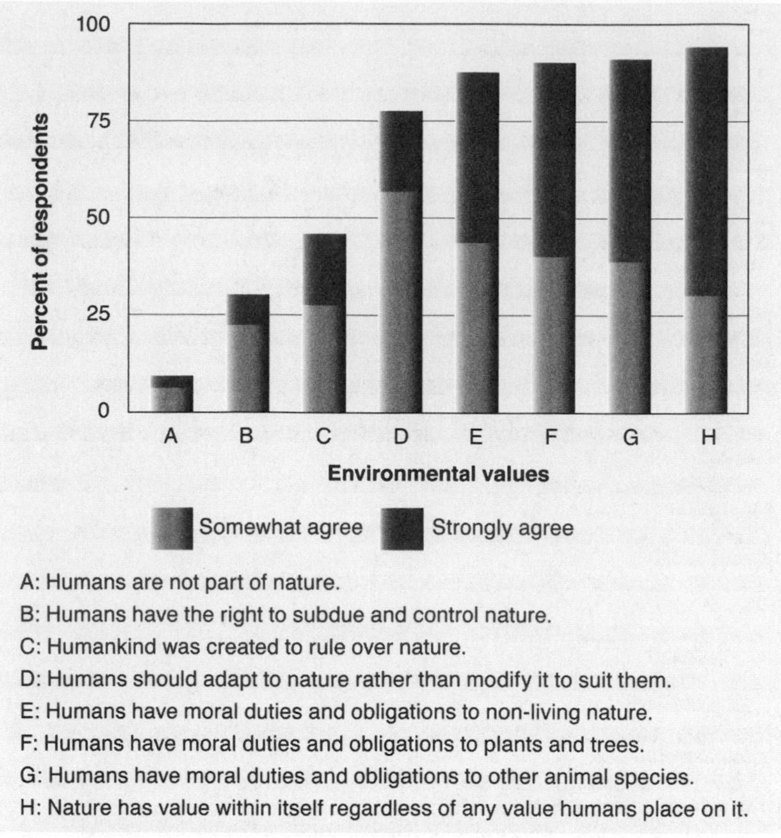

Figure 3 American (U.S.) Environmental Values.
Source: A. Leiserowitz, 2005.

and air pollution to global problems like ozone depletion and climate change.[23] Further, 52 percent of the global public said that if no action is taken, "species loss will seriously affect the planet's ability to sustain life" just 20 years from now.[24]

Environmental Protection versus Economic Growth

In two recent studies, 52 percent of respondents worldwide agreed that "protecting the environment should be given priority" over "economic growth and creating jobs," while 74 percent of respondents in the G7 countries prioritized environmental protection over economic growth, even if some jobs were lost.[25] Unfortunately, this now-standard survey question pits the environment against economic growth as an either/or dilemma. Rarely do surveys allow respondents to choose an alternative answer, that environmental protection can generate economic growth and create jobs (for example, in new energy system development, tourism, and manufacturing).

Attitudes toward Environmental Policies

In 1995, a large majority (62 percent) worldwide said they "would agree to an increase in taxes if the extra money were used to prevent environmental damage," while 33 percent said they would oppose them.[26] In 2000, there was widespread global support for stronger environmental protection laws and regulations, with 69 percent saying that, at the time of the survey, their national laws and regulations did not go at all far enough.[27] The 1992 Health of the Planet survey found that a very large majority (78 percent) favored the idea of their own national government "contributing money to an international agency to work on solving global environmental problems." Attitudes toward international agreements in this survey, however, were less favorable. In 1992, 47 percent worldwide agreed that "our nation's environmental problems can be solved without any international agreements," with respondents from low-income countries more likely to strongly agree (23 percent) than individuals from middle-income (17 percent) or high-income (12 percent) countries.[28] In 2001, however, 79 percent of respondents from the G8 countries said that international negotiations and progress on climate change was either "not good enough" (39 percent) or "not acceptable" (40 percent) and needed faster action. Surprisingly, this latter 40 percent supported giving the United Nations "the power to impose legally-binding actions on national governments to protect the Earth's climate."[29]

Environmental Behavior

Material consumption is one of the primary means by which environmental values and attitudes get translated into behavior. (For attitudes toward consumption per se, see the following section on affluence, poverty, and consumerism.)

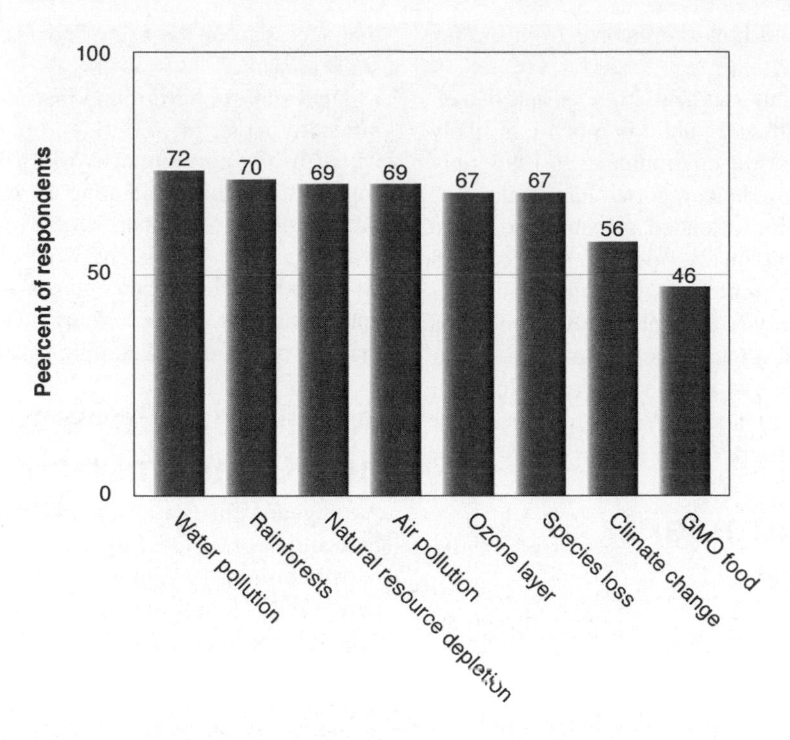

Figure 4 Percent of Global Public Calling Environmental Issues a "Very Serious Problem."

Source: A. Leiserowitz, 2005. Data from Environics International (Globe Scan), *Environics International Environmental Monitor Survey Dataset* (Kingston, Canada: Environics international, 2000), http://jeff-lab.queensu.ca/poadata/info/iem/iemlist.shtml (accessed 5 October 2004).

In 2002, Environics International (GlobeScan) found that 36 percent of respondents from 20 developed and developing countries stated that they had avoided a product or brand for environmental reasons, while 27 percent had refused packaging, and 25 percent had gathered environmental information.[30] Recycling was highly popular, with 6 in 10 people setting aside garbage for reuse, recycling, or safe disposal. These rates, however, reached 91 percent in North America versus only 36–38 percent in Latin America, Eastern Europe, and Central Asia,[31] which may be the result of structural barriers in these societies (for example, inadequate infrastructures, regulations, or markets). There is less survey data regarding international attitudes toward energy consumption, but among Europeans, large majorities said they had reduced or intended to reduce their use of heating, air conditioning, lighting, and domestic electrical appliances.[32]

In 1995, 46 percent of respondents worldwide reported having chosen products thought to be better for the environment, 50 percent of respondents said they had tried to reduce their own water consumption, and 48 percent reported that in the 12 months prior to the survey, they reused or recycled something rather than throwing it away. There was a clear distinction between richer and poorer societies: 67 percent of respondents from high-income countries reported that they had chosen "green" products, while only 30 percent had done so in low-income countries. Likewise, 75 percent of respondents from high-income countries said that they had reused or recycled something, while only 27 percent in low-income countries said this.[33] However, the latter results contradict the observations of researchers who have noted that many people in developing countries reuse things as part of everyday life (for example, converting oil barrels into water containers) and that millions eke out an existence by reusing and recycling items from landfills and garbage dumps.[34] This disparity could be the result of inadequate survey representation of the very poor, who are the most likely to reuse and recycle as part of survival, or, alternatively, different cultural interpretations of the concepts "reuse" and "recycle."

In 2002, 44 percent of respondents in high-income countries were very willing to pay 10 percent more for an environmentally friendly car, compared to 41 percent from low-income countries and 29 percent from middle-income countries.[35] These findings clearly mark the emergence of a global market for more energy-efficient and less-polluting automobiles. However, while many people appear willing to spend more to buy an environmentally friendly car, most do not appear willing to pay more for gasoline to reduce air pollution. The same 2002 survey found that among high-income countries, only 28 percent of respondents were very willing to pay 10 percent more for gasoline if the money was used to reduce air pollution, compared to 23 percent in medium-income countries and 36 percent in low-income countries.[36] People appear to generally oppose higher gasoline prices, although public attitudes are probably affected, at least in part, by the prices extant at the time of a given survey, the rationale given for the tax, and how the income from the tax will be spent.

Despite the generally pro-environment attitudes and behaviors outlined above, the worldwide public is much less likely to engage in political action for the environment. In 1995, only 13 percent of worldwide respondents reported having donated to an environmental organization, attended a meeting, or signed a petition for the environment in the prior 12 months, with more doing so in high-income countries than in low-income countries.[37] Finally, in 2000, only 10 percent worldwide reported having written a letter or made a telephone call to express their concern about an environmental issue in the past year, 18 percent had based a vote on green issues, and 11 percent belonged to or supported an environmental group.[38]

Drivers of Development and Environment

Many analyses of the human impact on life-support systems focus on three driving forces: population, affluence or income, and technology—the so-called I = PAT identity.[39] In other words, environmental impact is considered a function of these three drivers. In a similar example, carbon dioxide (CO_2) emissions from the energy sector are often considered a function of population, affluence (gross domestic product (GDP) per capita), energy intensity (units of energy per GDP), and technology (CO_2 emissions per unit of energy).[40] While useful, most analysts also recognize that these variables are not fundamental driving forces in and of themselves and are not independent from one another.[41] A similar approach has also been applied to human development (D = PAE), in which development is considered a function of population, affluence, and entitlements and equity.[42] What follows is a review of empirical trends in attitudes and behavior related to population, affluence, technology, and equity and entitlements.

Population

Global population continues to grow, but the rate of growth continues to decline almost everywhere. Recurrent Demographic and Health Surveys (DHS) have found that the ideal number of children desired is declining worldwide. Globally, attitudes toward family planning and contraception are very positive, with 67 percent worldwide and large majorities in 38 out of 40 countries agreeing that birth control and family planning have been a change for the better.[43] Worldwide, these positive attitudes toward family planning are reflected in the behavior of more than 62 percent of married women of reproductive age who are currently using contraception. Within the developing world, the United Nations reports that from 1990 to 2000, contraceptive use among married women in Asia increased from 52 percent to 66 percent, in Latin American and the Caribbean from 57 percent to 69 percent, but in Africa from only 15 percent to 25 percent.[44] Notwithstanding these positive attitudes toward contraception, in 1997, approximately 20 percent to 25 percent of births in the developing world were unwanted, indicating that access to or the use of contraceptives remains limited in some areas.[45]

DHS surveys have found that ideal family size remains significantly larger in western and middle Africa (5.2) than elsewhere in the developing world (2.9).[46] They also found that support for family planning is much lower in sub-Saharan Africa (44 percent) than in the rest of the developing world (74 percent).[47] Consistent with these attitudes, sub-Saharan Africa exhibits lower percentages of married women using birth control as well as lower rates of growth in contraceptive use than the rest of the developing world.[48]

Affluence, Poverty, and Consumerism

Aggregate affluence and related consumption have risen dramatically worldwide with GDP per capita (purchasing-power parity, constant 1995 international dollars) more than doubling between 1975 and 2002.[49] However, the rising tide has not lifted all boats. Worldwide in 2001, more than 1.1 billion people lived on less than $1 per day, and 2.7 billion people lived on less than $2 per day—with little overall change from 1990. However, the World Bank projects these numbers to decline dramatically by 2015—to 622 million living on less than $1 per day and 1.9 billion living on less than $2 per day. There are also large regional differences, with sub-Saharan Africa the most notable exception: There, the number of people living on less than $1 per day rose from an estimated 227 million in 1990 to 313 million in 2001 and is projected to increase to 340 million by 2015.[50]

Poverty

Poverty reduction is an essential objective of sustainable development.[51] In 1995, 65 percent of respondents worldwide said that more people were living in poverty than had been 10 years prior. Regarding the root causes of poverty, 63 percent blamed unfair treatment by society, while 26 percent blamed the laziness of the poor themselves. Majorities blamed poverty on the laziness and lack of willpower of the poor only in the United States (61 percent), Puerto Rico (72 percent), Japan (57 percent), China (59 percent), Taiwan (69 percent), and the Philippines (63 percent) (see Figure 5).[52] Worldwide, 68 percent said their own government was doing too little to help people in poverty within their own country, while only 4 percent said their government was doing too much. At the national level, only in the United States (33 percent) and the Philippines (21 percent) did significant proportions say their own government was doing too much to help people in poverty.[53]

Consumerism

Different surveys paint a complicated and contradictory picture of attitudes toward consumption. On the one hand, majorities around the world agree that, at the societal level, material and status-related consumption are threats to human cultures and the environment. Worldwide, 54 percent thought "less emphasis on money and material possessions" would be a good thing, while

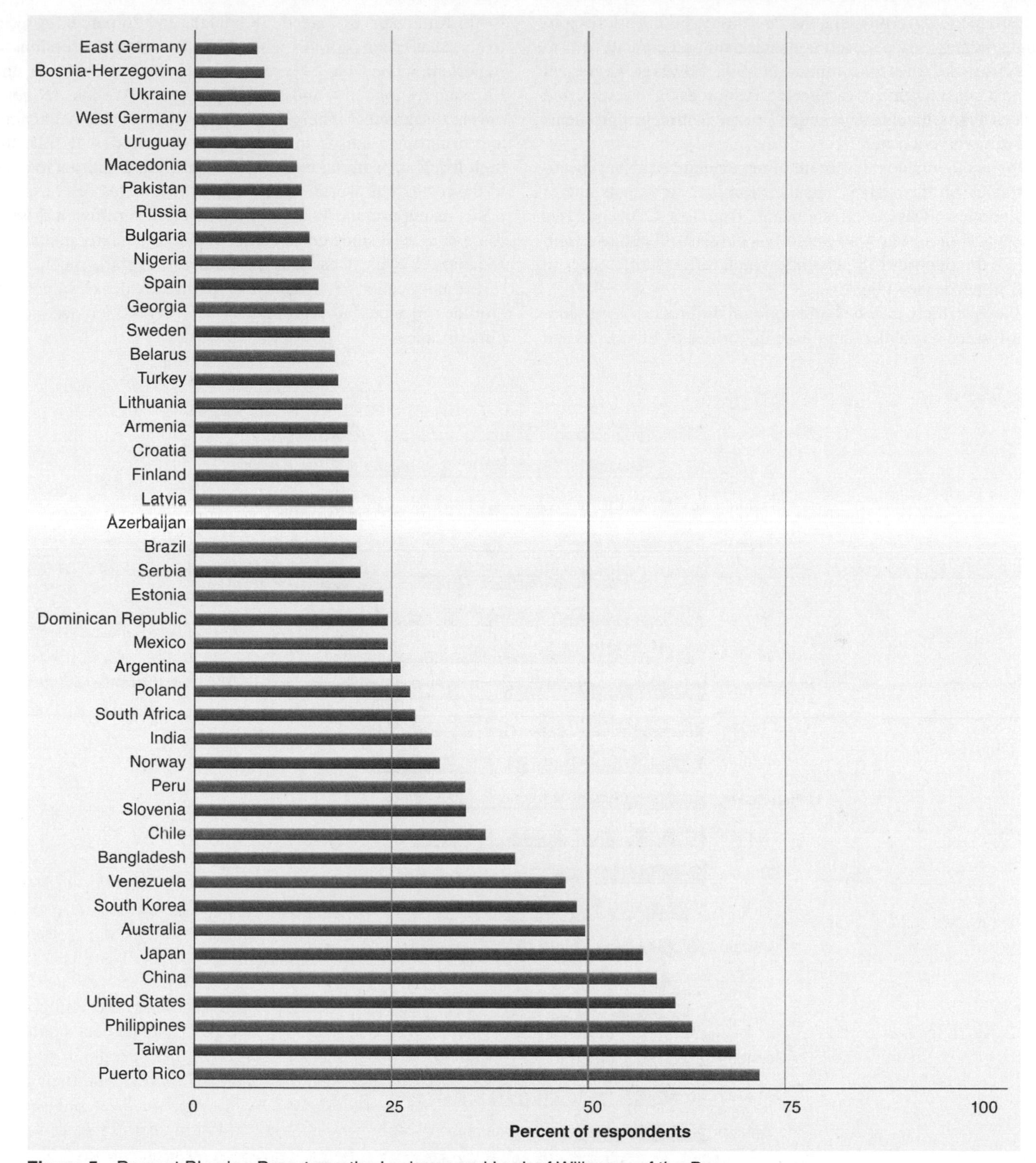

Figure 5 Percent Blaming Poverty on the Laziness and Lack of Willpower of the Poor.

Source: A. Leiserowitz, 2005. Data from R. Inglehart, et al., *World Values Surveys and European Values Surveys, 1981–1984, 1990–1993, and 1995–1997* [computer file], Inter-university Consortium for Political and Social Research (ICPSR) version (Ann Arbor, MI: Institute for Social Research [producer], 2000; Ann Arbor, MI: ICPSR [distributor], 2000).

only 21 percent thought this would be a bad thing.[54] Further, large majorities agreed that gaining more time for leisure activities or family life is their biggest goal in life.[55]

More broadly, in 2002 a global study sponsored by the Pew Research Center for the People & the Press found that 45 percent worldwide saw consumerism and commercialism as a threat to their own culture. Interestingly, more respondents from high-income and upper middle-income countries (approximately 51 percent) perceived consumerism as a threat than low-middle- and low-income countries (approximately

43 percent).[56] Unfortunately, the Pew study did not ask respondents whether they believed consumerism and commercialism were a threat to the environment. In 1992, however, 41 percent said that consumption of the world's resources by industrialized countries contributed "a great deal" to environmental problems in developing countries."[57]

On the other hand, 65 percent of respondents said that spending money on themselves and their families represents one of life's greatest pleasures. Respondents from low-GDP countries were much more likely to agree (74 percent) than those from high-GDP countries (58 percent), which reflects differences in material needs (see Figure 6).[58]

Likewise, there may be large regional differences in attitudes toward status consumerism. Large majorities of Europeans and North Americans disagreed (78 percent and 76 percent respectively) that other people's admiration for one's possessions is important, while 54 to 59 percent of Latin American, Asian, and Eurasian respondents, and only 19 percent of Africans (Nigeria only), disagreed.[59] There are strong cultural norms against appearing materialistic in many Western societies, despite the high levels of material consumption in these countries relative to the rest of the world. At the same time, status or conspicuous consumption has long been posited as a significant driving force in at least some consumer behavior, especially in affluent societies.[60] While these studies are a useful start, much more research is needed to unpack and explain the roles of values and attitudes in material consumption in different socioeconomic circumstances.

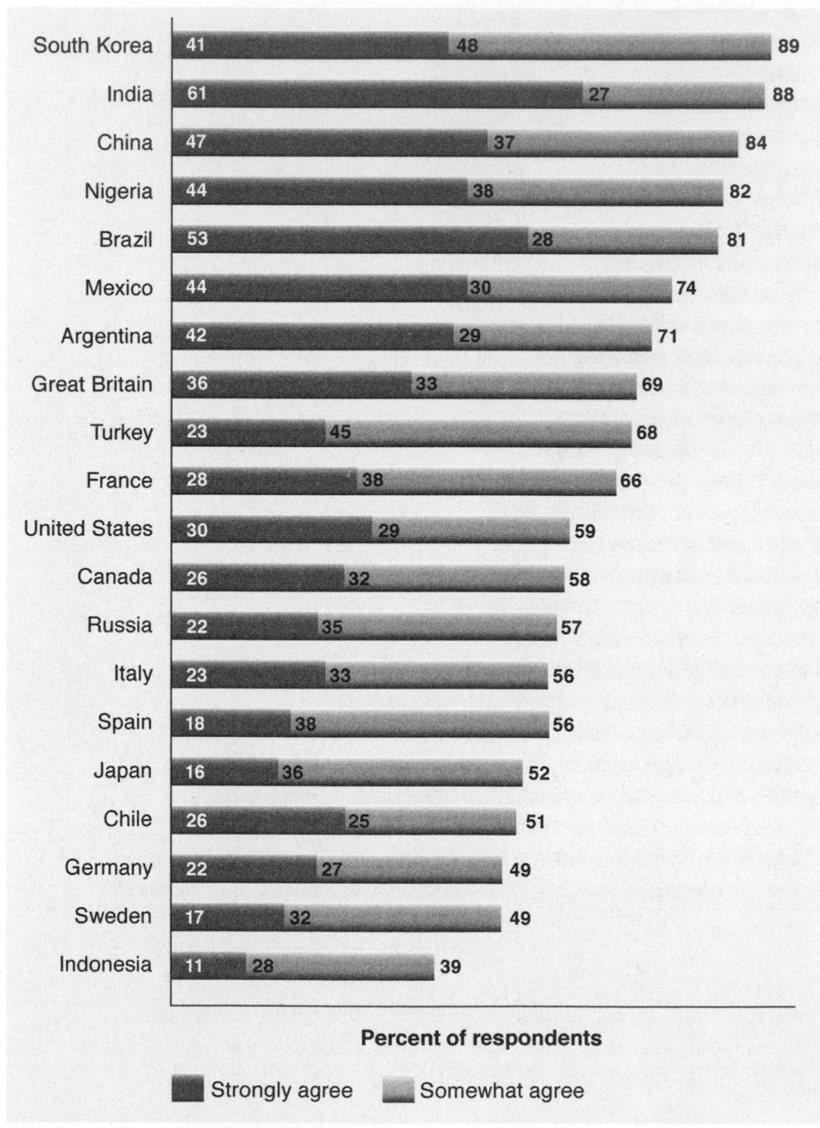

Figure 6 Purchasing for Self and Family Gives Greatest Pleasure ("Strongly" and "Somewhat" agree).

Note. The question was, "To spend money, to buy something new for myself or my family, is one of the greatest pleasures in my life."

Source: Environics International (GlobeScan), *Consumerism: A Special Report* (Toronto: Environics International, 2002). 6.

Science and Technology

Successful deployment of new and more efficient technologies is an important component of most sustainability strategies, even though it is often difficult to assess all the environmental, social, and public health consequences of these technologies in advance. Overall, the global public has very positive attitudes toward science and technology. The 1995 World Values Survey asked respondents, "In the long run, do you think the scientific advances we are making will help or harm mankind?" Worldwide, 56 percent of respondents thought science will help mankind, while 26 percent thought it will harm mankind. Further, 67 percent said an increased emphasis on technological development would be a good thing, while only 9 percent said it would be bad.[61] Likewise, in 2002, GlobeScan found large majorities worldwide believed that the benefits of modern technology outweigh the risks.[62] The support for technology, however, was significantly higher in countries with low GDPs (69 percent) than in high-GDP countries (56 percent), indicating more skepticism among people in technologically advanced societies. Further, this survey found dramatic differences in technological optimism between richer and poorer countries. Asked whether "new technologies will resolve most of our environmental challenges, requiring only minor changes in human thinking and individual behavior," 62 percent of respondents from low-GDP countries agreed, while 55 percent from high-GDP countries disagreed (see Figure 7).

But what about specific technologies with sustainability implications? Do these also enjoy strong public support? What follows is a summary of global-scale data on attitudes toward renewable energy, nuclear power, the agricultural use of chemical pesticides, and biotechnology.

Europeans strongly preferred several renewable energy technologies (solar, wind, and biomass) over all other energy sources, including solid fuels (such as coal and peat), oil, natural gas, nuclear fission, nuclear fusion, and hydroelectric power. Also, Europeans believed that by the year 2050, these energy sources will be best for the environment (67 percent), be the least expensive (40 percent), and will provide the greatest amount of useful energy (27 percent).[63] Further, 37 percent of Europeans and approximately 33 percent of respondents in 16 developed and developing countries were willing to pay 10 percent more for electricity derived from renewable energy sources.[64]

Nuclear power, however, remains highly stigmatized throughout much of the developed world.[65] Among respondents from 18 countries (mostly developed), 62 percent considered nuclear power stations "very dangerous" to "extremely dangerous" for the environment.[66] Whatever its merits or demerits as an alternative energy source, public attitudes about nuclear power continue to constrain its political feasibility.

Regarding the use of chemical pesticides on food crops, a majority of people in poorer countries believed that the benefits are greater than the risks (54 percent), while respondents in high-GDP countries were more suspicious, with only 32 percent believing the benefits outweigh the risks.[67] Since 1998, however, support for the use of agricultural chemicals has dropped worldwide. Further, chemical pesticides are now one of the top food-related concerns expressed by respondents around the world.[68]

Additionally, the use of biotechnology in agriculture remains controversial worldwide, and views on the issue are divided between rich and poor countries. Across the G7 countries, 70 percent of respondents were opposed to scientifically altered fruits and vegetables because of health and environmental concerns,[69] while 62 percent of Europeans and 45 percent of Americans opposed the use of biotechnology in agriculture.[70] While majorities in poorer countries (65 percent) believed the benefits of using biotechnology on food crops are greater than the risks, majorities in high-GDP countries (51 percent) believed the risks outweigh the benefits.[71]

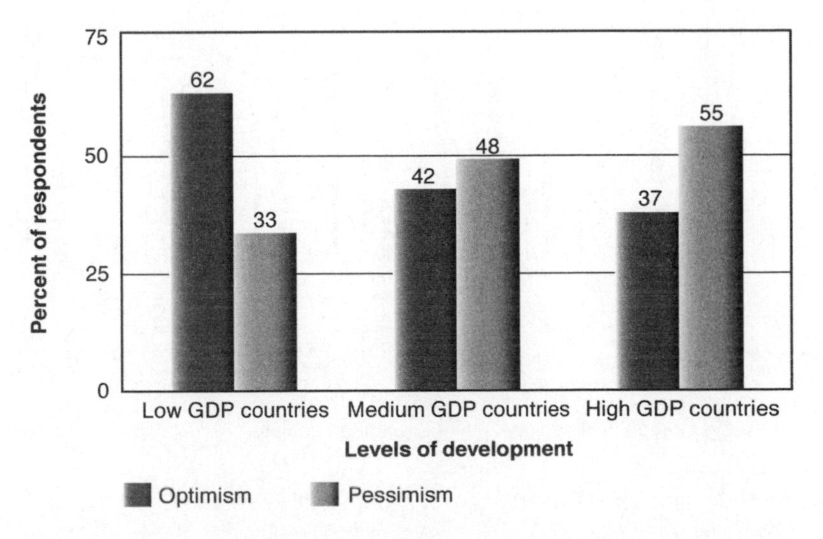

Figure 7 Technological Optimism Regarding Environmental Problems.

Source: A. Leiserowitz, 2005. Data from Environics International (GlobeScan), *International Environmental Monitor* (Toronto: Environics International, 2002), 135.

More broadly, public understanding of biotechnology is still limited, and slight variations in question wordings or framings can have significant impacts on support or opposition. For example, 56 percent worldwide thought that biotechnology will be good for society in the long term, yet 57 percent also agreed that "any attempt to modify the genes of plants or animals is ethically and morally wrong."[72] Particular applications of biotechnology also garnered widely different degrees of support. While 78 percent worldwide favored the use of biotechnology to develop new medicines, only 34 percent supported its use in the development of genetically modified food. Yet, when asked whether they supported the use of biotechnology to produce more nutritious crops, 61 percent agreed.[73]

Income Equity and Entitlements

Equity and entitlements strongly determine the degree to which rising population and affluence affect human development, particularly for the poor. For example, as global population and affluence have grown, income inequality between rich and poor countries has also increased over time, with the notable exceptions of East and Southeast Asia—where incomes are on the rise on a par with (or even faster than) the wealthier nations of the world.[74] Inequality within countries has also grown in many rich and poor countries. Similarly, access to entitlements—the bundle of income, natural resources, familial and social connections, and societal assistance that are key determinants of hunger and poverty[75]—has recently declined with the emergence of market-oriented economies in Eastern and Central Europe, Russia, and China; the rising costs of entitlement programs in the industrialized countries, including access to and quality of health care, education, housing, and employment; and structural adjustment programs in developing countries that were recommended by the International Monetary Fund. Critically, it appears there is no comparative data on global attitudes toward specific entitlements; however, there is much concern that living conditions for the elderly, unemployed, and the sick and injured are deteriorating, as cited above in the discussion on human development.

In 2002, large majorities said that the gap between rich and poor in their country had gotten wider over the previous 5 years. This was true across geographic regions and levels of economic development, with majorities ranging from 66 percent in Asia, 72 percent in North America, and 88 percent in Eastern Europe (excepting Ukraine) stating that the gap had gotten worse.[76] Nonetheless, 48 percent of respondents from 13 countries preferred a "competitive society, where wealth is distributed according to one's achievement," while 34 percent preferred an "egalitarian society, where the gap between rich and poor is small, regardless of achievement" (see Figure 8).[77]

More broadly, 47 percent of respondents from 72 countries preferred "larger income differences as incentives for individual effort," while 33 percent preferred that "incomes should be made more equal."[78] These results suggest that despite public perceptions of growing economic inequality, many accept it as an important incentive in a more individualistic and competitive economic system. These global results, however, are limited to just a few variables and gloss over many countries that strongly prefer more egalitarian distributions of wealth (such as India). Much more research is needed to understand how important the principles of income equality and equal economic opportunity are considered globally, either as global goals or as means to achieve other sustainability goals.

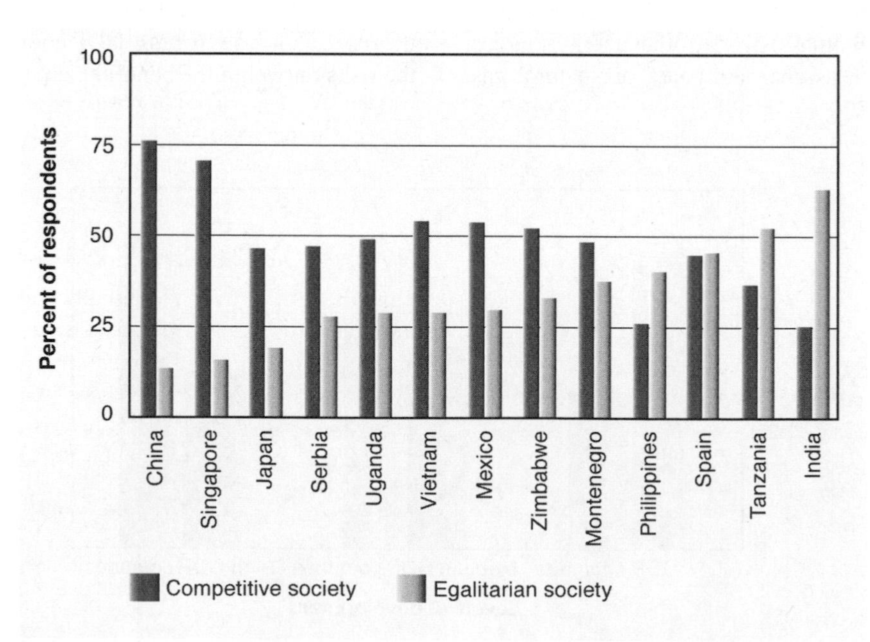

Figure 8 Multinational Preferences for a Competitive Versus Egalitarian Society.

Source: A. Leiserowitz, 2005. Data from World Values Survey. *The 1999–2002 Values Surveys Integrated Data File 1.0,* CD-ROM in R. Inglehart, M. Basanez, J. Diez-Medrano, L. Halman, and R. Luijkx, eds., *Human Beliefs and Values: A Cross-Cultural Sourcebook Based on the 1999–2002 Values Surveys*, first edition (Mexico City: Siglo XX[[]]I, 2004).

Does the Global Public Support Sustainable Development?

Surprisingly, the question of public support for sustainable development has never been asked directly, at least not globally. But two important themes emerge from the multinational data and analysis above. First, in general, the global public supports the main tenets of sustainable development. Second, however, there are many contradictions, including critical gaps between what people say and do—both as individuals and in aggregate. From these themes emerge a third finding: Diverse barriers stand between sustainability attitudes and action.

- *Large majorities worldwide appear to support environmental protection and economic and human development—the three pillars of sustainable development.* They express attitudes and have taken modest actions consonant with support for sustainable development, including support for environmental protection; economic growth; smaller populations; reduced poverty; improved technology; and care and concern for the poor, the marginal, the young, and the aged.
- *Amid the positive attitudes, however, are many contradictions.* Worldwide, all the components of the Human Development Index—life expectancy, adult literacy, and per capita income—have dramatically improved since World War II.[79] Despite the remarkable increases in human well-being, however, there appears to be a globally pervasive sense that human well-being has more recently been deteriorating. Meanwhile, levels of development assistance are consistently overestimated by lay publics, and the use of such aid is misunderstood, albeit strongly supported. Overall, there are very positive attitudes toward science and technology, but the most technologically sophisticated peoples are also the most pessimistic about the ability of technology to solve global problems. Likewise, attitudes toward biotechnology vary widely, depending on how the question is asked.
- Further, there are serious gaps between what people believe and what people do, both as individuals and as polities. Worldwide, the public strongly supports significantly larger levels of development assistance for poor countries, but national governments have yet to translate these attitudes into proportional action. Most people value the environment—for anthropocentric as well as ecocentric reasons—yet many ecological systems around the world continue to degrade, fragment, and lose resilience. Most favor smaller families, family planning, and contraception, but one-fifth to one-quarter of children born are not desired. Majorities are concerned with poverty and think more should be done to alleviate it, but important regions of the world think the poor themselves are to blame, and a majority worldwide accepts large gaps between rich and poor. Most people think that less emphasis on material possessions would be a good thing and that more time for leisure and family should be primary goals, but spending money often provides one of life's greatest pleasures. While many would pay more for fuel-efficient cars, fuel economy has either stagnated or even declined in many countries. Despite widespread public support for renewable energy, it still accounts for only a tiny proportion of global energy production.
- *There are diverse barriers standing between pro-sustainability attitudes and individual and collective behaviors.*[80] These include at least three types of barriers. First are the direction, strength, and priority of particular attitudes. Some sustainability attitudes may be widespread but not strongly or consistently enough relative to other, contradictory attitudes. A second type of barrier between attitudes and behavior relates to individual capabilities. Individuals often lack the time, money, access, literacy, knowledge, skills, power, or perceived efficacy to translate attitudes into action. Finally, a third type of barrier is structural and includes laws, regulations, perverse subsidies, infrastructure; available technology, social norms and expectations; and the broader social, economic, and political context (such as the price of oil, interest rates, special interest groups, and the election cycle).

Thus, each particular sustainability behavior may confront a unique set of barriers between attitudes and behaviors. Further, even the same behavior (such as contraceptive use) may confront different barriers across society, space, and scale—with different attitudes or individual and structural barriers operating in developed versus developing countries, in secular versus religious societies, or at different levels of decisionmaking (for example, individuals versus legislatures). Explaining unsustainable behavior is therefore "dauntingly complex, both in its variety and in the causal influences on it."[81] Yet bridging the gaps between what people believe and what people do will be an essential part of the transition to sustainability.

Promoting Sustainable Behavior

Our limited knowledge about global sustainability values, attitudes, and behaviors does suggest, however, that there are short and long-term strategies to promote sustainable behavior. We know that socially pervasive values and attitudes are often highly resistant to change. Thus, in the short term, leveraging the values and attitudes already dominant in particular cultures may be more practical than asking people to adopt new value orientations.[82] For example, economic values clearly influence and motivate many human behaviors, especially in the market and cash economies of the developed countries. Incorporating environmental and social "externalities" into prices or accounting for the monetary value of ecosystem services can thus encourage both individual and collective sustainable behavior.[83] Likewise, anthropocentric concerns about the impacts of environmental degradation and exploitative labor conditions on human health and social well-being remain strong motivators for action in both the developed and developing worlds.[84] Additionally, religious values are vital sources of meaning, motivation, and direction for much of the world, and many religions are actively re-evaluating and reinterpreting their traditions in support of sustainability.[85]

In the long term, however, more fundamental changes may be required, such as extending and accelerating the shift from materialist to post-materialist values, from anthropocentric to ecological worldviews, and a redefinition of "the good life."[86] These long term changes may be driven in part by impersonal forces, like changing economics (globalization) or technologies (for example, mass media and computer networks) or by broadly based social movements, like those that continue to challenge social attitudes about racism, environmental degradation, and human rights. Finally, sustainability science will play a critical role, at multiple scales and using multiple methodologies, as it works to identify and explain the key relationships between sustainability values, attitudes, and behaviors—and to apply this knowledge in support of sustainable development.

Notes

1. For example, see U.S. National Research Council, Policy Division, Board on Sustainable Development, *Our Common Journey: A Transition toward Sustainability* (Washington, DC: National Academy Press, 1999); and P. Raskin et al., *Great Transition: The Promise and Lure of the Times Ahead* (Boston: Stockholm Environment Institute, 2002).
2. R. W. Kates, T. M. Parris, and A. Leiserowitz, "What Is Sustainable Development? Goals, Indicators, Values, and Practice," *Environment,* April 2005, 8–21.
3. For simplicity, the words "global" and "worldwide" are used throughout this article to refer to survey results. Please note, however, that there has never been a truly representative global survey with either representative samples from every country in the world or in which all human beings worldwide had an equal probability of being selected. Additionally, some developing country results are taken from predominantly urban samples and are thus not fully representative.
4. For more detail about these surveys and the countries sampled, see Appendix A in A. Leiserowitz, R. W. Kates, and T. M. Parris, *Sustainability Values, Attitudes and Behaviors: A Review of Multi-national and Global Trends* (No. CID Working Paper No. 113) (Cambridge, MA: Science, Environment and Development Group, Center for International Development, Harvard University, 2004), http://www.cid.harvard.edu/cidwp/113.htm
5. Pew Research Center for the People & the Press, *Views of a Changing World* (Washington, DC: The Pew Research Center for the People & the Press, 2003), T72.
6. See R. Inglehart, "Globalization and Postmodern Values," *Washington Quarterly* 23, no. 1 (1999): 215–28.
7. Leiserowitz, Kates, and Parris, note 4 above, page 8.
8. Pew Research Center for the People & the Press, *The Pew Global Attitudes Project Dataset* (Washington, DC: The Pew Research Center for the People & the Press, 2004).
9. Gross national income (GNI) is "[t]he total market value of goods and services produced during a given period by labor and capital supplied by residents of a country, regardless of where the labor and capital are located. [GNI] differs from GDP primarily by including the capital income that residents earn from investments abroad and excluding the capital income that nonresidents earn from domestic investment." Official development assistance (ODA) is defined as "[t]hose flows to developing countries and multilateral institutions provided by official agencies, including state and local governments, or by their executive agencies, each transaction of which meets the following tests: (a) it is administered with the promotion of the economic development and welfare of developing countries as its main objective; and (b) it is concessional in character and conveys a grant element of at least 25 percent." UN Millennium Project, *The O. 7% Target: An In-Depth Look,* http://www.unmillenniumproject.org/involved/action07.htm (accessed 24 August 2005). Official development assistance (ODA) does not include aid flows from private voluntary organizations (such as churches, universities, or foundations). For example, it is estimated that in 2000, the United States provided more than $4 billion in private grants for development assistance, versus nearly $10 billion in ODA. U.S. Agency for International Development (USAID), *Foreign Aid in the National Interest* (Washington. DC, 2002), 134.
10. Organisation for Economic Co-operation and Development (OECD), *Official Development Assistance Increases Further—But 2006 Targets Still a Challenge* (Paris: OECD, 2005), http://www.oecd.org/document/3/0,2340,en_2649_34447_34700611_1_1_1_1,00.html (accessed 30 July 2005).
11. Environics International (GlobeScan), *The World Economic Forum Poll: Global Public Opinion on Globalization* (Toronto: Environics International, 2002), http://www.globescan.com/brochures/WEF_Poll_Brief.pdf (accessed 5 October 2004), 3. Note that Environics International changed its name to Globe Scan Incorporated in November 2003.
12. OECD, *Public Opinion and the Fight Against Poverty* (Paris: OECD Development Centre, 2003), 17.
13. Ibid, page 19.
14. Program on International Policy Attitudes (PIPA), *Americans on Foreign Aid and World Hunger: A Study of U.S. Public Attitudes* (Washington, DC: PIPA, 2001), http://www.pipa.org/OnlineReports/BFW (accessed 17 November 2004); and OECD, note 12 above, page 22.
15. See OECD Development Co-operation Directorate, *OECD-DAC Secretariat Simulation of DAC Members' Net ODA Volumes in 2006 and 2010,* http://www.oecd.org/dataoecd/57/30/35320618.pdf; and Central Intelligence Agency, The World Factbook, http://www.cia.gov/cia/publications/factbook/.
16. OECD, note 12 above, page 20.
17. I. Smillie and H. Helmich, eds., *Stakeholders: Government-NGO Partnerships for International Development* (London: Earthscan, 1999).

18. L. White Jr., "The Historical Roots of Our Ecologic Crisis," *Science,* 10 March 1967, 1203–07.
19. See C. Merchant, *The Death of Nature: Women, Ecology, and the Scientific Revolution* (1st ed.) (San Francisco: Harper & Row 1980); C. Merchant, *Radical Ecology: The Search for a Livable Worm* (New York: Routledge, 1992); and G. Sessions, *Deep Ecology for the Twenty-First Century* (1st ed.) (New York: Shambhala Press 1995).
20. World Values Survey, *The 1999–2002 Values Surveys Integrated Data File 1.0,* CD-ROM in R. Inglehart, M. Basanez, J. Diez-Medrano, L. Halman, and R. Luijkx, eds., *Human Beliefs and Values: A Cross-Cultural Sourcebook Based on the 1999–2002 Values Surveys,* first edition (Mexico City: Siglo XXI, 2004).
21. These results come from a representative national survey of American Climate change risk perceptions, policy preferences, and behaviors and broader environmental and cultural values. From November 2002 to February 2003, 673 adults (18 and older) completed a mail-out, mail-back questionnaire, for a response rate of 55 percent. The results are weighted to bring them in line with actual population proportions. See A. Leiserowitz, "American Risk Perceptions: Is Climate Change Dangerous?" *Risk Analysis,* in press; and A. Leiserowitz, "Climate Change Risk Perception and Policy Preferences: The Role of Affect, Imagery, and Values," *Climatic Change,* in press.
22. These results support the argument that concerns about the envi-ronment are not "a luxury affordable only by those who have enough economic security to pursue quality-of-life goals." See R. E. Dunlap, G. H. Gallup Jr., and A. M. Gallup, "Of Global Concern: Results of the Health of the Planet Survey," *Environment,* November 1993, 7–15, 33–39 (quote at 37); R. E. Dunlap, A. G. Mertig, "Global Concern for the Environment: Is Affluence a Prerequisite?" *Journal of Social Issues* 511, no. 4 (1995): 121–37; S. R. Brechin and W. Kempton, "Global Environmentalism: A Challenge to the Postmaterialism Thesis?" *Social Science Quarterly* 75, no. 2 (1994): 245–69.
23. Environics International (GlobeScan), *Environics International Environmental Monitor Survey Data-set* (Kingston, Canada: Environics International, 2000), http://jeff-lab.queensu.ca/poadata/info/iem/iemlist.shtml (accessed 5 October 2004). These multinational levels of concern and perceived seriousness of environmental problems remained roughly equivalent from 1992 to 2000, averaged across the countries sampled by the 1992 Health of the Planet and the Environics surveys, although some countries saw significant increases in perceived seriousness of environmental problems (India, the Netherlands, the Philippines, and South Korea), while others saw significant decreases (Turkey and Uruguay). See R. E. Dunlap, G. H. Gallup Jr., and A. M. Gallup, *Health of the Planet: Results of a 1992 International Environmental Opinion Survey of Citizens in 24 Nations* (Princeton, N J: The George H. Gallup International Institute, 1993); and R. E. Dunlap, G. H. Gallup Jr., and A. M. Gallup, "Of Global Concern: Results of the Health of the Planet Survey," *Environment,* November 1993, 7–15, 33–39.
24. GlobeScan, *Results of First-Ever Global Poll on Humanity's Relationship with Nature* (Toronto: GlobeScan Incorporated, 2004), http://www.globescan.com/news_archives/IUCN_PR.html (accessed 30 July 2005).
25. World Values Survey, note 20 above; and Pew Research Center for the People & the Press, *What the World Thinks in 2002* (Washington, DC: The Pew Research Center for the People & the Press, 2002), T-9. The G7 includes Canada, France, Germany, Great Britain, Italy, Japan and the United States. It expanded to the G8 with the addition of Russia in 1998.
26. R. Inglehart, et al., *World Values Surveys and European Values Surveys, 1981–1984, 1990–1993, and 1995–1997* [computer file], Inter-university Consortium for Political and Social Research (ICPSR) version (Ann Arbor, MI: Institute for Social Research [producer], 2000; Ann Arbor, MI: ICPSR [distributor], 2000).
27. Environics International (GlobeScan), note 23 above.
28. Dunlap, Gallup Jr., and Gallup, *Health of the Planet: Results of a 1992 International Environmental Opinion Survey of Citizens in 24 Nations,* note 23 above.
29. Environics International (GlobeScan), *New Poll Shows G8 Citizens Want Legally-Binding Climate Accord* (Toronto: Environics International, 2001), http://www.globescan.com/news_archives/IEM_climatechange.pdf (accessed 30 July 2005).
30. Environics International (GlobeScan), *International Environmental Monitor* (Toronto: Environics International, 2002), 44.
31. Ibid., page 49.
32. The European Opinion Research Group, *Eurobarometer: Energy: Issues, Options and Technologies, Science and Society,* EUR 20624 (Brussels: European Commission, 2002), 96–99.
33. Inglehart, note 26 above.
34. C. M. Rogerson, "The Waste Sector and Informal Entrepreneurship in Developing World Cities," *Urban Forum* 12, no. 2 (2001): 247–59.
35. Environics International (GlobeScan), note 30 above, page 63. These results are based on the sub-sample of those who own or have regular use of a car.
36. Environics International (GlobeScan), note 30 above, page 65.
37. Inglehart, note 26 above.
38. Environics International (GlobeScan), note 23 above.
39. P. A. Ehrlich and J. P. Holdren, review of *The Closing Circle,* by Barry Commoner, *Environment,* April 1972, 24, 26–39.
40. Y. Kaya, "Impact of Carbon Dioxide Emission Control on GNP Growth: Interpretation of Proposed Scenarios," paper presented at the Intergovernmental Panel on Climate Change (IPCC) Energy and Industry Subgroup, Response Strategies Working Group, Paris, France, 1990; and R. York, E. Rosa, and T. Dietz, "STIRPAT, IPAT and ImPACT: Analytic Tools for Unpacking the Driving Forces of Environmental Impacts," *Ecological Economics* 46, no. 3 (2003): 351.

41. IPCC, *Emissions Scenarios* (Cambridge: Cambridge University Press, 2000); and E. F. Lambin, et al., "The Causes of Land-Use and Land-Cover Change: Moving Beyond the Myths," *Global Environmental Change: Human and Policy Dimensions* 11, no. 4 (2001):
42. T. M. Parris and R. W. Kates, "Characterizing a Sustainability Transition: Goals, Targets, Trends, and Driving Forces," *Proceedings of the National Academy of Sciences of the United States of* America 100, no. 14 (2003): 6.
43. Pew Research Center for the People & the Press, note 8 above, page T17.
44. United Nations, *Majority of World's Couples Are Using Contraception* (New York: United Nations Population Division, 2001).
45. J. Bongaarts, "Trends in Unwanted Childbearing in the Developing World," *Studies in Family Planning* 28, no. 4 (1997): 267–77.
46. Demographic and Health Surveys (DHS), *STATCompiler* (Calverton, MD: Measure DHS, 2004), http://www.measuredhs.com/ (accessed 5 October, 2004).
47. Ibid.
48. U.S. Bureau of the Census, *World Population Profile: 1998,* WP/98 (Washington, DC, 1999), 45.
49. World Bank, *World Development Indicators CD-ROM 2004* [computer file] (Washington, DC: International Bank for Reconstruction and Development (IBRD) [producer], 2004).
50. World Bank, *Global Economic Prospects 2005: Trade, Regionalism, and Development* [computer file] (Washington, DC: IBRD [producer] 2005).
51. For more information on poverty reduction strategies, see T. Banuri, review of *Investing in Development: A Practical Plan to Achieve the Millennium Goals,* by UN Millennium Project, *Environment,* November 2005 (this issue), 37.
52. Inglehart, note 26 above.
53. Inglehart, note 26 above.
54. Inglehart, note 26 above.
55. Environics International (GlobeScan), *Consumerism: A Special Report* (Toronto: Environics International, 2002), 6.
56. Pew Research Center for the People & the Press, note 25 above.
57. Dunlap, Gallup Jr., and Gallup, *Health of the Planet: Results of a 1992 International Environmental Opinion Survey of Citizens in 24 Nations,* note 23 above, page 57.
58. Environics International (GlobeScan), note 55 above, pages 3–4.
59. Environics International (GlobeScan), note 55 above, pages 3–4.
60. T. Veblen, *The Theory of the Leisure Class: An Economic Study of Institutions* (New York: Macmillan 1899).
61. Inglehart, note 26 above.
62. Environics International (GlobeScan), note 30 above, page 133.
63. The European Opinion Research Group, note 32 above, page 70.
64. Environics International (GlobeScan), note 23 above.
65. For example, see J. Flynn, P. Slovic, and H. Kunreuther, *Risk, Media and Stigma: Understanding Public Challenges to Modern Science and Technology* (London: Earthscan, 2001).
66. International Social Science Program, *Environment II,* (No. 3440) (Cologne: Zentralarchiv für Empirische Sozialforschung, Universitaet zu Koeln (Central Archive for Empirical Social Research, University of Cologne), 2000), 114.
67. Environics International (GlobeScan), note 30 above, page 139.
68. Environics International (GlobeScan), note 30 above, page 141.
69. Pew Research Center for the People & the Press, note 25 above, page T20.
70. Chicago Council on Foreign Relations (CCFR), *Worldviews 2002* (Chicago: CCFR, 2002), 26.
71. Environics International (GlobeScan), note 30 above, page 163.
72. Environics International (GlobeScan), note 30 above, page 156–57.
73. Environics International (GlobeScan), note 30 above, page 57.
74. W. J. Baumol, R. R. Nelson, and E. N. Wolff, *Convergence of Productivity: Cross-National Studies and Historical Evidence* (New York: Oxford University Press, 1994).
75. A. K. Sen, *Poverty and Famines: An Essay on Entitlement and Deprivation* (Oxford: Oxford University Press, 1981).
76. Pew Research Center for the People & the Press, note 5 above, page 37.
77. World Values Survey, note 20 above.
78. World Values Survey, note 20 above.
79. The human development index (HDI) measures a country's average achievements in three basic aspects of human development: longevity, knowledge, and a decent standard of living. Longevity is measured by life expectancy at birth; knowledge is measured with the adult literacy rate and the combined primary, secondary, and tertiary gross enrollment ratio; and standard of living is measured by gross domestic product per capita (purchase-power parity US$). The UN Development Programme (UNDP) has used the HDI for its annual reports since 1993. UNDP, *Questions About the Human Development Index (HDI),* http://www.undp.org/hdr2003/faq.html#21 (accessed 25 August 2005).
80. See, for example, J. Blake, "Overcoming the 'Value-Action Gap' in Environmental Policy: Tensions Between National Policy and Local Experience," *Local Environment* 4, no. 3 (1999): 257–78; A. Kollmuss and J. Agyeman, "Mind the Gap: Why Do People Act Environmentally and What Are the Barriers to Pro-EnvironmentalBehavior?" *Environmental Education Research* 8, no. 3 (2002): 239–60; and E C. Stem, "Toward a Coherent Theory of Environmentally Significant Behavior,"*Journal of Social Issues* 56, no. 3 (2000): 407–24.
81. Stern, ibid., page 421.
82. See, for example, P. W. Schultz and L. Zelezny, "Reframing Environmental Messages to Be Congruent with American Values," *Human Ecology Review* 10, no. 2 (2003): 126–36.

83. Millennium Ecosystem Assessment, *Ecosystems and Human Well-Being: Synthesis* (Washington, DC: Island Press, 2005).

84. Dunlap, Gallup Jr., and Gallup, *Health of the Planet: Results of a 1992 International Environmental Opinion Survey of Citizens in 24 Nations,* note 23 above, page 36.

85. See *The Harvard Forum on Religion and Ecology,* http://environment.harvard.edu/religion/main.html; R. S. Gottlieb, *This Sacred Earth: Religion, Nature, Environment* (New York: Routledge, 1996); and G. Gardner, *Worldwatch Paper # 164: Invoking the Spirit: Religion and Spirituality in the Quest for a Sustainable World* (Washington, DC: Worldwatch Institute, 2002).

86. R. Inglehart, *Modernization and Postmodernization: Cultural, Economic and Political Change in 43 Societies* (Princeton: Princeton University Press, 1997); T. O'Riordan, "Frameworks for Choice: Core Beliefs and the Environment," *Environment,* October 1995, 4–9, 25–29; and E Raskin and Global Scenario Group, *Great Transition: The Promise and Lure of the Times Ahead* (Boston: Stockholm Environment Institute, 2002).

Anthony A. Leiserowitz is a research scientist at Decision Research and an adjunct professor of environmental studies at the University of Oregon, Eugene. He is also a principal investigator at the Center for Research on Environmental Decisions at Columbia University. Leiserowitz may be reached at (541) 485-2400 or by email at ecotone@uoregon.edu. **Robert W. Kates** is an independent scholar based in Trenton, Maine, and a professor emeritus at Brown University, where he served as director of the Feinstein World Hunger Program. He is also a former vice-chair of the Board of Sustainable Development of the U.S National Academy's National Research Council. In 1991, Kates was awarded the National Medal of Science for his work on hunger, environment, and natural hazards. He is an executive editor of *Environment* and may be contacted at rkates@acadia.net. **Thomas M. Parris** is a research scientist at and director of the New England office of ISCIENCES, LLC. He is a contributing editor of *Environment.* Parris may be reached at parris@isciences.com.

Paying for Climate Change

Governments must manage the incentives for households and firms to counter and adapt to climate change.

BENJAMIN JONES, MICHAEL KEEN, AND JON STRAND

Climate science tells that the earth is warming as a result of human activities. But considerable uncertainty regarding the precise nature and extent of the risks remains. Economists are needed to develop sensible policies to address these risks, which account for the uncertainties. In particular, the world needs public finance economists to consider what role fiscal instruments—notably, taxing and public spending—have to play in dealing with climate change.

Country efforts to adapt to and mitigate climate change are interrelated—broadly speaking, they are substitutes—but differ in important respects. Most adaptation, often involving relatively modest changes in behavior, will be carried out through private markets, though policy interventions may be needed to facilitate it—for example, by improving weather forecasting.

Mitigation, by contrast, generally needs to be driven by deliberate policy to a greater extent. Much adaptation can, and should, wait until the climate process has evolved: it makes little sense to adapt now to changes that will materialize mainly in, say, 30–100 years. However, mitigation needs to start well in advance of the damage it seeks to avoid because damage arises not from current emissions but from the slow-moving stock of greenhouse gases (GHGs) cumulated in the atmosphere.

This article argues that the role of fiscal instruments is central—indeed indispensable—for both mitigating and adapting to climate change. It looks at how efficient fiscal policies can help minimize the negative effects of climate change and examines the policy options available to governments. Fiscal instruments cannot provide a complete solution. But taxes and public spending are key to getting the incentives right for households and firms, as well as to ensuring a fair distribution of the associated costs and benefits. They can help ensure that those whose GHG emissions affect climate developments pay a proper price for doing so, and they can provide the resources needed to pay for dealing with it.

Adaptation—How Much Could It Cost?

Even with unchanged fiscal policies, climate change may have effects on both tax revenue (tax bases being eroded, perhaps, by declining agricultural productivity or by intensified extreme weather events, such as storms, flooding, and droughts) and public spending (perhaps to deal with increased prevalence of malaria). In some cases, the net effect might be beneficial, though the overall tendency is likely to amplify the problems faced by those countries—often among the poorest—most adversely affected in general by climate change.

The most likely negative effects of future climate change include sea-level rise, productivity losses in climate-exposed sectors such as tourism and agriculture, and more intense and perhaps more frequent and extreme weather events—all with potential adverse repercussions for fiscal positions and external stability.

Outside such catastrophic events as melting of the West Antarctic Ice Sheet, human societies are likely to adapt to most of these changes, although at a cost. How to minimize those costs, and how governments can best help, is not always clear. Typically, it will not be optimal to adapt so fully as to eliminate the entire climate effect: averting all damage may simply be too expensive. And difficult choices arise between taking early precautions and waiting for better information to become available. For example, whereas sinking costs into strengthening coastal defenses will seem a wise decision if future storm surge problems worsen, it will look like a white elephant if they do not.

Very little is known about the aggregate extent of the costs of adaptation, but there are some rough estimates. One survey concludes that these costs typically make up at most 25 percent of total climate impact costs (Tol, 2005). So if doubling GHG concentrations (a prospect under "business-as-usual" assumptions in this century) leads to an overall climate cost of 1–2 percent of world GDP, adaptation costs would be about 0.2–0.5 percent of world GDP, or about $70–150 billion a year. The World Bank

(2006) also estimates adaptation costs for lower-income countries in the tens of billions of dollars annually.

Given the importance for adaptation of such public goods as coastal defenses and health provision, a substantial proportion of these costs can be expected to fall on the public sector, but how much that is likely to be is even less clear: the World Bank, for example, roughly estimated that about a third of adaptations costs could be public. However, a better understanding of the likely fiscal costs of adapting to climate change, at the country level, is urgently needed if the fiscal risks that it poses are to be properly prepared for.

Mitigation—Dealing with Market Failures

Effective mitigation of GHG emissions is likely to require the use of fiscal instruments to overcome a deep market failure—a classic free-rider problem. The problem is simply that individuals, firms, and governments have insufficient incentives to limit their GHG emissions: whereas they incur the full costs of doing so, the benefits (from less global warming) accrue to the entire global community. The consequences are excessive emissions and too little effort in developing alternatives to fossil fuels.

At the local or national level, there may be some co-benefits from reduced burning of fossil fuels in the form of less local and regional pollution, but these do not eliminate the basic difficulty: everyone would prefer that others take the pain of reducing global emissions. Moreover, the benefits of current mitigation will accrue largely to future generations—so the extent to which the current costs are worth incurring depends on the weight one attaches to the well-being of future generations, and how much allowance to make for the likelihood that they will be better off than we are. The discount rate used to compare current costs and future benefits then proves critical in evaluating and forming climate-related policies—more so than in most other cost-benefit analyses because of the unusually long time horizons involved.

The second market failure relates to the development of new energy technologies that will permit substantial reductions in GHG emissions. Most such research and development (R&D) activity will—and, from efficiency considerations, probably should—be undertaken by individuals and businesses in pursuit of commercial gain. But they will typically not be able to appropriate all the social benefit of their innovations, so there is a risk of underinvestment in climate- related R&D.

The same considerations also apply, of course, to much R&D that has nothing to do with climate change, and many governments already offer generous tax breaks and other forms of fiscal support for commercial R&D. However, the importance now attached to climate- and energy-related research, including energy security considerations, and the particularly high risks for individual developers (in particular, related to developing more fundamental, "breakthrough" technologies—such as the capture and storage of CO_2 emissions and geo-engineering techniques for offsetting climate modifications), may argue for additional support of climate-related research. In practice, however, energy-related R&D remains well below its peak in the 1970s.

Pricing Emissions

Another complex issue is how to price emissions. In principle, optimal policy to reduce GHG emissions is simple: every emitter should be charged a price per unit of emissions, equal to the (net present value of) damage it causes (this in addition to the price paid for the coal or other underlying resource). That is, to ensure that the cost of reducing emissions is minimized, the charge should be the same for all emissions, wherever and however they arise. The use of fossil fuels, for example, should be charged at a rate—a "carbon price"—that reflects the carbon content of each and, hence, the CO_2 that they emit when burned.

Though the principle is simple, its application is complex. Deciding the "correct" value of marginal damage from emissions—we shall speak of the carbon price, although the same principles apply to all GHGs—requires taking a view on matters ranging from the highly speculative (such as the likely pace and nature of technical progress) to the philosophical (in the choice of discount rate).

And it is not just (or even mainly) today's carbon price that matters. Investments decisions made today in risky R&D, or in developing power stations that will last decades, require some view on future fossil fuel prices, including carbon prices.

The carbon price is likely to increase over time in real terms, at least for the foreseeable future: as the time of most intense damage comes nearer, the carbon price rises in present value and, hence, so too does the charge. It may not be wise for the carbon price to increase too fast, however, because that could create an incentive for owners of fossil fuels to extract more rapidly now, when the charge is low, making future problems worse (Sinn, 2007). Although the appropriate rate of increase remains an open question, a key challenge for policymakers, which they are far from solving, is to find ways of making credible the expectation of reasonably rising carbon prices.

Figure 1 illustrates some of the ambiguities and uncertainties related to the correct emissions price path and is based on simulations under the U.S. Climate Change Science Program. It uses two integrated assessment models applied in that work: the IGSM model developed by the Massachusetts Institute of Technology, and the MiniCAM model, developed by teams at the Pacific Northwest National Laboratories and University of Maryland.

The assessment of future emissions prices varies widely—both for a given year and by time frame, across models and long-run GHG concentration targets. (In 2040, for example, the price will range from $13 per ton of carbon (tC) for MiniCAM given a long-run target of 650 parts per million (ppm) for atmospheric carbon to $562/tC for IGSM under a 450 ppm target.)

Differences between models represent uncertainties about such factors as mitigation costs and baseline energy use; the "correct" emissions target is also uncertain. Some types of uncertainty are not modeled: assumptions about discounting, for example, are the same in all these calculations (4 percent a year). For comparison, most assessments of the current "correct" emissions price are in the range $15–$60/tC (with the value proposed by the Stern Review at about $330/tC, something of an outlier).

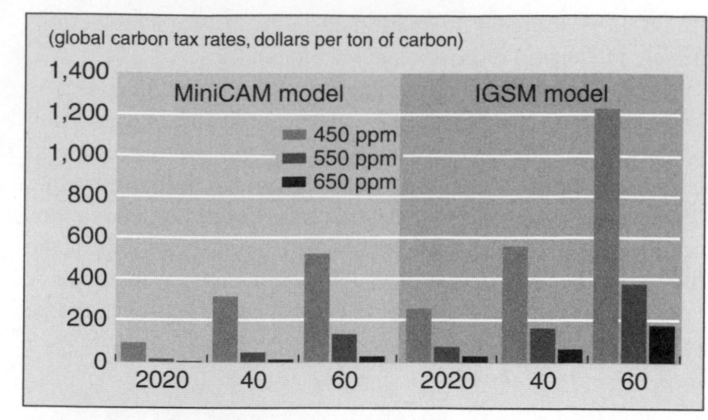

Figure 1 Different rates. The IGSM model, which assumes a higher baseline growth in emissions than the MiniCAM model, requires higher tax rates to achieve targeted emissions levels.

Source: IMF staff calculations using MiniCAM output.

Carbon Taxes, Cap-and-Trade, and All That

Further issues arise in implementing carbon prices. There are two archetypal market-based methods: carbon taxation and cap-and-trade schemes (under which rights to emit are issued—either sold or given away—up to some fixed amount and then bought by those who find abating relatively hard from those who find it relatively easy). Most schemes proposed in practice are hybrids: they may involve, for instance, permit trade but with the government ready to issue enough permits to keep the price above some floor. But these two polar forms illustrate many of the key choices to be made.

In the simplest case, no choice need be made. If all emission quotas under a cap-and-trade arrangement are auctioned to the highest bidders, and with full certainty about emissions (and the emissions price), the two mechanisms are equivalent: replacing a cap-and-trade scheme with a carbon tax at a rate equal to the market-clearing permit price, emissions, and government revenue will be exactly the same.

But in the presence of uncertainty, the equivalence breaks down. Cap-and-trade provides certainty on aggregate emissions; carbon taxes provide relative certainty on prices. In the face of uncertainty as to how costly reducing emissions will be, taxes may have some advantage as a mitigation device because they better match the marginal costs and benefits of mitigation.

Suppose, for example, that abatement proves much more costly than expected. Under cap-and-trade, emissions would be unaffected, but the necessary abatement would be very costly. Under a carbon tax, those costs would be avoided, but emissions would be higher than desired. Such a surge in emissions may be of relatively little concern, however, because emissions over any short period matter little to atmospheric concentrations, which are what really matter.

The equivalence will also fail if—as has often happened in practice—emissions rights under cap-and-trade are not auctioned, but given away. For example, under the current phase of the European Union Emissions Trading Scheme (EU-ETS), set up to help implement the EU's Kyoto Protocol commitments, no more than 10 percent of emissions quotas may be auctioned. This leads to an implicit revenue loss of about €40 billion a year and to a substantial and opaque measure of redistribution.

Such "grandfathering" of emissions rights can have other adverse effects too. Firms may expect future allocations to depend on current emissions, thereby blunting their incentive to abate now. Entry and exit rules also matter. If exiting firms lose their rights, for instance (rather than being able to sell them), they may be less likely to exit, making abatement more difficult. Grandfathering may have been reasonable for investments sunk before carbon pricing was even imaginable. But that is no longer the case. And, indeed, the European Commission proposes to eliminate grandfathering during the third phase of the EU-ETS, from 2013 to 2020—a firm step in the right direction, and an example for others.

What to Do with the Revenue?

How much money optimally imposed emissions taxes will raise for governments is an important fiscal issue. Figure 2 shows projected revenues from charges on carbon emissions in percent of world GDP by 2020, 2040, and 2060, with projected tax rates and emissions calculated by the integrated assessment models used in that exercise. We see that these numbers range from totally insignificant (0.1 percent of income under MiniCAM in 2020 with a 650 ppm target), to substantial (more than 3 percent of income under IGSM in 2060 with a 450 ppm target). While regional distributions are not given here, the share of total emissions for lower-income countries is projected to increase gradually (more so under the MiniCAM model), implying that these countries also will collect a greater share of overall tax revenue (exceeding 65 percent for non-OECD countries by 2060 under MiniCAM).

When equivalence of the kind described above holds, the same total revenue could also be achieved under a cap-and-trade arrangement with full auctioning of emissions rights. But the revenue distribution across countries could be quite different.

The widespread presumption under carbon taxation is that revenues would accrue to the country in which the carbon is used (although this would not prevent subsequent international transfers). Under cap-and-trade, however, some rule must be adopted for allocating the total emissions rights across countries. And how that is done—in proportion to emissions under business as usual, for instance, or in proportion to population—can have powerful implications for the direction and extent of international trade in permits.

Different exercises give somewhat different results but tend to agree that Africa and India would likely be sellers of permits

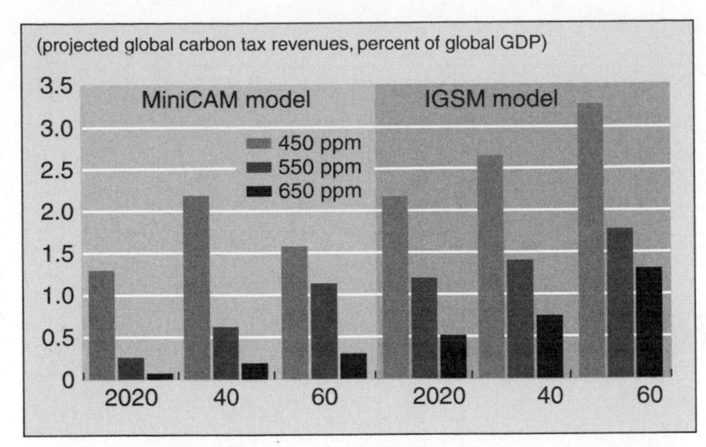

Figure 2 Money maker. The IGSM model produces higher revenues than the MiniCAM model, mainly due to higher tax rates on emissions.

Source: IMF staff calculations using MiniCAM output.

(forcing them with an incentive to participate in the scheme), whereas the industrial countries would be buyers. Such schemes, on top of having deterrent effects on emissions, would imply an effective resource transfer from high-income to lower-income countries. Clearly, the implementation of such transfers would raise difficulties: it would need, in particular, some agreed system by which each country can be assured that others are indeed emitting no more than allowed by the permits they hold.

For cash-strapped governments, the potential revenue from carbon pricing would seem to provide at least one benefit from climate change. And, indeed, it would enable them to make less use of more distortionary taxes and deal more confidently with the potential revenue challenges arising from trade liberalization and globalization. (Some, of course, will worry that they will instead simply waste this additional revenue.)

But carbon pricing may well worsen the distortions caused by the existing tax systems, tending, broadly, to reduce the level of economic activity and so exacerbate marginal disincentives caused by the tax system. So it is generally a good use of the revenue from carbon pricing to shift away from more distorting tax instruments. Exactly what those other instruments are may vary across countries. Several European countries, for example, have sought to alleviate labor market problems by using increased energy taxes to reduce social contributions. Others might see the corporate tax as a prime candidate for reduction.

Participation and Fairness

Not the least of the potential roles for fiscal design is to encourage wide participation in mitigation—to limit emissions at least possible cost—and, a related challenge, to help spread the burden of climate change in ways that are perceived to be fair. This means, for example, using other instruments to soften the distributional impact of carbon pricing within countries (which can be particularly difficult when it comes to raising unduly low energy prices in lowincome countries) and addressing such controversial issues as the potential use of border tax adjustments if neighboring countries do not have similar carbon tax rates.

Thus, it is increasingly clear that fiscal design issues will be central to any effective response to the difficulties posed by climate change.

References

Sinn, Hans-Werner, 2007, "Public Policies Against Global Warming," CESifo Working Paper No. 2087 (www.csfinfo.de).

Stern, Nicholas, and others, 2007, *The Economics of Climate Change* ("The Stern Review") (Cambridge: Cambridge University Press).

Tol, Richard S. J., 2005, "Adaptation and Mitigation: Trade-Offs in Substance and Methods," *Environmental Science and Policy,* Vol. 8, pp. 395–418.

Weitzman, Martin L., 1974, "Prices Versus Quantities," *Review of Economic Studies,* Vol. 41, pp. 477–91.

World Bank, 2006, *Clean Energy and Development: Towards an Investment Framework* (Washington).

Michael Keen is an Advisor, **Benjamin Jones** is an Economist, and **Jon Strand** is a Technical Assistance Advisor in the IMF's Fiscal Affairs Department.

Article 7

High-Tech Trash

Will Your Discarded TV or Computer End up in a Ditch in Ghana?

CHRIS CARROLL

June is the wet season in Ghana, but here in Accra, the capital, the morning rain has ceased. As the sun heats the humid air, pillars of black smoke begin to rise above the vast Agbogbloshie Market. I follow one plume toward its source, past lettuce and plantain vendors, past stalls of used tires, and through a clanging scrap market where hunched men bash on old alternators and engine blocks. Soon the muddy track is flanked by piles of old TVs, gutted computer cases, and smashed monitors heaped ten feet high. Beyond lies a field of fine ash speckled with glints of amber and green—the sharp broken bits of circuit boards. I can see now that the smoke issues not from one fire, but from many small blazes. Dozens of indistinct figures move among the acrid haze, some stirring flames with sticks, others carrying armfuls of brightly colored computer wire. Most are children.

Choking, I pull my shirt over my nose and approach a boy of about 15, his thin frame wreathed in smoke. Karim says he has been tending such fires for two years. He pokes at one meditatively, and then his top half disappears as he bends into the billowing soot. He hoists a tangle of copper wire off the old tire he's using for fuel and douses the hissing mass in a puddle. With the flame retardant insulation burned away—a process that has released a bouquet of carcinogens and other toxics—the wire may fetch a dollar from a scrap-metal buyer.

Another day in the market, on a similar ash heap above an inlet that flushes to the Atlantic after a downpour, Israel Mensah, an incongruously stylish young man of about 20, adjusts his designer glasses and explains how he makes his living. Each day scrap sellers bring loads of old electronics—from where he doesn't know. Mensah and his partners—friends and family, including two shoeless boys raptly listening to us talk—buy a few computers or TVs. They break copper yokes off picture tubes, littering the ground with shards containing lead, a neurotoxin, and cadmium, a carcinogen that damages lungs and kidneys. They strip resalable parts such as drives and memory chips. Then they rip out wiring and burn the plastic. He sells copper stripped from one scrap load to buy another. The key to making money is speed, not safety. "The gas goes to your nose and you feel something in your head," Mensah says, knocking his fist against the back of his skull for effect. "Then you get sick in your head and your chest." Nearby, hulls of broken monitors float in the lagoon. Tomorrow the rain will wash them into the ocean.

People have always been proficient at making trash. Future archaeologists will note that at the tail end of the 20th century, a new, noxious kind of clutter exploded across the landscape: the digital detritus that has come to be called e-waste.

More than 40 years ago, Gordon Moore, co-founder of the computer-chip maker Intel, observed that computer processing power roughly doubles every two years. An unstated corollary to "Moore's law" is that at any given time, all the machines considered state-of-the-art are simultaneously on the verge of obsolescence. At this very moment, heavily caffeinated software engineers are designing programs that will overtax and befuddle your new turbo-powered PC when you try running them a few years from now. The memory and graphics requirements of Microsoft's recent Vista operating system, for instance, spell doom for aging machines that were still able to squeak by a year ago. According to the U.S. Environmental Protection Agency, an estimated 30 to 40 million PCs will be ready for "end-of-life management" in each of the next few years.

Computers are hardly the only electronic hardware hounded by obsolescence. A switchover to digital high-definition television broadcasts is scheduled to be complete by 2009, rendering inoperable TVs that function perfectly today but receive only an analog signal. As viewers prepare for the switch, about 25 million TVs are taken out of service yearly. In the fashion-conscious mobile market, 98 million U.S. cell phones took their last call in 2005. All told, the EPA estimates that in the U.S. that year, between 1.5 and 1.9 million tons of computers, TVs, VCRs, monitors, cell phones, and other equipment were discarded. If all sources of electronic waste are tallied, it could total 50 million tons a year worldwide, according to the UN Environment Programme.

So what happens to all this junk?

In the United States, it is estimated that more than 70 percent of discarded computers and monitors, and well over 80 percent of TVs, eventually end up in landfills, despite a growing number of state laws that prohibit dumping of e-waste, which may leak lead, mercury, arsenic, cadmium, beryllium, and other toxics into the ground. Meanwhile, a staggering volume of unused electronic gear sits in storage—about 180 million TVs, desktop PCs, and other components as of 2005, according to the EPA. Even if this obsolete equipment remains in attics and basements indefinitely, never reaching a landfill, this solution has its own, indirect impact on the environment. In addition to toxics, e-waste contains goodly amounts of silver, gold, and other valuable metals that are highly efficient conductors of electricity. In theory, recycling gold from old computer motherboards is far more efficient and less environmentally destructive than ripping it from the earth, often by surface-mining that imperils pristine rain forests.

Currently, less than 20 percent of e-waste entering the solid waste stream is channeled through companies that advertise themselves as recyclers, though the number is likely to rise as states like California crack down on landfill dumping. Yet recycling, under the current system, is less benign than it sounds. Dropping your old electronic gear off with a recycling company or at a municipal collection point does not guarantee that it will be safely disposed of. While some recyclers process the material with an eye toward minimizing pollution and health risks, many more sell it to brokers who ship it to the developing world, where environmental enforcement is weak. For people in countries on the front end of this arrangement, it's a handy out-of-sight, out-of-mind solution.

Many governments, conscious that electronic waste wrongly handled damages the environment and human health, have tried to weave an international regulatory net. The 1989 Basel Convention, a 170-nation accord, requires that developed nations notify developing nations of incoming hazardous waste shipments. Environmental groups and many undeveloped nations called the terms too weak, and in 1995 protests led to an amendment known as the Basel Ban, which forbids hazardous waste shipments to poor countries. Though the ban has yet to take effect, the European Union has written the requirements into its laws.

The EU also requires manufacturers to shoulder the burden of safe disposal. Recently a new EU directive encourages "green design" of electronics, setting limits for allowable levels of lead, mercury, fire retardants, and other substances. Another directive requires manufacturers to set up infrastructure to collect e-waste and ensure responsible recycling—a strategy called take-back. In spite of these safeguards, untold tons of e-waste still slip out of European ports, on their way to the developing world.

In the United States, electronic waste has been less of a legislative priority. One of only three countries to sign but not ratify the Basel Convention (the other two are Haiti and Afghanistan), it does not require green design or take-back programs of manufacturers, though a few states have stepped in with their own laws. The U.S. approach, says Matthew Hale, EPA solid waste program director, is instead to encourage responsible recycling by working with industry—for instance, with a ratings system that rewards environmentally sound products with a seal of approval. "We're definitely trying to channel market forces, and look for cooperative approaches and consensus standards," Hale says.

The result of the federal hands-off policy is that the greater part of e-waste sent to domestic recyclers is shunted overseas.

"We in the developed world get the benefit from these devices," says Jim Puckett, head of Basel Action Network, or BAN, a group that opposes hazardous waste shipments to developing nations. "But when our equipment becomes unusable, we externalize the real environmental costs and liabilities to the developing world."

Asia is the center of much of the world's high-tech manufacturing, and it is here the devices often return when they die. China in particular has long been the world's electronics graveyard. With explosive growth in its manufacturing sector fueling demand, China's ports have become conduits for recyclable scrap of every sort: Steel, aluminum, plastic, even paper. By the mid-1980s, electronic waste began freely pouring into China as well, carrying the lucrative promise of the precious metals embedded in circuit boards.

Vandell Norwood, owner of Corona Visions, a recycling company in San Antonio, Texas, remembers when foreign scrap brokers began trolling for electronics to ship to China. Today he opposes the practice, but then it struck him and many other recyclers as a win-win situation. "They said this stuff was all going to get recycled and put back into use," Norwood remembers brokers assuring him. "It seemed environmentally responsible. And it was profitable, because I was getting paid to have it taken off my hands." Huge volumes of scrap electronics were shipped out, and the profits rolled in.

Any illusion of responsibility was shattered in 2002, the year Puckett's group, BAN, released a documentary film that showed the reality of e-waste recycling in China. *Exporting Harm* focused on the town of Guiyu in Guangdong Province, adjacent to Hong Kong. Guiyu had become the dumping ground for massive quantities of electronic junk. BAN documented thousands of people—entire families, from young to old—engaged in dangerous practices like burning computer wire to expose copper, melting circuit boards in pots to extract lead and other metals, or dousing the boards in powerful acid to remove gold.

China had specifically prohibited the import of electronic waste in 2000, but that had not stopped the trade. After the worldwide publicity BAN's film generated, however, the government lengthened the list of forbidden e-wastes and began pushing local governments to enforce the ban in earnest.

On a recent trip to Taizhou, a city in Zhejiang Province south of Shanghai that was another center of e-waste processing, I saw evidence of both the crackdown and its limits. Until a few years ago, the hill country outside Taizhou was the center of a huge but informal electronics disassembly industry that rivaled Guiyu's. But these days, customs officials at the nearby Haimen and Ningbo ports—clearinghouses for massive volumes of metal scrap—are sniffing around incoming shipments for illegal hazardous waste.

High-tech scrap "imports here started in the 1990s and reached a peak in 2003," says a high school teacher whose students tested the environment around Taizhou for toxics from e-waste. He requested anonymity from fear of local recyclers angry about the drop in business. "It has been falling since 2005 and now is hard to find."

Today the salvagers operate in the shadows. Inside the open door of a house in a hillside village, a homeowner uses pliers to rip microchips and metal parts off a computer motherboard. A buyer will burn these pieces to recover copper. The man won't reveal his name. "This business is illegal," he admits, offering a cigarette. In the same village, several men huddle inside a shed, heating circuit boards over a flame to extract metal. Outside the door lies a pile of scorched boards. In another village a few miles away, a woman stacks up bags of circuit boards in her house. She shoos my translator and me away. Continuing through the hills, I see people tearing apart car batteries, alternators, and high-voltage cable for recycling, and others hauling aluminum scrap to an aging smelter. But I find no one else working with electronics. In Taizhou, at least, the e-waste business seems to be waning.

In China the air near some electronics salvage operations contains the highest amounts of dioxin measured anywhere in the world. Soils are saturated with the chemical.

Yet for some people it is likely too late; a cycle of disease or disability is already in motion. In a spate of studies released last year, Chinese scientists documented the environmental plight of Guiyu, the site of the original BAN film. The air near some electronics salvage operations that remain open contains the highest amounts of dioxin measured anywhere in the world. Soils are saturated with the chemical, a probable carcinogen that may disrupt endocrine and immune function. High levels of flame retardants called PBDEs—common in electronics, and potentially damaging to fetal development even at very low levels—turned up in the blood of the electronics workers. The high school teacher in Taizhou says his students found high levels of PBDEs in plants and animals. Humans were also tested, but he was not at liberty to discuss the results.

China may someday succeed in curtailing electronic waste imports. But e-waste flows like water. Shipments that a few years ago might have gone to ports in Guangdong or Zhejiang Provinces can easily be diverted to friendlier environs in Thailand, Pakistan, or elsewhere. "It doesn't help in a global sense for one place like China, or India, to become restrictive," says David N. Pellow, an ethnic studies professor at the University of California, San Diego, who studies electronic waste from a social justice perspective. "The flow simply shifts as it takes the path of least resistance to the bottom."

It is next to impossible to gauge how much e-waste is still being smuggled into China, diverted to other parts of Asia, or—increasingly—dumped in West African countries like Ghana, Nigeria, and Ivory Coast. At ground level, however, one can pick out single threads from this global toxic tapestry and follow them back to their source.

In Accra, Mike Anane, a local environmental journalist, takes me down to the seaport. Guards block us at the gate. But some truck drivers at a nearby gas station point us toward a shipment facility just up the street, where they say computers are often unloaded. There, in a storage yard, locals are opening a shipping container from Germany. Shoes, clothes, and handbags pour out onto the tarmac. Among the clutter: some battered Pentium 2 and 3 computers and monitors with cracked cases and missing knobs, all sitting in the rain. A man hears us asking questions. "You want computers?" he asks. "How many containers?"

Near the port I enter a garage-like building with a sign over the door: "Importers of British Used Goods." Inside: more age-encrusted PCs, TVs, and audio components. According to the manager, the owner of the facility imports a 40-foot container every week. Working items go up for sale. Broken ones are sold for a pittance to scrap collectors.

All around the city, the sidewalks are choked with used electronics shops. In a suburb called Darkuman, a dim stall is stacked front to back with CRT monitors. These are valueless relics in wealthy countries, particularly hard to dispose of because of their high levels of lead and other toxics. Apparently no one wants them here, either. Some are monochrome, with tiny screens. Boys will soon be smashing them up in a scrap market.

A price tag on one of the monitors bears the label of a chain of Goodwill stores headquartered in Frederick, Maryland, a 45-minute drive from my house. A lot of people donate their old computers to charity organizations, believing they're doing the right thing. I might well have done the same. I ask the proprietor of the shop where he got the monitors. He tells me his brother in Alexandria, Virginia, sent them. He sees no reason not to give me his brother's phone number.

When his brother Baah finally returns my calls, he turns out not to be some shady character trying to avoid the press, but a maintenance man in an apartment complex, working 15-hour days fixing toilets and lights. To make ends meet, he tells me, he works nights and weekends exporting used computers to Ghana through his brother. A Pentium 3 brings $150 in Accra, and he can sometimes buy the machines for less than $10 on Internet liquidation websites—he favors private ones, but the U.S. General Services Administration runs one as well. Or he buys bulk loads from charity stores. (Managers of the Goodwill store whose monitor ended up in Ghana denied selling large quantities of computers to dealers.) Whatever the source, the profit margin on a working computer is substantial.

The catch: Nothing is guaranteed to work, and companies always try to unload junk. CRT monitors, though useless, are often part of the deal. Baah has neither time nor space to unpack and test his monthly loads. "You take it over there and half of them don't work," he says disgustedly. All you can do then is sell it to scrap people, he says. "What they do with it from that point, I don't know nothing about it."

In the long run, the only way to prevent e-waste from flooding Accra, Taizhou, or a hundred other places is to carve a new, more responsible direction for it to flow in.

Baah's little exporting business is just one trickle in the cataract of e-waste flowing out of the U.S. and the rest of the developed world. In the long run, the only way to prevent it from flooding Accra, Taizhou, or a hundred other places is to carve a new, more responsible direction for it to flow in. A Tampa, Florida, company called Creative Recycling Systems has already begun.

The key to the company's business model rumbles away at one end of a warehouse—a building-size machine operating not unlike an assembly line in reverse. "David" was what company president Jon Yob called the more than three-million-dollar investment in machines and processes when they were installed in 2006; Goliath is the towering stockpile of U.S. e-scrap. Today the machine's steel teeth are chomping up audio and video components. Vacuum pressure and filters capture dust from the process. "The air that comes out is cleaner than the ambient air in the building," vice president Joe Yob (Jon's brother) bellows over the roar. A conveyor belt transports material from the shredder through a series of sorting stations: vibrating screens of varying finenesses, magnets, a device to extract leaded glass, and an eddy current separator—akin to a reverse magnet, Yob says—that propels nonferrous metals like copper and aluminum into a bin, along with precious metals like gold, silver, and palladium. The most valuable product, shredded circuit boards, is shipped to a state-of-the-art smelter in Belgium specializing in precious-metals recycling. According to Yob, a four-foot-square box of the stuff can be worth as much as $10,000.

In Europe, where the recycling infrastructure is more developed, plant-size recycling machines like David are fairly common. So far, only three other American companies have such equipment. David can handle some 150 million pounds of electronics a year; it wouldn't take many more machines like it to process the entire country's output of high-tech trash. But under current policies, pound for pound it is still more profitable to ship waste abroad than to process it safely at home. "We can't compete economically with people who do it wrong, who ship it overseas," Joe Yob says. Creative Recycling's investment in David thus represents a gamble—one that could pay off if the EPA institutes a certification process for recyclers that would define minimum standards for the industry. Companies that rely mainly on export would have difficulty meeting such standards. The EPA is exploring certification options.

Ultimately, shipping e-waste overseas may be no bargain even for the developed world. In 2006, Jeffrey Weidenhamer, a chemist at Ashland University in Ohio, bought some cheap, Chinese-made jewelry at a local dollar store for his class to analyze. That the jewelry contained high amounts of lead was distressing, but hardly a surprise; Chinese-made leaded jewelry is all too commonly marketed in the U.S. More revealing were the amounts of copper and tin alloyed with the lead. As Weidenhamer and his colleague Michael Clement argued in a scientific paper published this past July, the proportions of these metals in some samples suggest their source was leaded solder used in the manufacture of electronic circuit boards.

"The U.S. right now is shipping large quantities of leaded materials to China, and China is the world's major manufacturing center," Weidenhamer says. "It's not all that surprising things are coming full circle and now we're getting contaminated products back." In a global economy, out of sight will not stay out of mind for long.

Down with Carbon: Scientists Work to Put the Greenhouse Gas in Its Place

SID PERKINS

One morning each week, a scientist takes a stroll on the barren upper slopes of Hawaii's Mauna Loa volcano, a basketball-sized glass sphere in hand. At some point, the researcher faces the wind, takes a deep breath, holds it and strides forward while twisting open a stopcock. With a whoosh lasting no more than a few seconds, 5 liters of the most pristine air on the planet replaces the vacuum inside the thick-walled orb.

Once every couple of weeks, a parkaclad researcher at the South Pole conducts the same ritual. At these remote sites and dozens of others, instruments also sniff the air, adding measurements of atmospheric chemistry to a dataset that stretches back more than 50 years. The nearly continuous record results from one of the longest-running most comprehensive earth science experiments in history, says Ralph F. Keeling, a climate scientist at Scripps Institution of Oceanography in La Jolla, Calif. He carries on the effort his father, Charles Keeling, began as a graduate student in the 1950s.

Several trends pop out of the data, says Ralph Keeling. First, in the Northern Hemisphere the atmospheric concentration of carbon dioxide rises and falls about 7 parts per million over the course of a year. The concentration typically reaches a peak each May, then starts to drop as the hemisphere's flush of new plant growth converts the gas into sprouts, vegetation and wood. In October, the decomposition of newly fallen leaves again boosts CO_2 levels. Populations of algae at the base of the ocean's food chain follow the same trend, waxing each spring and waning each autumn.

A second trend is that each year's 7-ppm, saw-tooth variation in CO_2 is superimposed on an average concentration that is steadily rising. Today's average is more than 380 ppm, compared with 315 ppm 50 years ago. And it's still rising, about 2 ppm each year, mainly from burning fossil fuels.

Largely because CO_2 traps heat, Earth's average temperature has climbed about 0.74°C over the past century (SN: 2/10/07, p. 83), a trend that scientists expect will accelerate. In the next 20 years, the average global temperature is projected to rise another 0.4°C.

Squelching additional temperature increases depends on limiting, if not eliminating, the rise in CO_2 levels, many scientists say. And, Keeling says, "It's clear that if we want to stabilize CO_2 concentrations in the atmosphere, we need to stop the rise in fossil fuel emissions."

But halting the increase in amounts of CO_2 in the air doesn't necessarily mean doing away with fossil fuels. Many experts suggest that capturing CO_2 emissions, rather than only reducing them, could ultimately provide climate relief.

Possible solutions range from boosting natural forms of carbon capture and storage, or sequestration—fertilizing the oceans to enhance algal blooms, say, or somehow augmenting the soil's ability to hold organic matter—to schemes for snatching CO_2 from smokestacks and disposing of it deep underground or in seafloor sediments.

Success in sequestering carbon comes down to meeting two challenges: How to remove CO_2 from the air (or prevent it from getting there in the first place) and what to do with it once it has been collected.

Doing It Naturally

Organisms that dominate the base of the world's food chains soak up quite a bit of CO_2—currently about 2 percent of the atmosphere's stockpile each growing season. That gas, plus sunlight and other nutrients, is converted into carbon-rich sugars and biological tissues that nourish humans and all other animals. Unfortunately, most of that carbon makes its way back to the atmosphere rather quickly: Animals metabolize their food, breathing out CO_2. Decomposition of dead plants and animals likewise produces the greenhouse gas.

Over the long haul, though, ecosystems can sequester significant amounts of carbon. About 30 percent of the carbon in the world's soil is locked in peat lands of the Northern Hemisphere, for instance, with most of that accumulating since the end of the last ice age about 10,000 years ago (SN: 2/10/01, p. 95).

Recent data suggest that North American ecosystems sequester, on average, 505 million metric tons of carbon each year. Some accumulates as organic material in soil, wetlands or the carbon-rich sediments deposited in the continent's rivers and lakes. More is stored in woody plants that have invaded grasslands or trees that have taken over shrublands. Most of the sequestered carbon, about 301 million tons, is locked away in

North American forests or in the wood products harvested from them, notes Anthony W. King, an ecosystem scientist at Oak Ridge National Laboratory in Tennessee. He and his colleagues reported their analysis of these carbon sinks last November in an assessment issued by a consortium of U.S. government agencies.

"New, vigorously growing forests are where most carbon sequestration takes place," King says.

Some researchers, including Ning Zeng, an atmospheric scientist at the University of Maryland, College Park, seek to harness the prodigious carbon-storing power of forests. Right now, forest floors worldwide are lined with coarse wood—everything from twigs and limbs shed during growth to entire fallen trees—containing about 65 billion tons of carbon, says Zeng. Left undisturbed, that material would return its carbon to the atmosphere via decomposition or wildfire. Bury that wood in an oxygen-poor environment, however, and the carbon could be locked away for centuries.

Furthermore, Zeng notes, each year the world's forests naturally produce enough coarse wood to lock away about 10 billion tons of carbon. Burying just half of that amount would significantly counteract the estimated 6.9 billion tons of carbon released into the atmosphere each year via fossil fuel emissions.

While the price tag for this technique would be relatively reasonable—photosynthesis is free, and burying the wood would cost about $14 per ton—the environmental toll could be substantial. Coarse wood collected from the average square kilometer of forest could contain about 500 tons of carbon, Zeng reported in December in San Francisco at a meeting of the American Geophysical Union. That volume of wood would fill a trench 10 meters wide, 10 meters deep and 25 meters long. To sequester 5 billion tons of carbon each year, logging crews would need to dig and fill 10 million such trenches, about one every three seconds.

"This is not an environmentally friendly method" of carbon sequestration, Zeng admits.

Life at Sea

In certain parts of the oceans, especially along the western coasts of large continents, nutrient-rich waters fuel the growth of algae and other phytoplankton. Their growth pulls CO_2 from the atmosphere. Many parts of the ocean, however, lack one or more vital nutrients, particularly dissolved iron, and are therefore nearly devoid of life (SN: 8/4/07, p. 77).

Adding iron to the surface waters in some seas could help reduce CO_2 build-up in the atmosphere and forestall climate change, some scientists suggest. In the late 1980s, oceanographer John Martin, an early proponent of this idea, boasted: "Give me half a tanker of iron, and I'll give you the next ice age."

Or maybe not. Recent studies in the North Atlantic and North Pacific confirm that natural algal blooms can indeed sequester CO_2, but in many cases the phenomenon maybe only temporary, with little if any carbon making its way into deep water or seafloor sediments (SLY: 5/19/07, p. 307). In late 2004 and early 2005, a similar study near the Crozet Islands southeast of South Africa further demonstrated that natural algal blooms result in only modest carbon sequestration.

Peter Statham, a marine biogeochemist at the National Oceanography Centre in Southampton, England, and his colleagues installed sediment traps at a depth of 2,000 meters at several spots near the islands. South of the islands, particles drifting down through a 1-square-meter area together carry only 0.087 grams of carbon each year, the researchers estimate. North of the islands, where ocean currents have carried dissolved iron and other minerals eroded from the islands, the carbon flux to deep water is almost five times higher, Statham and his colleagues reported in Orlando, Fla., in March at the Ocean Sciences Meeting.

Many uncertainties remain about how effective any artificial attempts to boost algal growth might be, says Statham. First of all, he notes, scientists aren't sure which forms of iron are the ones that marine phytoplankton find most nutritious. And the long-term effects of adding the wrong type of iron—or maybe even the right one, he adds—could damage marine ecosystems for years. "There's a huge gap in our understanding of these phenomena," he says.

Finally, fertilizing the seas to sequester carbon, even with no bad side effects, may have little if any effect on climate. "Even in the most favorable circumstances, oceans would sequester only a small fraction of the carbon dioxide that humans are emitting," Statham argues.

Down and Away

Today, coal and petroleum each account for about 40 percent of global CO_2 emissions. Of the two, however, coal poses by far the larger threat to future climate. For one thing, coal produces more CO_2 per unit of energy than any other fossil fuel—about twice that generated by burning natural gas, for example. Also, coal is abundant and therefore relatively cheap: The amount of carbon found in the world's coal reserves is about triple that locked away in petroleum and natural gas deposits.

Worldwide, coal-fired power plants each year generate about 8 billion tons of CO_2, an amount that contains about 2.2 billion tons of carbon. And, says Daniel Schrag, a geochemist at Harvard University, emissions are poised to get even worse: About 150 power plants fueled by pulverized coal are now at various stages in the permitting process in the United States, and China reportedly cuts the ribbon on one such plant every week or so.

All told, the coal-fired power plants built in the next 25 years will, during their projected 50- to 50-year lifetimes, generate about 660 billion tons of CO_2, says George Peridas, an analyst with the Natural Resources Defense Council office in San Francisco. That's about 25 percent more than all the CO_2 that humans have produced by burning coal since 1751, a period that encompasses the entire Industrial Revolution.

Because coal-fired power plants are point sources of immense volumes of CO_2, they're tempting targets for sequestration efforts, says Tom Feeley, an environmental scientist at the National Energy Technology Laboratory in Pittsburgh. He and his colleagues are studying ways to capture emissions, ranging from using CO_2-hungry materials to sop CO_2 from smokestacks to building new types of plants that burn coal altogether differently. The goal is to develop techniques for large-scale field tests by 2012 that can capture at least 90 percent of a power plant's CO_2 emissions but boost the price of its electricity by no more than 20 percent.

In current power plants, CO_2-absorbing materials would be placed in a stream of 200°C emissions, mostly nitrogen with between 3 and 15 percent CO_2. The active materials could either absorb the gas, just as a sponge sops up water, or chemically bind to it.

Materials called metal-organic frameworks (SN: 1/7/06, p. 4) fall into the category of CO_2 sponges. In their gaseous state, CO_2 molecules fly about at great speeds and keep a considerable distance from each other, but inside the pores of some of these crystalline sieves, the molecules line up and cram close together, says Rahul Banerjee, a chemist at the University of California, Los Angeles.

Discovering the reactions that produce a substance that effectively captures CO_2 takes time. So, Banerjee and his colleagues recently adopted a technique common in the pharmaceutical industry: They used a computer-controlled device to automatically dispense various combinations and concentrations of reactants into each of 96 tiny wells on a single plate—each well, in essence, its own 300-microliter beaker—which was then heated. The researchers then assessed the CO_2-sopping ability of the resulting crystals.

In less than three months, the researchers generated 16 new zeolites, a type of metal-organic framework composed of aluminum silicates, Banerjee and his colleagues reported in the Feb. 15 Science. Three of the zeolites are highly porous, with each gram of the material having a large surface area—where CO_2 molecules can attach—of between 1,000 and 2,000 square meters. A 1-liter sample of one of those supersponges, a substance dubbed ZIF-69, could hold up to 83 liters of CO_2 under normal atmospheric pressure.

Another team of scientists has produced a CO_2-absorbing substance—one that binds the gas via a chemical reaction—by painting an organic compound called aziridine on a wafer of silica. Unlike previously developed aminosilica materials, the new substance has a high storage capacity for CO_2, says Christopher W. Jones, a chemical engineer at Georgia Institute of Technology in Atlanta. The chemical reaction can be reversed by heating the CO_2-saturated material, enabling researchers to capture the gas and dispose of it. A series of lab tests indicates that the material, whose amine-rich coating is tightly bound to the silica substrate, retains its capacity to soak up CO_2 after nearly a dozen cycles, the researchers reported in the March 12 *Journal of the American Chemical Society.*

Dump Sites

Capturing vast amounts of power plant emissions is just half the task. The next step is storage. Many scientists propose locking CO_2 underground or in the deep ocean.

Under high pressure, as in ocean depths below 500 meters, CO_2 is a dense liquid, not a gas, and doesn't mix well with water. Therefore it's possible to deep six CO_2 on the ocean floor, but many researchers have concerns about how large pools of concentrated CO_2 might affect ecosystems there (SN: 6/19/99, p. 392). The CO_2 might slowly dissolve into the surrounding water, creating acidic conditions.

A new and relatively simple twist on the deep-ocean technique may address many such concerns. If liquid CO_2 is blended with a mixture of seawater and pulverized limestone, the CO_2 breaks up into globules that are 200 to 500 micrometers in diameter and coated with limestone powder, says Dan Golomb, a physical chemist at the University of Massachusetts, Lowell. The resulting emulsion has a consistency between that of milk and mayonnaise. Injected into the deep sea, the limestone-veneered droplets sink about 200 meters per day, lab tests suggest. As the droplets dissolve into the surrounding water or break up as they jostle about on the seafloor, the limestone's carbonate dissolves too, buffering much of the resulting acidity, like a tiny Tums. Golomb and his colleagues described their carbon-dumping process last July in *Environmental Science & Technology.*

Immense volumes of subterranean strata are a tempting dumping ground, too. Some types of rock formations are naturally impervious to the flow of gases and liquids. In fact, some of these geological reservoirs have already proven themselves by sequestering naturally formed CO_2 for millions of years. Oil companies have been mining that CO_2, transporting it through pipelines and pumping it into the ground to enhance the recovery of petroleum from faltering oil fields for decades—an irony indeed to think that CO_2 is being pumped into the ground so that petroleum, a raw material for even more CO_2, can be extracted.

In many regions of the world, saline aquifers lie deep beneath the ground. Because that salty water isn't suitable for drinking, some of those strata, especially those sandwiched between or capped by impervious rocks, could be used to store CO_2. Scientists estimate that such reservoirs might hold hundreds of years' worth of captured emissions.

Disposal of CO_2 in ancient volcanic rocks may provide an even more secure sequestration technique. A multimillion-dollar field test soon to be under way in southeastern Washington state is designed to find out.

Lab tests suggest that liquid CO_2 will chemically react with basalt to produce various minerals, including calcium carbonate, in a matter of months, says Pete McGrail, an environmental engineer at Pacific Northwest National Laboratory in Richland, Wash. Therefore, concerns about the CO_2 escaping from its underground prison are minimized. Thick layers of basalt, the result of widespread volcanic activity in the region between 6 million and 17 million years ago, underlie the tristate area surrounding McGrail's lab. Although most think of basalt as impervious, many of these deposits are porous because they were frothy when they cooled or they cracked extensively when subsequent flows of lava heated them up or weighed them down.

Later this year, McGrail and his colleagues will inject between 1,000 and 3,000 tons of liquid CO_2—enough, give or take, to fill an Olympic-sized swimming pool—into the porous rocks at a depth of about 1 kilometer. Then, researchers will assess the effectiveness of their sequestration by occasionally collecting fluid samples at the injection site. Analyses suggest that this volume of CO_2 will react to form carbonate minerals within five years, says McGrail.

If this sequestration technique is deemed suitable, the region's ancient basalts could hold a volume of CO_2 approaching that emitted by every coalfired power plant in the United States over a 20- to 50-year period, McGrail and his colleagues estimate. Across the nation, deep geologic formations such as saline aquifers and coal layers could sequester 150 years' worth of worldwide power-plant emissions, possibly providing a rocksolid solution to one of the world's most pressing problems.

The United States and the world need carbon sequestration—not right now, says Harvard's Schrag, but soon, and on an enormous scale. The challenge, he notes, is to ensure that carbon capture and sequestration technologies are ready when serious political action on climate change is finally taken.

And that time may be coming soon, says Oak Ridge's King. "It's beginning to dawn on people," he says, "that they can change the planet in ways larger than the planet can change itself."

UNIT 3

Energy: Present and Future Problems

Unit Selections

9. **Gassing up with Hydrogen,** Sunita Stayapal, John Petrovic, and George Thomas
10. **Wind Power: Obstacles and Opportunities,** Martin J. Pasqualetti
11. **A Solar Grand Plan,** Ken Zweibel, James Mason, and Vasilis Fthenakis
12. **The Rise of Renewable Energy,** Daniel M. Kammen
13. **What Nuclear Renaissance?,** Christian Parenti
14. **The Biofuel Future: Scientists Seek Ways to Make Green Energy Pay Off,** Rachel Ehrenberg
15. **Putting Your Home on an Energy Diet,** Marianne Lavelle

Key Points to Consider

- What are our alternative energy sources and how feasible are they? Can we afford to switch to alternative energies such as solar and wind?
- Where does nuclear power fit in? Much was made of nuclear in the last presidential election, and many European countries generate the vast majority of their electricity from nuclear. What are the benefits and downsides for additional development in the United States?
- How can conservation measures actually increase the economic benefits of energy production? What impact would improving energy efficiency have on fossil fuel consumption and on potential climate change?
- What are some of the major benefits of such alternate energy sources as solar power and wind power? Do these energy alternatives really have a chance at competing with fossil fuels for a share of the global energy market?
- Where does conservation fit in? In 2001 then–Vice President Dick Cheney said "Conservation may be a sign of personal virtue but it is not a sufficient basis for a sound, comprehensive energy policy." How important is conservation?

Student Website

www.mhhe.com/cls

Internet References

Alliance for Global Sustainability (AGS)
http://globalsustainability.org

Alternative Energy Institute, Inc.
http://www.altenergy.org

Department of Energy—Energy efficiency and renewable energy
http://www.eere.energy.gov

Energy and the Environment: Resources for a Networked World
http://zebu.uoregon.edu/energy.html

Fuel economy (Department of Energy)
http://www.fueleconomy.gov

Institute for Global Communication/EcoNet
http://www.igc.org

Nuclear Power Introduction
http://library.thinkquest.org/17658/pdfs/nucintro.pdf

U.S. Department of Energy
http://www.energy.gov

Energy: What a rollercoaster ride we've been on! Until recently, one of the certainties was that ridiculously cheap fuel was always with us, except for several erratic departures in the 1970s and 1980s related to OPEC oil embargos. Then oil prices shot up to over $4/gallon, and now, with the ongoing economic crisis, plunged down again. All those thinking that the price spike was an aberration are misleading themselves. The world's oil supply is limited, and we have reached the point where demand has caught up with supply. As China and India continue to grow economically, overall world demand rises, while oil supplies decrease. We must accept that the price of this commodity will increase. Our heretofore cheap energy allowed for our explosive population growth over the last several centuries and is responsible for the increased levels of greenhouse gases in our atmosphere. We are presently adding a staggering 6,000 million metric tons of CO_2 to the atmosphere annually, which is expected to increase to 8,000 million metric tons by 2030 (Energy Information Administration—http://www.eia.doe.gov). Promising advances have been made in extracting fossil fuels from oil shales, but at what cost? If we continue to "successfully" burn carbon-based fuels, CO_2 levels will continue to increase. In contrast to the climate skeptics who stick their head in the sand and ignore the dire warnings of climate scientists, we must actively pursue alternatives, whether they are wind, solar, nuclear, biofuels, or something else on the horizon.

Unit 3 explores these alternatives. All have benefits, and all have limitations. If a magic bullet existed, we would have exploited it already. The fact is that there's no simple answer to our energy demands. What is needed is a creative, flexible, and intelligent energy policy for the 21st century. Energy requirements for homes and buildings consume the largest fraction of our energy needs, but transportation—particularly automobiles—is not far behind. The first article "Gassing up with Hydrogen," by Sunita Satyapal, John Petrovic, and George Thomas, explores the future of hydrogen fuel cells as an alterative to the traditional internal combustion engine. Hydrogen is a super-clean fuel, with no harmful emissions. The obstacles to a hydrogen-based economy are many, however. Producing hydrogen is costly. A nationwide network of hydrogen fueling stations will have to be implemented. Advances in other technologies, such as electric cars, must necessarily be minimal to keep hydrogen at an economic advantage. The cost of producing a hydrogen car would have to come down by orders of magnitude from the present price tag of close to $1 million. And finally, as discussed in the article, a safe and economical method of storage would have to be developed. Unfortunately, a nation driven by hydrogen cars is not in the immediate future.

The next article explores an energy source that is already firmly taking hold, namely electricity generation from wind. Once expensive compared to traditional energy sources, wind power is now cheaper than any other form except for hydroelectric. The negatives to wind power certainly exist, but compared to other energy sources, they are relatively minor. Initial installation costs are high, but at the same time, wind turbines can be erected quickly and require very little maintenance. Wind is fickle. It doesn't always blow. Wind power is noisy and

kills birds (although modern large turbines incur far fewer bird deaths). As prices drop, wind energy will become an ever more important part of our overall energy mix. Wind farms are growing exponentially, and prices are coming down just as fast. Not the silver bullet, but a wonderful supplement to our energy needs.

Zweibel, Mason, and Fthenakis in "A Solar Grand Plan" outline strategies for generating electricity from solar an important part of our energy mix. They explain that solar should be considered a major, rather than small, part of our future energy mix. The next article "The Rise of Renewable Energy" provides a general overview of different renewable energy sources, written by Daniel Kammen and first published in *Scientific American.*

"What Nuclear Renaissance?" debunks much of the hype about nuclear power. In last year's presidential elections, there was much discussion of nuclear power as our energy savior. So what has kept this energy source from taking off? We have not started the construction of a new nuclear power plant in over 30 years. Why? Nuclear power has zero CO_2 emissions and does not require importation of the fuel source. It is essentially "renewable" because our nuclear resources are extensive. But the negatives to nuclear power weigh heavy. Christian Parenti distills the problem down to one simple factor—cost. Given the concerns about accidents, cost overruns, and potential terrorist attacks, nuclear power simply does not hold up as a competitive energy source. First, there are huge upfront costs that make nuclear power viable only with large federal subsides. Second, nuclear power plants will always be vulnerable to terrorist attack. Third, the fuel is extremely dangerous, and great care is required to ensure that it is handled correctly. And finally, there is the political thorny issue of disposal. In 100 years, we will probably have answers to the problems of nuclear disposal, but for now, serious questions remain.

How about using nature as a fuel source? "The Biofuel Future" discusses possible biofuels as an energy source. Both fossil fuels and biofuels are based on the same principle. You burn them and they generate heat that is used for electricity generation. The profound difference is that fossil fuels in essence mine carbon from the ground and send it into the atmosphere. Biofuels also release CO_2, but the released CO_2 had been *withdrawn* from the atmosphere during the growth of the plants. In theory, biofuels are carbon neutral—they take CO_2 from the atmosphere and return it upon burning and electricity generation. The question, explored in the article, is whether biofuel technology can be improved to the point where it become a significant part of our energy mix.

This section ends with a much more modest proposal. "Putting Your Home on an Energy Diet" highlights the multipronged attack to global warming. While biofuels or wind power may satiate our appetite for electricity, they cannot solve the problem alone. Modest, individual action can make a difference. The energy we use to heat and power our homes and businesses is larger than any other sector. Better insulation, changing to fluorescent light bulbs, turning down the thermostat, and turning off vampire electronics makes a huge difference in our energy consumption. We as individuals are not just sitting on the sidelines. There's lots we can do right now.

Gassing up with Hydrogen

Researchers are working on ways for fuel-cell vehicles to hold the hydrogen gas they need for long-distance travel.

SUNITA SATYAPAL, JOHN PETROVIC, AND GEORGE THOMAS

On a late summer day in Paris in 1783, Jacques Charles did something astonishing. He soared 3,000 feet above the ground in a balloon of rubber-coated silk bags filled with lighter-than-air hydrogen gas. Terrified peasants destroyed the balloon soon after it returned to earth, but Charles had launched a quest that researchers two centuries later are still pursuing: to harness the power of hydrogen, the lightest element in the universe, for transportation.

Burned or used in fuel cells, hydrogen is an appealing option for powering future automotive vehicles for several reasons. Domestic industries can make it from a range of chemical feedstocks and energy sources (for instance, from renewable, nuclear and fossil-fuel sources), and the nontoxic gas could serve as a virtually pollution-free energy carrier for machines of many kinds. When it burns, it releases no carbon dioxide, a potent greenhouse gas. And if hydrogen is fed into a fuel-cell stack—a battery-like device that generates electricity from hydrogen and oxygen—it can propel an electric car or truck with only water and heat as by-products [see "On the Road to Fuel-Cell Cars," by Steven Ashley; *Scientific American,* March 2005]. Fuel-cell-powered vehicles could offer more than twice the efficiency of today's autos. Hydrogen could therefore help ease pressing environmental and societal problems, including air pollution and its health hazards, global climate change and dependence on foreign oil imports.

Yet barriers to gassing up cars with hydrogen are significant. Kilogram for kilogram, hydrogen contains three times the energy of gasoline, but today it is impossible to store hydrogen gas as compactly and simply as the conventional liquid fuel. One of the most challenging technical issues is how to efficiently and safely store enough hydrogen onboard to provide the driving range and performance that motorists demand. Researchers must find the "Goldilocks" storage solutions that are "just right." Storage devices should hold sufficient hydrogen to support today's minimum acceptable travel range—300 miles—on a tank of fuel in a volume of space that does not compromise passenger or luggage room. They should release it at the required flow rates for acceleration on the highway and operate at practical temperatures. They should be refilled or recharged in a few minutes and come with a competitive price tag. Current hydrogen storage technologies fall far short of these goals.

Researchers worldwide in the auto industry, government and academia are expending considerable effort to overcome these limitations. The International Energy Agency's Hydrogen Implementing Agreement, signed in 1977, is now the largest international group focusing on hydrogen storage, with more than 35 researchers from 13 countries. The International Partnership for the Hydrogen Economy, formed in 2003, now includes 17 governments committed to advancing hydrogen and fuel-cell technologies. And in 2005 the U.S. Department of Energy set up a National Hydrogen Storage Project with three Centers of Excellence and many industry, university and federal laboratory efforts in both basic and applied research. Last year alone this project provided more than $30 million to fund about 80 research projects.

Infrastructural Hurdles

One obstacle to the wide adoption of hydrogen fuel-cell cars and trucks is the sheer size of the problem. U.S. vehicles alone consume 383 million gallons of gasoline a day (about 140 billion

Overview/Hydrogen Storage

- One of the biggest obstacles to future fuel-cell vehicles is how engineers will manage to stuff enough hydrogen onboard to provide the 300-mile minimum driving range that motorists demand.
- Typically hydrogen is stored in pressurized tanks as a highly compressed gas at ambient temperature, but the tanks do not hold enough gas. Liquid-hydrogen systems, which operate at cryogenic temperatures, also suffer from significant drawbacks.
- Several alternative high-density storage technologies are under development, but none is yet up to the challenge.

gallons annually), which accounts for about two thirds of the total national oil consumption. More than half of that petroleum comes from overseas. Clearly, the nation would need to invest considerable capital to convert today's domestic auto industry to fuel-cell vehicle production and the nation's extensive gasoline refining and distribution network to one that handles vast quantities of hydrogen. The fuel-cell vehicles themselves would have to become cheap and durable enough to compete with current technology while offering equivalent performance. They also must address safety concerns and a lingering negative public perception—people still remember the 1937 *Hindenburg* airship tragedy and associate it with hydrogen, despite some credible evidence that the airship's flammable skin was the crucial factor in the ignition of the blaze.

Why is it so difficult to store enough hydrogen onboard a vehicle? At room temperature and atmospheric pressure (one atmosphere is about 14.5 pounds per square inch, or psi), hydrogen exists as a gas with an energy density about $\frac{1}{3,000}$ that of liquid gasoline. A 20-gallon tank containing hydrogen gas at atmospheric pressure would propel a standard car only about 500 feet. So engineers must increase the density of stored hydrogen in any useful onboard hydrogen containment system.

A 300-mile minimum driving range is one of the principal operational aims of an industry-government effort—the FreedomCAR and Fuel Partnership—to develop advanced technology for future automobiles. Engineers employ a useful rule of thumb in making such calculations: a gallon of gasoline is equal, on an energy basis, to one kilogram (2.2 pounds) of hydrogen. Whereas today's average automobile needs about 20 gallons of gasoline to travel at least 300 miles, the typical fuel-cell vehicle would need only about eight kilograms of hydrogen because of its greater operational efficiency. Depending on the vehicle type and size, some models would require less hydrogen to go that far, some more. Tests of about 60 hydrogen-fueled prototypes from several automakers have so far demonstrated driving ranges of 100 to 190 miles.

Aiming for a practical goal that could be achievable by 2010 (when some companies expect the first production fuel-cell cars to hit the road), researchers compare the performance of various storage technologies against the "6 weight percent" benchmark. That is, a fuel storage system in which 6 percent of its total weight is hydrogen. For a system weighing a total of 100 kilograms (a reasonable size for a vehicle), six kilograms would be stored hydrogen. Although 6 percent may not seem like much, achieving that level will be extremely tough; less than 2 percent is the best possible today—using storage materials that operate at relatively low pressures. Further, keeping the system's total volume to about that of a standard automotive gasoline tank will be even more difficult, given that much of its allotted space will be taken up by the tanks, valves, tubing, regulators, sensors, insulation and anything else that is required to hold the six kilograms of hydrogen. Finally, a useful system must release hydrogen at rates fast enough for the fuel-cell and electric motor combination to provide the power and acceleration that drivers expect.

Containing Hydrogen

At present, most of the several hundred prototype fuel-cell vehicles store hydrogen gas in high-pressure cylinders, similar to scuba tanks. Advanced filament-wound, carbon-fiber composite technology has yielded strong, lightweight tanks that can safely contain hydrogen at pressures of 5,000 psi (350 times atmospheric pressure) to 10,000 psi (700 times atmospheric pressure). Simply raising the pressure does not proportionally increase the hydrogen density, however. Even at 10,000 psi, the best achievable energy density with current high-pressure tanks (39 grams per liter) is about 15 percent of the energy content of gasoline in the same given volume. Today's high-pressure tanks can contain only about 3.5 to 4.5 percent of hydrogen by weight. Ford recently introduced a prototype "crossover SUV" called Edge that is powered by a combination plug-in hybrid/fuel-cell system that stores 4.5 kilograms of hydrogen fuel in a 5,000-psi tank to achieve a total maximum range of 200 miles.

High-pressure tanks would be acceptable in certain transportation applications, such as transit buses and other large vehicles that have the physical size necessary to accommodate storage for sufficient hydrogen, but it would be difficult to manage in cars. Also, the current cost of such tanks is 10 or more times higher than what is competitive for autos.

Liquefying stored hydrogen can improve its energy density, packing the most hydrogen into a given volume of any existing option. Like any gas, hydrogen that is cooled sufficiently condenses into a liquid, which at atmospheric pressure occurs around –253 degrees Celsius. Liquid hydrogen exhibits a density of 71 grams per liter, or about 30 percent of the energy density of gasoline. The hydrogen weight densities achievable by these systems depend on the containment and insulation equipment they use.

Liquefied hydrogen has important drawbacks, though. First, its very low boiling point necessitates cryogenic equipment and special precautions for safe handling. In addition, because it operates at low temperature, the containers have to be insulated extremely well. Finally, liquefying hydrogen takes more energy than compressing the gas to high pressures. This requirement drives up the cost of the fuel and reduces the overall energy efficiency of the cryocooling process.

Nevertheless, one carmaker is pushing this technology onto the road. BMW plans to introduce a vehicle this year called Hydrogen 7, which will incorporate an internal-combustion engine capable of running on either gasoline (for 300 miles) or on liquid hydrogen for 125 miles. Hydrogen 7 will be sold on a limited basis to selected customers in the U.S. and other countries with local access to hydrogen refueling stations.

Chemical Compaction

Searching for promising ways to raise energy density, scientists may be able to take advantage of the chemistry of hydrogen itself. In their pure gas and liquid phases, hydrogen molecules contain two bound atoms each. But when hydrogen atoms are chemically bound to certain other elements, they can be packed even closer together than in liquid hydrogen. The principal aim of hydrogen storage research now is finding the materials that can pull off this trick.

Some researchers are focusing on a class of substances called reversible metal hydrides, which were discovered by accident in 1969 at the Philips Eindhoven Labs in the Netherlands. Investigators found that a samarium-cobalt alloy exposed to pressurized hydrogen gas would absorb hydrogen, somewhat like a sponge soaks up water. When the pressure was then removed, the hydrogen within the alloy reemerged; in other words, the process was reversible.

Intensive research followed this discovery. In the U.S., scientists James Reilly of Brookhaven National Laboratory and Gary Sandrock of Inco Research and Development Center in Suffern, N.Y., pioneered the development of hydride alloys with finely tuned hydrogen absorption properties. This early work formed the basis for today's widely used nickel-metal hydride batteries. The density of hydrogen in these alloys can be very high: 150 percent more than liquid hydrogen, because the hydrogen atoms are constrained between the metal atoms in their crystal lattices.

Many properties of metal hydrides are well suited to automobiles. Densities surpassing that of liquid hydrogen can be achieved at relatively low pressures, in the range of 10 to 100 times atmospheric pressure. Metal hydrides are also inherently stable, so they require no extra energy to maintain storage, although heat is required to release the stored gas. But their Achilles' heel is mass. They weigh too much for practical onboard storage. Metal hydride researchers have so far attained a maximum hydrogen capacity of 2 percent of the total material weight (2 weight percent). This level translates into a 1,000-pound hydrogen storage system (for a 300-mile driving range), which is clearly too heavy for today's 3,000-pound car.

Metal hydride studies currently concentrate on materials with inherently high hydrogen content, which researchers then modify to meet the hydrogen storage system requirements of operating temperatures in the neighborhood of 100 degrees C, pressures from 10 to 100 atmospheres and delivery rates sufficient to support rapid vehicle acceleration. In many cases, materials that contain useful proportions of hydrogen are a bit too stable in that they require substantially higher temperatures to release the hydrogen. Magnesium, for example, forms magnesium hydride with 7.6 weight percent hydrogen but must be heated to above 300 degrees C for release to occur. If a practical system is to rely on waste heat from a fuel-cell stack (about 80 degrees C) to serve as the "switch" to liberate hydrogen from a metal hydride, then the trigger temperature must be lower.

Destabilized Hydrides

Chemists John J. Vajo and Gregory L. Olson of HRL Laboratories in Malibu, Calif., as well as researchers elsewhere are exploring a clever approach to overcoming the temperature problem. Their "destabilized hydrides" combine several substances to alter the reaction pathway so that the resulting compounds release the gas at lower temperatures.

Destabilized hydrides are part of a class of hydrogen-containing materials called complex hydrides. Chemists long thought that many of these compounds were not optimal for refueling a vehicle, because they were irreversible—once the hydrogen was freed by decomposition of the compounds, the materials would require reprocessing to return them to a hydrogenated state. Chemists Borislav Bogdanovic and Manfred Schwickardi of the Max Planck Institute of Coal Research in Mulheim, Germany, however, stunned the hydride research community in 1996 when they demonstrated that the complex hydride sodium alanate becomes reversible when a small amount of titanium is added. This work triggered a flurry of activity during the past decade. HRL's lithium borohydride destabilized with magnesium hydride, for example, holds around 9 percent of hydrogen by weight reversibly and features a 200 degree C operating temperature. This improvement is notable, but its operating temperature is still too high and its hydrogen release rate too slow for automotive applications. Nevertheless, the work is promising.

Although current metal hydrides have limitations, many automakers see them as the most viable low-pressure approach in the near- to mid-term. Toyota and Honda engineers, for example, are planning a so-called hybrid approach in a system that combines a solid metal hydride with moderate pressure (significantly lower than 10,000 psi), which they predict could achieve a driving range of more than 300 miles. General Motors has teams of storage experts, including Scott Jorgensen, who are supporting research on a wide range of metal hydride systems worldwide (including in Russia, Canada and Singapore). GM is also collaborating with Sandia National Laboratories on a four-year, $10-million effort to produce a prototype complex metal hydride system.

Hydrogen Carriers

Other hydrogen storage options have the potential to work well in cars, but they suffer a penalty in the refueling step. In general, these chemical hydride substances need industrial processing to reconstitute the spent material. The step requires off-board regeneration; that is, once hydrogen stored onboard a vehicle is released, a leftover by-product must be reclaimed at a service station and regenerated in a chemical plant.

More than 20 years ago Japanese researchers studied this approach using, for example, the decalin-naphthalene system. When decalin (C10H18) is heated, it converts chemically to naphthalene (a pungent-smelling compound with the formula C10H8) by changing the nature of its chemical bonds, which liberates five hydrogen molecules. Hydrogen gas thus bubbles out of the liquid decalin as it transforms into naphthalene. Exposing naphthalene to moderate hydrogen gas pressures reverses the process; it absorbs hydrogen and changes back to decalin (6.2 weight percent for the material alone). Research chemists Alan Cooper and Guido Pez of Air Products and Chemicals in Allentown, Pa., are investigating a similar technique using organic (hydrocarbon-based) liquids. Other scientists, including S. Thomas Autrey and his co-workers at the Pacific Northwest National Laboratory and chemistry professor Larry G. Sneddon of the University of Pennsylvania, are working on new liquid carriers, such as aminoboranes, that can store large amounts of hydrogen and release it at moderate temperatures.

Designer Materials

Yet another approach to the hydrogen storage problem centers on lightweight materials with very high surface areas to which hydrogen molecules stick (or adsorb). As one might expect, the amount of hydrogen retained on any surface correlates with the material's surface area. Recent developments in nanoscale engineering have yielded a host of new high-surface-area materials, some with more than 5,000 square meters of surface area per gram of material. (This amount equates to about three acres of surface area within just a teaspoon of powder.) Carbon-based materials are particularly interesting because they are lightweight, can be low cost and can form a variety of nanosize structures: carbon nanotubes, nanohorns (hornlike tubes), fullerenes (ball-shaped molecules) and aerogels (ultraporous solids). One relatively cheap material, activated carbon, can store up to about 5 weight percent hydrogen.

These carbon structures all share a common limitation, however. Hydrogen molecules bond very weakly with the carbon atoms, which means that the high-surface-area materials must be kept at or near the temperature of liquid nitrogen, −196 degrees C. In contrast to hydride research, in which scientists are struggling to lower the hydrogen binding energy, carbon researchers are exploring ways to raise the binding energy by modifying the surfaces of materials or by adding metal dopants that may alter their properties. These investigators employ theoretical modeling of carbon structures to discover promising systems for further study.

Beyond the carbon-based approaches, another fascinating nanoscale engineering concept is a category of substances called metal-organic materials. A few years ago Omar Yaghi, then a chemistry professor at the University of Michigan at Ann Arbor and now at the University of California, Los Angeles, invented these so-called metal-organic frameworks, or MOFs. Yaghi and his co-workers showed that this new class of highly porous, crystalline materials could be produced by linking inorganic compounds together with organic "struts". The resulting MOFs are synthetic compounds with elegant-looking structures and physical characteristics that can be controlled to provide various desired functions. These heterogeneous structures can have very large surface areas (as high as 5,500 square meters per gram), and researchers can tailor chemical sites on them for optimal binding to hydrogen. To date, investigators have demonstrated MOFs that exhibit hydrogen capacities of 7 weight percent at −196 degrees C. They continue to work on boosting this performance.

Although current progress on hydrogen storage methods is encouraging, finding the "just right" approach may take time, requiring sustained, innovative research and development efforts. Over the centuries, the basic promise—and challenge—of using hydrogen for transportation has remained fundamentally unchanged: Holding onto hydrogen in a practical, lightweight container allowed Jacques Charles to travel across the sky in his balloon during the last decades of the 18th century. Finding a similarly suitable container to store hydrogen in automobiles will permit people to travel across the globe in the coming decades of the 21st century without fouling the sky above.

Sunita Satyapal, John Petrovic and **George Thomas** work in the U.S. Department of Energy's applied research and development program in hydrogen storage technology. Satyapal, who has held various positions in academia and industry, serves as the team leader for the DOE's applied hydrogen storage R&D activities. Petrovic, a laboratory fellow (retired) of Los Alamos National Laboratory, is a consultant for the DOE and a fellow of both the American Ceramic Society and the American Society for Materials international. Thomas, currently a consultant with the DOE, has more than 30 years of experience studying the effects of hydrogen on metals at Sandia National Laboratories. These views are those of the authors only; they do not reflect the positions of the U.S. Department of Energy.

Wind Power
Obstacles and Opportunities

MARTIN J. PASQUALETTI

To know the wind is to respect nature. You ride with the wind when it fills your sails, but pay its power no heed and risk inconvenience, expense, even death. Drive through calm air in Los Angeles one moment only to encounter 30 minutes later Santa Anas whipping wildfires across mountaintops and pushing tractor-trailers into ditches. Lounge on the beach on Kauai one day, but find yourself huddling for protection the next day as a hurricane levels entire forests.[1] Sit on the porch during a quiet and muggy Oklahoma night when suddenly a mass of debris, once a house, swirls past, before dropping nearby as a pile of kindling and shattered dreams. More than any other force of nature, we have little defense against the wind. The wind keeps us on our toes.

If we cannot control the wind, perhaps we can put it to our use. It is a challenge with which we have had some success. Historically, we have used the wind to help us with work that would otherwise fall heavily upon our own backs. The wind helped humans explore the world; they had no other energy source. It continues to help us prepare foods and pump water. In some places, the wind is such a part of daily life that in its absence, silence blankets the landscape and puts us out of sorts. When it picks up again, flags flutter, well water rises, and grains are again ground to flour.

The wind machines humans developed were among the earliest icons of civilization. We can see them in early sketches from the Orient, scrolls from Persia, paintings from the Low Countries, photographs from the Dust Bowl, and even movies from Hollywood. Putting the wind to work was our first conscious use of solar power.

Perhaps the most widespread use of wind machines, at least in the United States, has been to pump water. Dotting the Great Plains by the hundreds of thousands, farm windmills—along with grain silos—were once as characteristic of the landscape as coal spoils were of Appalachia. Spinning whenever air moved, they brought to the surface the water that allowed ranches and settlements to flourish in an area otherwise too dry for either to exist for long. Most of these ingenious whirling devices eventually gave way to powerful compact motors that ran on fossil fuels, and as a result, wind energy landscapes largely disappeared. Before they all were removed, some folks preserved a few of them, drawn to their quaint beauty and the nostalgia they evoked as symbols of a Great Plains lifestyle. By then, however, most people considered the era of wind machines dead.

As it has happened, the epitaphs were premature. Today, wind machines are back. It has not been a quiet resurrection but rather one with substantial notoriety and publicity, plus a controversial mix of support and resistance. The new devices look and act little like their ancestors: Instead of the creaking, wooden machines of the past, those of the new species are made of metal and fiberglass—and are bigger, quieter, sleeker, and more powerful than ever. Instead of pumping water, the moving blades spin generators housed with an assemblage of gears in the nacelle, which is located behind the hub where all the blades meet. Instead of a stream of water, modern wind machines are pumping a stream of electrons, a product proving to be a valuable asset to farmers who are trying to address present day economic realities of living off the land.

The new appearance and mechanics of wind machines reflects their different role. Instead of producing mechanical power for the purposes of pumping and grinding, the new machines convert mechanical energy into electricity. Instead of being erected here and there in splendidly independent isolation, many are being clustered in symmetrically interdependent neighborhoods, designed to work together as parts of a larger organism. Nor are they just generating electricity: Unexpectedly, modern wind machines are prompting us to consider how best to weigh the energy we need against the environmental quality we want. All the while they are continuing their transformation from public indifference to public curiosity, from an overlooked energy supplier to alternative energy's "holy grail," one possible way to get most of what we want and little of which we do not.

An Old Resource with a New Mission

Compared to the variety of uses that stretch back millennia, converting wind energy to electricity is a recent application. Although a few people were trying to accomplish this at the same time Thomas Edison opened his coal-fired Pearl Street generating plant in the latter years of the nineteenth century, it would be another 80 years before such proof-of-concept machines would

Table 1 Historical Wind Turbines

Turbine, Country	Date in Service	Diameter (meters)	Swept Area (meters)	Power (kilowatts)	Specific Power (kilowatts per square meter)	Number of Blades	Tower Height (meters)
Poul la Cour, Denmark	1891	23	408	18	0.04	4	—
Smith-Putnam, United States	1941	53	2,231	1,250	0.56	2	34
F. L. Smidth, Denmark	1941	17	237	50	0.21	3	24
F. L. Smidth, Denmark	1942	24	456	70	0.15	3	24
Gedser, Denmark	1957	24	452	200	0.44	3	25
Hütter, Germany	1958	34	908	100	0.11	2	22

Source: P. Gipe, *Wind Energy Comes of Age* (New York: John Wiley & Sons, 1995), 78.

evolve into the commercial generators that started sprouting in the California landscape in 1981 (see Table 1). Indeed, the beginning of the modern era of wind power bore few similarities to earlier water pumping. The vision of modern wind power was much grander in scale, one that has evolved to row upon row of machines spreading over hundreds of acres, contributing enough electricity to power an entire city but—and this is the big difference—without undesirable side effects that accompanied the use of conventional resources.

Obviously, the machinery of the late twentieth century differs both in form and function from the equipment that nineteenth century ranchers and farmers developed to help them wrest a living from the dry lands that predominate west of the 100th meridian. For them, it was enough that machines were turning when the air was on the move. In the new era, such simple fulfillment is not enough; wind power today is viewed less from the living room and more from the boardroom. Wind power is big business, and the managers of that business must be sophisticated not just in the ways of making money, but in several disciplines that lead to success. Even something as seemingly innocent as turbine placement can no longer be considered just from the perspectives of convenience, necessity, or whim. Instead, the new wind barons must understand meteorology, metallurgy, physics, aerodynamics, capacity factors, land ownership, planning, zoning, and the influence of public perception.

Wind power's popularity is widespread and growing, a result of its increasing profitability and the perceived environmental benefits it engenders. It also results from the simple fact that, unlike fossil and nuclear fuels, wind is a widely available, familiar element of the environment. The first step is seemingly the easiest: Finding it.

Wind and the Family Farm

The initial step in developing any resource, be it gold or wind, is locating it. While it is often an uncomplicated step, not every place is attractive. Just like gold, wind is not evenly distributed in the richness of its product. Subtropical deserts, such as the Sonoran Desert surrounding Phoenix, are created by persistent high pressure and are often unsuitable for wind development. For obvious reasons, forested areas are unattractive for wind turbines, as are equatorial areas with their characteristically light and variable winds. The rest of the world is more promising, although detailed data collection must precede full-fledged capital investment.

Finding windy places is a relatively easy step: Unlike fossil or nuclear fuels, it does not require drilling rigs, seismic gear, or Geiger counters. Wind power, in most places, is simply "out there." This means that the most obvious early task of wind prospectors is to determine where winds are strong enough. Once identified, such areas reveal several common characteristics, including exposed terrain, colliding air masses, and, in particular circumstances, topographic funneling, as through mountain passes. In the United States, several areas meet such criteria, including sites in California, southeastern Washington, central Wyoming, east-central New Mexico, and most notably the Great Plains (see Figure 1). In Europe, strong winds are significant along the west coast of Ireland, Great Britain, the eastern North Sea, the southern Baltic Sea, the Pyrénées Mountains, and the Rhone Valley (see Figure 2). Many of these areas are "nuggets" that we are plucking first, and they have been stimulating further prospecting and the development of grand plans for the future. By the end of 2003, the total installed capacity in the European Union was 28,440 megawatts (MW).[2]

Like the gold rush of the 1850s, the modern wind rush started in California. California still leads the way, with 2,042 MW installed by January 2004, principally in four locations: Altamont Pass, on the edge of the Central Valley east of San Francisco; Tehachapi Pass at the southern end of the Sierra Nevada; San Gorgonio Pass, near Palm Springs; and in the rolling hills between San Francisco and Sacramento. Although first, the primacy of California is certainly temporary: With a potential for 6,770 MW, it ranks only seventeenth among the 50 states in potential (see Table 2).[3]

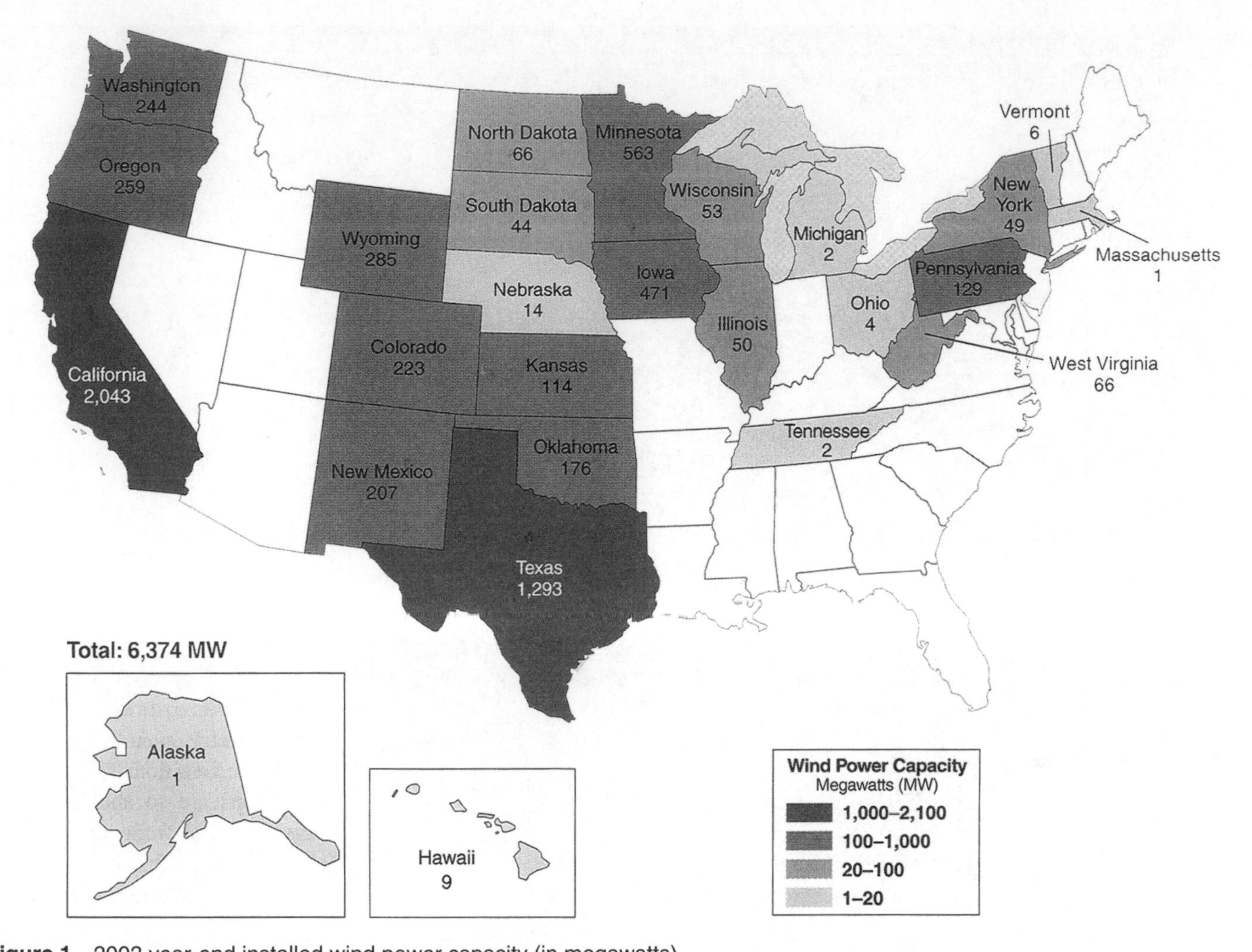

Figure 1 2003 year-end installed wind power capacity (in megawatts).

Source: American Wind Energy Association, *Wind Energy Projects Throughout the United States of America,* http://www.awea.org/projects (accessed 30 June 2004).

Among other states attracting interest, Texas has been the pacesetter with more than $1 billion in new wind investment and 1,293 MW installed, mostly in the western part of the state near such communities as Big Springs and McCamey. Coincidentally, many of these developments are positioned among the now derelict oil equipment that helped bring great wealth to this part of the state and has underpinned many of its towns and cities. At the end of January 2004, California, Texas, and 24 additional states held within their borders an installed capacity of 6,374 MW.[4]

Another 2,000 MW has been proposed for development in the near future, with some of the largest projects planned for the states of Washington and Massachusetts. However, the greatest potential remains where wind machines once so dominated the landscape—midway between these extremes in the Great Plains.

The wind of the Great Plains is as obvious as is its treeless expanse. When Francisco Vásquez de Coronado and his men crossed this region searching for the Seven Cities of Cíbola in 1540, they found no gold but two other resources instead. The most obvious and most useful to Coronado were the great herds of bison—totaling perhaps 50 million head—that were scattered across a million square miles of grassland. Not only did they provide food, but their droppings helped guide the expedition across otherwise indistinct landscapes. The other resource was the wind, but three centuries would pass before it was appreciated.

By the late 1800s, the general pattern had reversed; bison were being decimated for sport, and wind power was lifting water for irrigation. Today, only this second resource remains, yet it is being used for a different mission: To generate electricity and make money.[5] It is a realistic ambition: The winds of the Great Plains are so abundant that the energy potential from just three states (North Dakota, Texas, and Kansas), were they fully developed, would match the electrical needs of the entire country. These and several other Great Plains states hold the largest expanse of class 4 (400–500 watts per square meter) lands in the country (see the box).[6]

Although weather on the Great Plains is often viewed as being inhospitable—farming families there endure swirling snow in winter and blowing dust in summer—attitudes toward the frequent tempests are lately bending in a new direction.

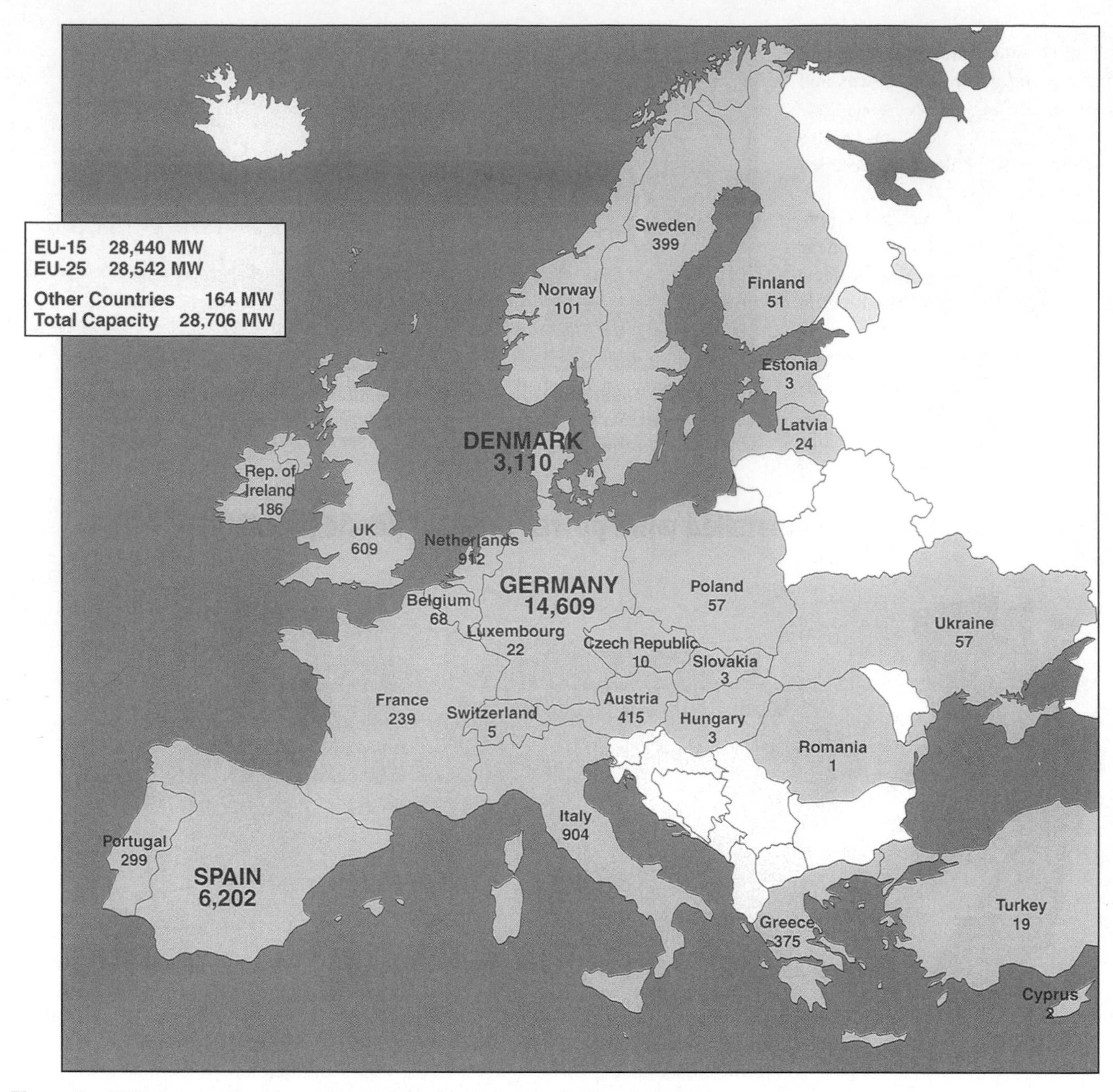

Figure 2 2003 year-end european installed wind power capacity (in megawatts).

Source: Adapted from a map compiled by the European Wind Energy Association, http://www.ewea.org, 3 February 2004.

Always alert for new sources of income to ease their financial volatility, locals are turning to wind developers with equanimity and even enthusiasm. They are finding that the same winds that strip soil from the fields and bury houses in snow can fuel rural economic development.[7]

Construction of a typical 100 MW wind farm produces more than 50,000 days (approximately 419,020 manhours) of employment. In Prowers County, Colorado, the recently completed wind development is each year providing $764,000 in new revenues, $917,000 in school general funds, $203,000 in school bond funds, $189,000 to the Prowers Medical Center, and $189,000 in additional revenue to the county tax base.[8] Meanwhile, a 250 MW project in Iowa is providing $2 million in property taxes and $5.5 million in operation and maintenance income. The leases on offer to farmers in this area commonly provide yearly royalties of more than $2,000 per turbine: For a single Iowa project, local farmers are receiving $640,000 annually. Other projects return $4,000–$5,000 per turbine. In some cases, a one-megawatt turbine could generate revenues for the owner of $150,000 per year once the debt for purchase is repaid.[9] Though appreciable at the individual level, such royalties are only a small part of the variable costs for wind developers (see Figure 3). To them, such largesse seems good business. To farmers, such revenues can mean the difference between bankruptcy and prosperity.

Table 2 Top 20 States for Wind Energy Potential (measured by annual energy potential in billions of kilowatt hours)

1	North Dakota	1,200
2	Texas	1,190
3	Kansas	1,070
4	South Dakota	1,030
5	Montana	1,020
6	Nebraska	868
7	Wyoming	747
8	Oklahoma	725
9	Minnesota	657
10	Iowa	551
11	Colorado	481
12	New Mexico	435
13	Idaho	73
14	Michigan	65
15	New York	62
16	Illinois	61
17	California	59
18	Wisconsin	58
19	Maine	56
20	Missouri	52

Note. As of July 2004, reevaluation of the wind potential has been done for 28 states by the U.S. Department of Energy's Windpowering America program; reevaluation of additional states is still in process. Once complete, the numbers for each state might change, as might the relative rankings. In many cases, the potential will increase. The Great Plains will continue to dominate the rest of the country in terms of potential.

Source: Pacific Northwest Laboratory, *An Assessment of the Available Wind Land Area and Wind Energy Potential in the Contiguous United States* (Richland, WA: Pacific Northwest National Laboratory, 1991).

In spite of a weighty list of environmental attributes, wind power carries some unexpectedly heavy baggage.

In spite of promising trends, it would be an overstatement to claim that wind power offers an energy panacea or a reversal of the national trend toward increasingly concentrated generation of electricity: It is, at least so far, a relatively small enterprise. Taken together, all the wind developments in the country contribute less than 1 percent of our current needs.[10] However, there is a real attraction to wind power's promise for the future: Its estimated generating potential in the United States alone is 10,777 billion kilowatts per hour (kWh) annually, or three times the electricity generated in the entire country today.[11] In recognition of such potential for pollution-free electricity, the U.S. government is sponsoring a program called Wind Powering America (WPA) to tap more deeply this vast natural resource. WPA's new goal is to increase to 30 the number of states with more than 100 megawatts of wind-generating capacity by 2010.[12] The program also aims to increase rural economic development and, to some degree, local energy independence.

The Environmental Irony

Wind power attracts many adherents from the environmental community. These organizations focus on its solar roots, emphasizing that it requires no mining, drilling, or pumping, no pipelines, port facilities, or supply trains. It produces no air pollution or radioactive waste, and it neither dirties water nor requires water for cooling. Wind power is relatively benign, simple, modular, affordable, and domestic. It is, in short, an environmental golden goose.

However, in spite of such a weighty list of attributes, wind power carries some unexpectedly heavy baggage. In England, anti-wind epithets have been particularly colorful: Developers have suffered their machines being called everything including "lavatory brushes in the air" for their busy top ends. This leads us to an irony of wind power: While we usually consider wind power environmentally friendly, most of the objections to its expansion have had environmental origins.

We can follow the thread of such reactions most clearly to Palm Springs, California, in the mid-1980s. Soon after installing thousands of turbines in windy San Gorgonio Pass just north of the city limits, developers were battered with complaints that the machines interfered with television reception, produced annoying and inconsistent noise, posed risks to wildlife and aircraft, and represented incompatible land-use practices.[13] The most troubling, bitter, and outraged complaint, however, was that the wind machines destroyed the aesthetic appeal of the landscape, thereby threatening the very attribute that most attracts tourists to the area's fancy resorts.[14]

Developers and bureaucrats of the day, all of whom had expected a warmer welcome, were startled by such reactions. It was apparent to everyone with hopes for contributions from wind power that any future success would have to rest on greater environmental compatibility and a more complete respect for public attitudes and opinions. The initial experience provoked industry musing as to what actions might better attract support. Soon, manufacturers began making improvements to design and engineering. Locally, the concerns over wind impacts led to stricter planning rules and more uniform standards. These adjustments softened the problems, but they could not eliminate them. Turbines remained unavoidably visible and the center of a classic example of incompatible land use. The very characteristic that had long kept residential development in the San Gorgonio Pass minimal—the strong wind—was the same characteristic that prompted developers to fill it with machines. There was very little compromise potential.

So strong was the backlash against wind development that the City of Palm Springs sued the U.S. Bureau of Land Management and the County of Riverside, claiming that developers had not

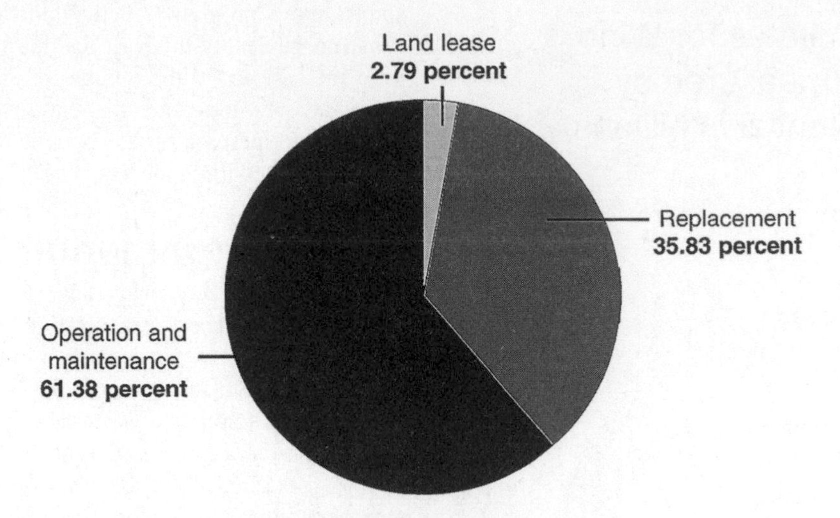

Figure 3 Variable costs of wind energy projects.

Source: E. DeMeo and B. Parsons, "Some Common Misconceptions about Wind Power," presented at the All States Wind Summit, Austin, TX, 22 May 2003. See U.S. Department of Energy, *State Wind Energy Handbook,* http://www.eere.energy.gov/windpoweringamerica/pdfs/wpa/34600_wind_handbook.pdf (accessed 30 June 2004), 90.

Wind Classes

Developers need to know the average wind speed at a particular site to design and build the most appropriate turbines. It would be no more prudent to size the turbines for the slowest speed than it would be to size them from the fastest, but infrequently occurring speed. One can get a good idea of this relationship by using the Weibull distribution, a plot of frequency against speed. This distribution helps identify various classes of wind (see the table). The higher the number, the stronger the average speed. A good wind speed is 7 meters per second (mps); 20 mps may be excessive and cause damage to equipment. Turbine manufacturers have a "rated wind speed" for all models and sizes of turbines they sell. Typical rated wind speed requirements are in the range of 8–13 mps, but many machines will produce some power with much slower speeds.[1] Currently, developers are concentrating on class 4 and above as the most promising areas.

1. For an excellent and detailed description of these and other principles, see P. Gipe, *Wind Energy* Comes of Age (New York: John Wiley & Sons, 1995).

Wind Power Classification

Wind Power Class	Resource Potential	Wind Power Density at 50 Meters (in Watts per square meter)	Wind Speed at 50 Meters (in meters per second)	Wind Speed at 50 Meters (in miles per hour)
2	Marginal	200–300	5.6–6.4	12.5–14.3
3	Fair	300–400	6.4–7.0	14.3–15.7
4	Good	400–500	7.0–7.5	15.7–16.8
5	Excellent	500–600	7.5–8.0	16.8–17.9
6	Outstanding	600–800	8.0–8.8	17.9–19.7
7	Superb	800–1600	8.8–11.1	19.7–24.8

followed proper environmental procedures. Although the suit was eventually abandoned, it was not before the local jurisdictions, including Palm Springs and Riverside County, enacted a long list of required adjustments, stipulating (for example) height limitations, the use of nonglinting paint, reporting mechanisms for endangered species, and the establishment of decommissioning bonds.

A few years later, in an unpredictable turnaround, attitudes changed. This happened once Palm Springs, led by its mayor Sonny Bono, began eyeing wind machines as generators of tax revenue, as well as electricity. With a financial windfall in mind, the city annexed several square miles of land in the middle of the windiest part of the pass, thereby enlarging the city limits and sweeping additional tax revenues into the municipal treasury. Also, counter to the early intuition and opposition of city officials, the wind turbines have become something of a tourist attraction. Organized tours are available, images of wind farms adorn many local postcards, and brochures advertise the Palm Springs wind industry. Even Hollywood producers have incorporated the striking wind energy landscapes in movies and advertisements. These changes reflect the progression in public attitudes toward greater acceptance, although a closer look still finds disgruntled residents who have the original objections. As they point out, we can paint them, size them, sculpt them, and engineer them to a fine edge, but we cannot make them disappear.

The Aesthetic Core

Reactions to wind power tend to be both quick and subjective. While one group fights intrusion, another is organizing visits for enthusiastic tourists. Where one person loathes turbines, that person's neighbors find them fascinating. Whichever reaction prevails in any given location, wind turbines cannot be ignored, for they do not fit naturally upon the land. They are, to apply Massachusetts Institute of Technology historian Leo Marx's famous phrase, "machines in the garden."[15]

The spread of wind power encounters the most strident opposition where it interferes with local land use.

Wind power's development contained a surprise: Among its corps of supporters, no one anticipated the need to defend wind projects. Why did no one foresee objections? We can only speculate, but it seems that the advantages were considered by adherents to be so obvious, especially when compared to nuclear power, that developing a defensive strategy for this new technology seemed superfluous. Supporters failed to recognize how opposite is the signature between the two: Nuclear power is compact and quiet, whereas wind power is expansive and obvious. Reflecting on this difference, resistance to nuclear power accumulated slowly only after a long educational process that culminated with accidents at Three Mile Island and Chernobyl, while resistance to wind power was immediate and instinctive.

Although this difference suggests the heft of visual aesthetics in shaping public opinion, it masks two other ingredients of equal importance. One is the immobility of the resource: Wind moves, windy sites do not. In this way, wind differs from coal and most other fuels, because its nature does not allow it to be extracted and transported for use at a distant site. For wind power to be successful, turbines must be installed where sufficient wind resources exist or not at all. Thus, just like two other resources—geothermal energy and hydropower—the site-specific nature of wind developments intrinsically invites conflicts with existing or planned land uses. This is true even in deserts, the common dumping ground of society.[16]

The second ingredient in helping form public attitudes toward wind power is the landscape itself. Simply put, some landscapes are more valued than others. Place turbines in sensitive areas—perhaps along the coast or in a national park—and prepare for an uproar. Place them out of view or in low-value areas—sanitary landfills, for example—and opposition diminishes.

These characteristics produce wind power's most intractable challenges. First, owing to resource immobility and the subjectivity of its aesthetic impact, total mitigation is impossible. Second, because environmental competition changes from place to place and from one time to another, generic solutions are few and elusive. Third, because nothing can make turbines invisible, little we do will make them more acceptable to those perceiving land-use interference. There is no escaping the essence of wind turbines: They will always be spinning, pulsing, exoskeletal contraptions that naturally attract the eye.

The foregoing notwithstanding, the future of wind power remains both promising and substantial, if we can identify and follow the appropriate path. Two general strategies suggest themselves: Work to bend public opinion in favor of wind power, or install the turbines out of view. The first approach is under way but slow. The second approach can be quicker and would seem to hold promise, but it is being met with mixed results, especially when projects are proposed for offshore locations—the newest tactic to avoid public criticism and maximize profits.

Larger and Larger Turbines

Wind turbines are getting larger and larger. What is driving this trend? To answer this question, we need to know that movement obtains its impetus from the sun; as solar energy strikes the surface of the Earth, it creates differences in pressure. The wind, in turn, moves "downhill" along the pressure gradients that are formed, from higher to lower pressure. Speed increases as the horizontal distance between different pressures is shortened. Wind also typically accelerates when it is constricted, as when it moves through a mountain pass. The faster the wind moves, the more energy it carries, but it is not in a linear function. Rather, it increases with the cube of the wind speed, usually written x^3. This means that a wind speed of 8 meters per second (mps), yields 314 (8^3) watts for every square meter exposed to the wind, while at 16 mps, we get 2,509 (16^3) watts per square meter, again eight times as much. This relationship puts a premium on sites having the strongest winds. This relationship also explains why the area "swept" by the turbine blades is so important and why the wind industry has been striving fervently to increase the scale of the equipment it installs. A one-megawatt turbine at a typical European site would produce enough electricity annually to meet the needs of 700 typical European households.

Moving Offshore

The spread of wind power encounters the most strident opposition where it interferes with local land use. Tourism, recreation, entertainment, resorts, and a host of other outdoor activities create most of the challenge because their function is to help people escape reality. For relief from this dilemma, developers are looking for sites offshore, and they have been finding them, especially in the shallow, wind-swept waters of the Irish Sea and North Sea. Denmark, which is characteristically leading the way with this strategy, has already installed and activated several fields of this type. Other projects are in place or planned off Ireland, the United Kingdom, the Netherlands, and several other countries.

In addition to prohibiting wind projects in populated areas, positioning them offshore offers several operational advantages. For example, winds passing over water tend to be stronger than those passing over land; offshore placement removes no land from existing or planned uses; any noise produced at sea is muffled by that of the surf; road use is largely a moot issue; and negotiation with multiple landowners is unnecessary. Nonetheless, offshore placement requires some tradeoffs. For example, offshore equipment is more costly to construct and maintain, and it inherently increases the potential for conflict with any recreational use of the seashore. It also tends to encourage the installation of larger turbines (see the box).

Strictly from a public perspective, offshore placement has the presumed advantage of mitigating complaints about aesthetic intrusion. It has not, however, turned out to be the expected universal remedy. Indeed, moving offshore is increasing rather than diminishing the enmity of wind power in some quarters, especially in the northeastern United States.

Tempted by the strong offshore winds of coastal Massachusetts and responding to the hostility to wind developments witnessed in California, several entrepreneurs advocated placing wind turbines on the shallow offshore banks. The proposal itself may go down as the most foolhardy miscalculation in renewable energy history. The problem, as usual, is incompatible use of space. Called Cape Wind, the project is proposed for Nantucket Sound, a site between the popular vacation spots on Cape Cod and the exclusive holiday retreats of Martha's Vineyard and Nantucket Island. Like development near Palm Springs, Cape Wind is colliding with the wishes of a prosperous and politically astute residential corps bent on protecting existing scenic and recreational qualities that it has come to cherish.

Riding a Roller Coaster

Wind energy has experienced a wild ride over the past 20 years, one where initial enthusiasm soured quickly with the perception that generous incentives and lax oversight were allowing

The Cape Wind Project: A Wind Power Lightning Rod

Cape Wind is a proposed $500–$750 million wind development project for Horseshoe Shoal in Nantucket Sound. If approved, the turbines will come within 5 miles of land, spread over an area of 24 square miles, and consist of 130, 417-foot wind turbines connected to a central service platform that includes a helicopter pad and crew quarters. Each turbine blade will be 164 feet long with a total diameter of 328 feet. Each turbine will have a base diameter of 16 feet and an above-water profile taller than the Statue of Liberty.

The proposal has become a lightning rod for the wind industry. The Alliance to Protect Nantucket Sound (the Alliance),[1] which strongly opposes the project, has been accumulating arguments against it. They point out that

- each turbine will have about 150 gallons of hydraulic oil, and the service platform will have at least 30,000 gallons of dielectric oil, and diesel fuel;
- the project will be within the flight path of thousands of small planes; and
- the turbines will pose a navigation hazard to the commercial ferry lines in the area.

These and other objections, however, take a secondary position to the Alliance's primary objection, that of aesthetic intrusion. The Alliance claims that the turbines will be visible for farther than 20 miles, that they will be lighted at night, and that they will flicker with changing sun angle. The Alliance has developed many computer visualizations of how they would appear. (Some proponents might point out that the Visualizations illustrate how inconsequential the turbines would appear from the beach.)

The pro-development side has not been idle. Cape Wind has its own website,[2] which identifies the many benefits of the project, including that it offsets the need for 113 million gallons of oil yearly and creates approximately 600–1,000 new jobs. They also refer to many studies attesting to the benefits of such projects as Cape Wind. One of the most recent references the positive impacts on sea creatures around the wind turbines off the southern Swedish coast.[3] Other websites provide many testimonials to the good sense of offshore wind power.[4] The controversy over Cape Wind's offshore proposal is just the beginning of many other anticipated projects along the East Coast, such as off Long Island.[5]

1. http://www.saveoursound.org
2. http://www.capewind.org
3. L. Nordstrom, "Windmills off Swedish Coast are Providing Unexpected Benefit for Marine Life, Scientists Say," *Environmental News Network,* 11 February 2004, http://www.enn.com/news/200402-11/s_13011.asp
4. windfarm@cleanpowernow.org; http://www.safewind.info
5. See the Safe Wind Coalition's Web site, http://www.safewind.info/wind_farms_where.htm; and the Long Island Offshore Wind Initiative's site at http://www.lioffshorewindenergy.org/

virtually any wind farm development, no matter how carelessly designed or operated, to be financially tenable. An undertow that quickly started to pull against the early currents of promise was a perception that wind developments were being installed without sufficient public notification, due consideration, or individual benefit. By the late 1980s, it was clear that improved turbines and business situations were going to be necessary if wind power was to develop a significant position in the alternative energy mix in the United States or abroad.

Some of the earliest advances first came into view in Europe, where even casual inspection spotted substantial differences from early installations in California. For example, instead of large clusters of turbines spread haphazardly upon the land, deployment of European turbines was more sensitively organized into smaller groupings that were carefully integrated into the landscape. This was partly a result of a higher sensitivity to existing conditions and partly a measured response to the experiences in California that had dulled the promise of wind power and threatened its future.[17]

Reflecting improvements and continued support, wind power's trajectory is once again upward. Today it is the fastest-growing renewable energy resource in the world.[18] Wind power is especially popular outside the United States: In countries like Germany, it is welcomed, encouraged, and promoted as one way to reduce greenhouse gas emissions. In Denmark, the value of wind power to the economy now exceeds that of its economic mainstay, ham. Spain's development of wind power is currently growing at a faster pace than it is in any other country.

The roller coaster ride is not over, however: Even amid news of improvements and quickened growth, wind power continues to have its critics. The more determined of these opponents work to keep wind machines from their property and out of their view. They hire public relations experts, make abundant use of the Internet to promote their view and attract adherents, and invite the support of prominent citizens to their cause. The group Save Our Sound is perhaps the most visible example of such techniques.[19] Such determined resistance was never envisioned when the champions of wind power came calling more than two decades ago. Today, despite progress in assuaging public apprehensions, a measure of uncertainty still hangs over wind's future.

From Incentives to Independence

What is to be made of the many incentives that wind power enjoys? Tax incentives, utility portfolio standards, feed-in laws,[20] and many other aids currently help make it an economically viable alternative energy provider. Some would say that the requirement for these incentives demonstrates that wind power is not a legitimate competitor for our energy dollars. Others might argue that the mere existence of these aids suggests how narrow the economic gap is between a present need for subsidies and independent viability. While its increasingly competitive status results partly from a rising cost of conventional energy, it also reflects the declining costs of all alternatives, including wind. The message is this: Even without incentives, wind power has been moving toward economic independence, and it seems destined to reach parity with conventional sources soon.

It is often the smallest margin of help that wins the day for an emerging technology. One way to demonstrate this is to examine the impact of higher conventional energy cost. In one study of 12 Midwestern states, where electricity sold at 4.5 cents per kWh, the regional potential for cost-effective wind power was about 7 percent of current total generation in the United States.[21] If the market would support a price of 5.0 cents per kWh, however, the potential would grow to 177 percent of current generation. If one additional penny is added to the price, the potential blossoms to 14 times current levels.[22]

Until conventional energy makes this inevitable jump, wind operators need another way to bridge the gap. This brings us to the U.S. Production Tax Credit (PTC). This credit originally provided for an inflation-adjusted 1.5 cents per kilowatt-hour for electricity generated with wind turbines. With PTC now at about 1.9 cents, wind projects are economically favored. In its absence, however, development of new projects virtually ceases. This occurred, for instance, when the credit expired at the end of 2001, before it was reinstated some months later. PTC lapsed again on 31 December 2003, and discussions in Congress are once again under way as to whether to extend it for a five-year period.[23]

Without such a credit, the U.S. wind industry will suffer. According to Craig Cox, executive director of the Interwest Energy Alliance, "The lapse of the PTC has created uncertainty in the wind energy marketplace, and interest in new developments has slowed."[24] Renewal of the credit is part of the $31 billion energy bill that stalled in Congress at the end of 2003, again putting the wind energy industry back on its "roller coaster" in the United States. The world's major wind turbine manufacturer, Vestas Group, delayed its decision to build a wind turbine plant in Oregon because of the uncertainty of the credit. Ultimately, such uncertainty spreads to all phases of wind energy development, not just deployment of turbines. "We've been looking to establish a manufacturing facility in the U.S. but have not done that only because of the boom and bust cycle of the wind energy industry in the U.S.," Scott Kringen of Vestas told Reuters.[25] Other spokespersons have made similar observations: "Today, a wide range of U.S. companies are interested in the wind industry, but many are staying on the sidelines because of the on-again, off-again nature of the market produced by frequent expirations of the PTC," said Randall Swisher, executive director of the Washington, DC-based American Wind Energy Association.[26] Most countries offer more stable, longer-term policy support for wind than does the United States, and they use mechanisms that are inherently more pluralistic and egalitarian. This helps explain why wind power is on such a fast track in countries such as Germany, Denmark, and Spain.

Also playing an important role in helping wind power gain a competitive advantage are Renewable Energy Credits, or "green tags."[27] These tags result from laws currently in force in 13 states that require electricity providers to include a prescribed amount

Wind Power and Bird Mortality

There is a persistent public impression that birds and windmills don't mix very well. Particularly for the smaller turbines, the spinning blades are hard to see during the day and are invisible at night. Many of those who campaign against wind power expansion cite this concern as part of their argument.

Concerns about turbine-related bird mortality stem largely from the experience at Altamont Pass, California, where approximately 7,000 wind turbines are located on rolling grassland 50 miles east of San Francisco Bay.[1] Between 1989 and 1991, 182 dead birds were found in study plots associated with wind turbines, including approximately 39 golden eagles killed per year by the turbines.[2] Golden eagles, red-tailed hawks, and American kestrels had higher mortality than more common American ravens and turkey vultures.[3] Deaths of eagles and potential danger to endangered California condors are the biggest issues at Altamont Pass. Bird mortality at comparably sized wind facilities has been reported as being similar or lower than those at Altamont Pass.[4]

While such fatalities are regrettable, there is serious question as to whether they are sufficient to slow or halt the use of wind power. One environmental group, The Defenders of Wildlife, recommends that bird mortality should be "kept in perspective."[5] For comparison glass windows kill 100–900 million birds per year; house cats, 100 million; cars and trucks, 50–100 million; transmission line collisions, up to 175 million; agriculture, 67 million; and hunting, more than 100 million.[6] Clean Power Now, an advocacy group encouraging wind development in Nantucket Sound, answers the question "Do wind turbines kill birds?" by stating "Very few and not always."[7] Altamont Pass, where much of the concern for avian safety originated, appears to be more of the exception than the rule. Data show the actual numbers killed in the pass do not exceed one bird per turbine per year, and for raptors, reported kill rates are 0.05 per turbine per year.[8] Nevertheless, the wind power industry has made several adjustments. For example, perch guards are being installed and a program to replace the old machines with modern turbines on high monopoles is ongoing (One modern turbine replaces seven older machines).[9] More study on this matter would be welcome.

1. W. G. Hunt, R. E. Jackman, T. L. Hunt, D. E. Driscoll, and L. Culp, *A Population Study of Golden Eagles in the Altamont Pass Wind Resource Area: Population Trend Analysis 1997,* report prepared for the National Renewable Energy Laboratory (NREL), Subcontract XAT-6-16459-01 (Santa Cruz, CA: Predatory Bird Research Group, University of California, 1998).
2. S. Orloff and A. Flannery, *Wind Turbine Effects on Avian Activity, Habitat Use, and Mortality in Altamont Pass and Solano County WRAs,* report prepared by BioSystems Analysis, Inc. for the California Energy Commission, 1992.
3. C. G. Thelander and L. Rugge, *Avian Risk Behavior and Fatalities at the Altamont Pass Wind Resource Area,* report prepared for the National Renewable Energy Laboratory: SR-500-27545, (Santa Cruz, CA: Predatory Bird Research Group, University of California, 2000).
4. M. D. McCrary et al.,*Summary of Southern California Edison's Bird Monitoring Studies in the San Gorgonio Pass,* unpublished data; and R. L. Anderson, J. Tom, N. Neumann, J. A. Cleckler, and J. A. Brownell, *Avian Monitoring and Risk Assessment at Tehachapi Pass Wind Resource Area, California* (Sacramento, CA: California Energy Commission, 1996).
5. Defenders of Wildlife, *Renewable Energy: Wind Energy Resources, Principles and Recommendations,* http://www.defenders.org/habitat/renew/wind.html
6. Curry & Kerlinger, *What Kills Birds?* http://www.currykerlinger.com/birds.htm
7. Clean Power Now, *Do Wind Turbines Kill Birds?* http://www.cleanpowernow.org/birdkills.php
8. P. Kerlinger, *An Assessment of the Impacts of Green Mountain Power Corporation's Wind Power Facility on Breeding and Migrating Birds in Searsburg, Vermont, July 1996–July 1998,* report prepared for the Vermont Department of Public Service, NREL/SR-500-28591 (Golden, CO: NREL, March 2002), page 64.
9. R. C. Curry and P. Kerlinger, *Avian Mitigation Plan: Kenetech Model Wind Turbines, Altamont Pass WRA, California,* report presented at the National Avian Wind Power Planning Meeting III, San Diego, CA, May 1998, page 26.

of renewable electricity in the electric power-supply portfolio they offer to their customers. Electricity providers meet this requirement through several possible approaches. They can generate the necessary amount of renewable electricity themselves, purchase it from someone else, or buy credits from other providers who have excess. The green tags rely almost entirely on private market forces. Taken together with production tax credits and various industry improvements, they are helping wind power continue its trend toward independent profitability. Such status, coupled with reduced public resistance, will move wind power from the realm of alternative to the position of mainstream energy resource. This might be possible if we can move from NIMBY to PIMBY.

From NIMBY to PIMBY

When plans encounter resistance, developers usually make amended suggestions to attract greater support. This would seem to be a sensible tactic for wind developers, but it was not a part of planning for wind power in the mid-1980s. Instead, a naive impression prevailed that wind power would attract unquestioned support. However, the public resisted the blatant

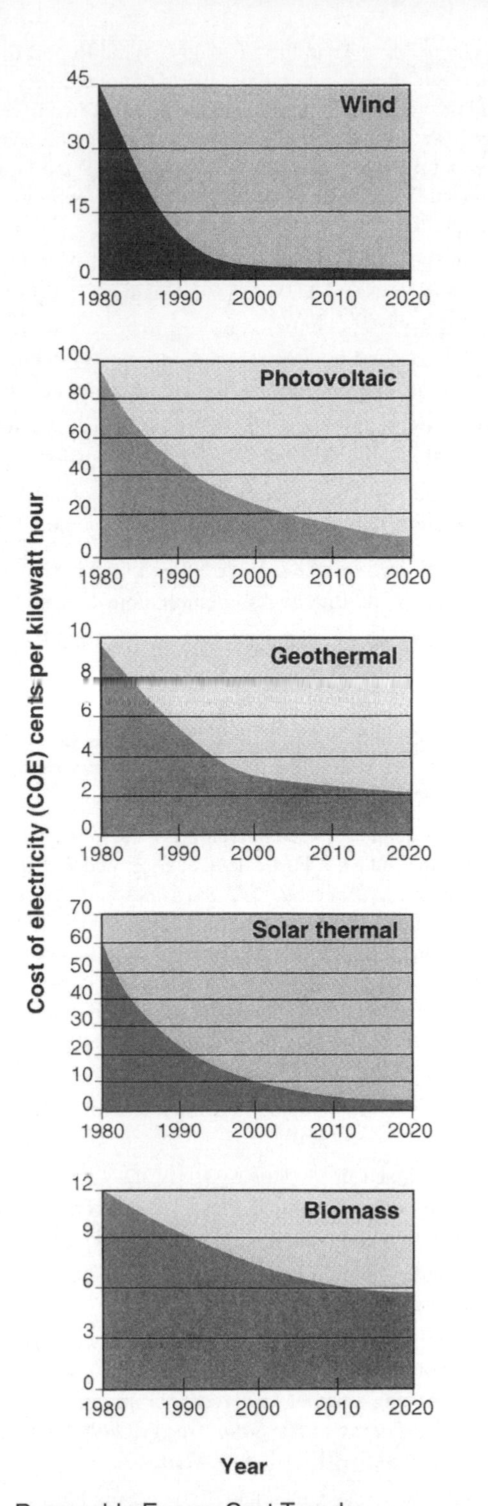

Figure 4 Renewable Energy Cost Trends.

Note. These graphs are reflections of historical cost trends, not precise annual historical data.
Source: National Renewable Energy Laboratory, 2002.

placement of wind turbines on the landscape. It was a particularly unexpected experience because California was known as a state where many of the most ardent environmentalists held forth. Instead of receiving congratulatory handshakes, wind developers (and various government officials) received notices of lawsuits for their trouble. NIMBY, even then a battlescarred acronym (for "Not In My Back Yard"), emerged in the headlines.

The ingredients mixed in the cauldron of subsequent wind power development made for a rich and complex brew. On the positive and promising side, developers learned to appreciate the power of public opinion and to work to inform it more completely. This also applied to regulators and policymakers. All parties began ascribing primacy to cooperation over imposition. By the mid-1990s, wind companies successfully improved efficiency and design, and jurisdictional authorities made zoning codes more appropriate if not more restrictive. The use of focus groups and public hearings became common elements in wind development planning procedures. As a result, controversy and press attention subsided, and projects continued to come on line with little fanfare or public notice in Iowa, Kansas, Minnesota, and Texas.

Then came Cape Wind, and much of the old debate began anew. Developers who had forgotten or never fully appreciated the power of public opinion started retreating. Despite many improvements and increasing experience, Cape Wind planners had devoted insufficient attention to considering the combination of factors that make one place unique from another. They reasoned that if offshore installations were meeting with success in Europe, why should they not find acceptance in "green" Massachusetts? However, they failed to realize the poor comparability between the mindset of people in the United States, who live in a spacious and largely post-industrial country, and their European contemporaries who have been living with industrial landscapes, greater population density, and much less personal space for centuries. In making their calculations, they neglected to note that the coastal areas of Massachusetts, heavily utilized for recreation, is not comparable with the lightly settled coastal areas of Europe.

In many ways, the Cape Wind episode is an East Coast version of the California experience 20 years earlier. Admittedly, the setting is different—desert versus ocean—but the underlying problem is the same: Wind turbines—immovable and numerous—interfering with the aesthetics of a valued recreational resource.

The experience of Cape Wind suggests the need for a fresh approach to wind power development. The key element of this new approach is simplicity itself: Avoid sites having a high potential for conflict. Making this assessment would involve two steps. The first step would be to assign sites "compatibility rankings," starting with the most compatible sites.

- Rank #1 properties would be those where it is not only suitable but overtly requested for wind development, such as farms in Iowa or Kansas.
- Rank #2 properties would likely be acceptable, such as in southeastern Washington.
- Rank #3 properties might be acceptable in certain circumstances, such as near Palm Springs.
- Rank #4 properties would be completely off-limits, for example, on the top of Mt. Rushmore.

Ranks would be determined according to points assigned to site-specific characteristics, including lines of site, type and color tone of terrain, ownership, bird flyways, endangered species, competitive economic value, transmission lines/corridors, protected status (such as national parks), economic development, energy security, and so forth. This should be part of the process of environmental impact assessment, and it should be initiated at any location with a strong, class 4 or above wind resource. Without such rankings, the current ad hoc and contentious approach will continue. It would be akin to a general plan for a city: Variances could be granted, but there would be a broad guidance document in place.

Part two of this plan would be to concentrate our attention on Rank #1 sites. In the United States, this means the Great Plains. There are two simple reasons to emphasize this region. First, the United States' greatest wind resource is there. Second, the small-scale farmers in the area generally welcome the turbines. The message is this: When contentious sites breed contempt, avoid them, at least for now, even if the resource base is attractive and the load centers are nearby. Admit that the wind power alternative is uniquely visible and interferes with scenic vistas and cease trying to force-feed developments down the throats of a resistant public. This is not good for the future of wind power.

On the other hand, there are places where wind power development is welcome. Small farms of the Great Plains have been losing ground for decades to consolidation and the vagaries of weather; they need an economic boost to stay viable. The owners of these farms have put out the welcome mat for wind developers in places such as along Buffalo Ridge, on the border between northwest Iowa and Minnesota, and even farther west in places like Lamar, Colorado. As Chris Rundell, a local rancher, phrased it: "The wind farm has installed a new spirit of community in Lamar . . . it's intangible but very real." They are embracing a new acronym, PIMBY—Please In My Back Yard.

Seeing the wind development in the Great Plains in recent years is a continuation of history, if in a slightly different form: Where a century ago hundreds of thousands of farm windmills made the local agricultural life possible, wind power is again proving its worth to those who would live there. It is bringing needed cash into the local economy and slowing a multiyear trend of farm abandonment and consolidation. The same lands that early wind machines helped develop, new wind machines are helping preserve.[28]

Notes

1. See http://www.state.hi.us/dbedt/ert/wwg/windy.html#molokai
2. European Wind Energy Association, "Wind Power Expands 23% in Europe but Still Is Only a 3-Member State Story," press release, 3 February 2004, www.ewea.org/documents/0203_ EU2003 figures&x005F; final6.pdf
3. Pacific Northwest Laboratory, *An Assessment of the Available Windy Land Area and Wind Energy Potential in the Contiguous United States* (Washington, DC: U.S. Department of Energy (DOE), 1991).
4. American Wind Energy Association, "Wind Energy: An Untapped Resource," fact sheet, 13 January 2004, http://www.awea.org/pubs/factsheets/WindEnergyAnUntappedResource.pdf
5. P. Gipe, "More Than First Thought? Wind Report Stirs Minor Tempest," *Renewable Energy World* 6, no. 5 (2003), available at http://www.jxj.com/magsandj/rew/2003_05/wind_report.html. See also C. L. Archer and M. Z. Jacobson, "The Spatial and Temporal Distributions of U.S. Winds and Wind Power at 80 M Derived from Measurements," *Journal of Geophysical Research* 108, no. D9 (2003): 4289.
6. North Dakota has the capacity to produce 138,000 MW; Texas, 136,000; Kansas, 122,000; South Dakota, 117,000 MW; and Montana, 116,000 MW.
7. American Wind Energy Association; see also the National Wind Technology Center Web site (http://www.nrel.gov/wind/).
8. Craig Cox, senior associate, Interwest Energy Alliance, personal communication with author, 27 April 2004.
9. Paul Gipe, executive director, Ontario Sustainable Energy Association, personal communication with author, 20 April 2004.
10. American Wind Energy Association, note 4 above.
11. American Wind Energy Association, note 4 above.
12. Lawrence Flowers, technical director, National Renewable Energy Laboratory, personal communication with author, 22 June 2004.
13. M. J. Pasqualetti and E. Butler, "Public Reaction to Wind Development in California," *International Journal of Ambient Energy* 8, no. 3 (1987): 83–90.
14. M. J. Pasqualetti, "Accommodating Wind Power in a Hostile Landscape," in M. J. Pasqualetti, P. Gipe, and R. Righter, eds., *Wind Power in View: Energy Landscapes in a Crowded World,* (San Diego, CA: Academic Press, 2002), 153–71; M. J. Pasqualetti, "Morality, Space, and the Power of Wind-Energy Landscapes," *The Geographical Review* 90, no. 3 (2001): 381–94; and M. J. Pasqualetti, "Wind Energy Landscapes: Society and Technology in the California Desert," *Society and Natural Resources* 14, no. 8 (2001): 689–99.
15. L. Marx, *The Machine in the Garden* (Oxford, UK: Oxford University Press, 1964).
16. C. C. Reith and B. M. Thomson, eds., *Deserts As Dumps? The Disposal of Hazardous Materials in Arid Ecosystems* (Albuquerque: University of New Mexico Press, 1992).
17. Pasqualetti, "Wind Energy Landscapes: Society and Technology in the California Desert," note 14 above.
18. Total worldwide wind energy installations were 1,000 MW in 1985, 18,000 MW in 2000, and nearly 40,000 MW in 2003, growing at about 35 percent per annum. See Solarbuzz, *Fast Solar Energy Facts: Solar Energy Global,* http://www.solarbuzz.com/FastFactslndustry.htm
19. See http://www.saveoursound.org/windspin.html
20. Today's support started with Public Utility Regulatory Policies Act (PURPA), the 1978 law that promoted alternative energy sources and energy efficiency by requiring utilities to buy power from independent companies that could produce power for less than what it would have cost for the utility to generate the power, the so-called "avoided cost." In the past 20 years, electricity feed-in laws have been popular in Denmark, Germany, Italy, France, Portugal, and Spain. Private generators, or producers, charge a feed-in tariff for the price-per-unit of electricity the suppliers or utility buy. The rate of the tariff is determined by the federal government. In other words, the government sets the price for electricity in the country.

Because the producer is guaranteed a price for the electricity, if he or she meets certain criteria, feed-in laws help attract new generation capacity. During the past decade, Germany became the world leader in wind development. Much of this success is due to *Stromeinspeisungsgesetz,* literally meaning the "Law on Feeding Electricity from Renewable Sources into the Public Network." The original Electricity Feed Law set the price for renewable electricity sources at 90 percent the retail residential price. In 2001, the German feed law was modified to a simple, fixed price for each renewable technology. See http://www.geni.org/globalenergy/policy/renewableenergy/electricityfeed-inlaws/germany/index.shtml

21. Union of Concerned Scientists, "How Wind Power Works," briefing, http://www.ucsusa.org/CoalvsWind/brief.wind.html
22. Ibid.
23. The U.S. Production Tax Cut, by its emphasis on the actual generation and transmission of electricity, not just the construction of equipment, provides additional incentives for greater technical efficiency.
24. See http://www.interwestenergy.org
25. Reuters went on to report that the "Wind industry backers say the gaps have created a roller coaster in U.S. wind production growth because companies become fearful of investing in the alternative energy source. They say the tax-break gaps hamper wind power growth in the United States which grew last year at a rate of only 10 percent compared to global growth of 28 percent." See "Wind Power Tax Credit Expires in December," *Reuters World Environment News,* 27 November 2003, http://www.planetark.com/dailynewsstory.cfm/newsid/22956/newsDate/27-Nov2003/story.htm
26. *First Quarter Report: Wind Industry Trade Group Sees Little To No Growth In 2004, Following Near-Record Expansion In 2003,* http://www.awea.org/news/news0405121qt.html
27. For more information about green tags for wind power, see http://www.sustainablemarketing.com/wind.php?google
28. For several economic summaries, see S. Clemmer, *The Economic Development Benefits of Wind Power,* presentation at Harvesting Clean Energy Conference, Boise, Idaho, 10 February 2003, available at http://www.eere.energy.gov/windpoweringamerica/pdfs/wpa/34600_wind_handbook.pdf

Martin J. Pasqualetti is a professor of geography at Arizona State University in Tempe. His research interests include renewable energy and the landscape impacts of energy development and use. He is a coeditor of and contributor to *Wind Power in View: Energy Landscapes in a Crowded World* (Academic Press, 2002). He also contributed articles on wind power to the *Encyclopedia of Energy* (Academic Press, 2004). His research has appeared in *The Geographical Review and Society and Natural Resources.* He thanks Paul Gipe and Robert Righter for reading the manuscript for this article and offering many helpful suggestions. Pasqualetti may be reached at (480) 965-7533 or via e-mail at pasqualetti@asu.edu.

Article 11

A Solar Grand Plan

By 2050 solar power could end U.S. dependence on foreign oil and slash greenhouse gas emissions.

KEN ZWEIBEL, JAMES MASON, AND VASILIS FTHENAKIS

High prices for gasoline and home heating oil are here to stay. The U.S. is at war in the Middle East at least in part to protect its foreign oil interests. And as China, India and other nations rapidly increase their demand for fossil fuels, future fighting over energy looms large. In the meantime, power plants that burn coal, oil and natural gas, as well as vehicles everywhere, continue to pour millions of tons of pollutants and greenhouse gases into the atmosphere annually, threatening the planet.

Well-meaning scientists, engineers, economists and politicians have proposed various steps that could slightly reduce fossil-fuel use and emissions. These steps are not enough. The U.S. needs a bold plan to free itself from fossil fuels. Our analysis convinces us that a massive switch to solar power is the logical answer.

Solar energy's potential is off the chart. The energy in sunlight striking the earth for 40 minutes is equivalent to global energy consumption for a year. The U.S. is lucky to be endowed with a vast resource; at least 250,000 square miles of land in the Southwest alone are suitable for constructing solar power plants, and that land receives more than 4,500 quadrillion British thermal units (Btu) of solar radiation a year. Converting only 2.5 percent of that radiation into electricity would match the nation's total energy consumption in 2006.

To convert the country to solar power, huge tracts of land would have to be covered with photovoltaic panels and solar heating troughs. A direct-current (DC) transmission backbone would also have to be erected to send that energy efficiently across the nation.

The technology is ready. On the following pages we present a grand plan that could provide 69 percent of the U.S.'s electricity and 35 percent of its total energy (which includes transportation) with solar power by 2050. We project that this energy could be sold to consumers at rates equivalent to today's rates for conventional power sources, about five cents per kilowatt-hour (kWh). If wind, bio-mass and geothermal sources were also developed, renewable energy could provide 100 percent of the nation's electricity and 90 percent of its energy by 2100.

Key Concepts

- A massive switch from coal, oil, natural gas and nuclear power plants to solar power plants could supply 69 percent of the U.S.'s electricity and 35 percent of its total energy by 2050.
- A vast area of photovoltaic cells would have to be erected in the Southwest. Excess daytime energy would be stored as compressed air in underground caverns to be tapped during nighttime hours.
- Large solar concentrator power plants would be built as well.
- A new direct-current power transmission backbone would deliver solar electricity across the country.
- But $420 billion in subsidies from 2011 to 2050 would be required to fund the infrastructure and make it cost-competitive.

—The Editors

The federal government would have to invest more than $400 billion over the next 40 years to complete the 2050 plan. That investment is substantial, but the payoff is greater. Solar plants consume little or no fuel, saving billions of dollars year after year. The infrastructure would displace 300 large coal-fired power plants and 300 more large natural gas plants and all the fuels they consume. The plan would effectively eliminate all imported oil, fundamentally cutting U.S. trade deficits and easing political tension in the Middle East and elsewhere. Because solar technologies are almost pollution-free, the plan would also reduce greenhouse gas emissions from power plants by 1.7 billion tons a year, and another 1.9 billion tons from gasoline vehicles would be displaced by plug-in hybrids refueled by the solar power grid. In 2050 U.S. carbon dioxide emissions would be 62 percent below 2005 levels, putting a major brake on global warming.

U.S. Plan for 2050

Solar Power Provides . . .

69% of electricity | **35%** of total energy

By 2050 vast photovoltaic arrays in the Southwest would supply electricity instead of fossil-fueled power plants and would also power a widespread conversion to plug-in electric vehicles. Excess energy would be stored as compressed air in underground caverns. Large arrays that concentrate sunlight to heat water would also supply electricity. A new high-voltage, direct-current transmission backbone would carry power to regional markets nationwide. The technologies and factors critical to their success are summarized at the right, along with the extent to which the technologies must be deployed by 2050. The plan would substantially cut the country's consumption of fossil fuels and its emission of greenhouse gases (*below*). We have assumed a 1 percent annual growth in net energy demand. And we have anticipated improvements in solar technologies forecasted only until 2020, with no further gains beyond that date.

—*K.Z., J.M. and V.F.*

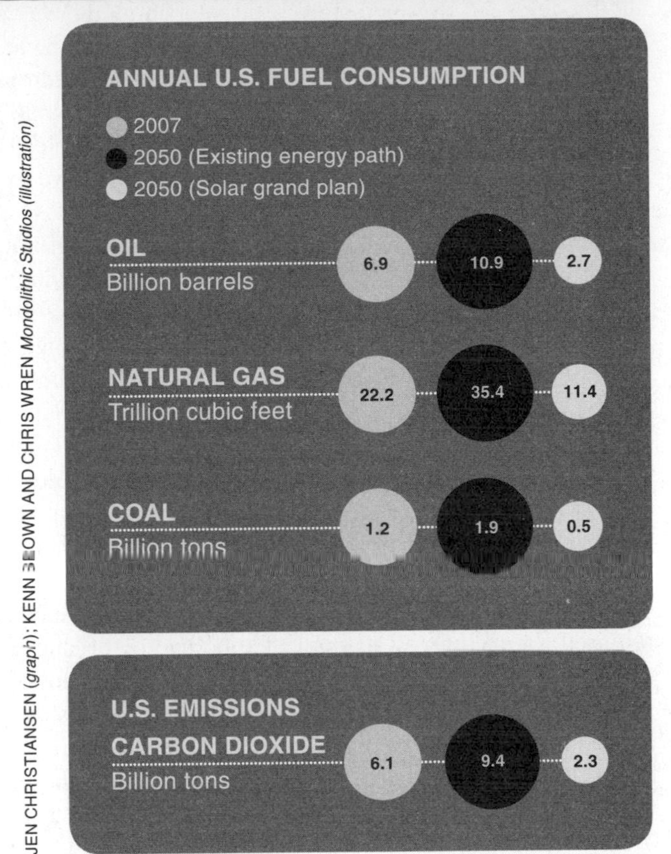

JEN CHRISTIANSEN (*graph*); KENN BROWN AND CHRIS WREN *Mondolithic Studios (illustration)*

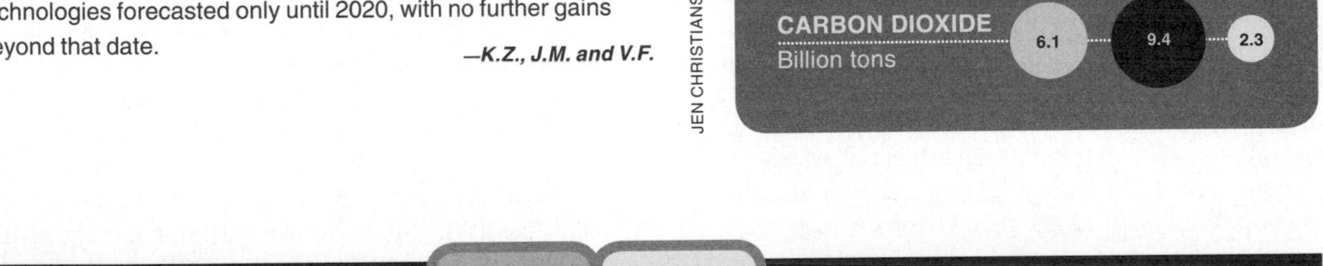

TECHNOLOGY	CRITICAL FACTOR	2007	2050	ADVANCES NEEDED
PHOTOVOLTAICS	Land area	10 sq miles	30,000 sq miles	Policies to develop large public land areas
	Thin-film module efficiency	10%	14%	More transparent materials to improve light transmission; more densely doped layers to increase voltage; larger modules to reduce inactive area
	Installed cost	$4/W	$1.20/W	Improvements in module efficiency; gains from volume production
	Electricity price	16¢/kWh	5¢/kWh	Follows from lower installed cost
	Total capacity	0.5 GW	2,940 GW	National energy plan built around solar power
COMPRESSED-AIR ENERGY STORAGE (with photovoltaic electricity)	Volume	0	535 billion cu ft	Coordination of site development with natural gas industry
	Installed cost	$5.80/W	$3.90/W	Economies of scale; decreasing photovoltaic electricity prices
	Electricity price	20¢/kWh	9¢/kWh	Follows from lower installed cost
	Total capacity	0.1 GW	558 GW	National energy plan
CONCENTRATED SOLAR POWER	Land area	10 sq miles	16,000 sq miles	Policies to develop large public land areas
	Solar-to-electric efficiency	13%	17%	Fluids that transfer heat more effectively
	Installed cost	$5.30/W	$3.70/W	Single-tank thermal storage systems; economies of scale
	Electricity price	18¢/kWh	9¢/kWh	Follows from lower installed cost
	Total capacity	0.5 GW	558 GW	National energy plan
DC TRANSMISSION	Length	500 miles	100,000–500,000 miles	New high-voltage DC grid from Southwest to rest of country

Photovoltaic Farms

In the past few years the cost to produce photo-voltaic cells and modules has dropped significantly, opening the way for large-scale deployment. Various cell types exist, but the least expensive modules today are thin films made of cadmium telluride. To provide electricity at six cents per kWh by 2020, cadmium telluride modules would have to convert electricity with 14 percent efficiency, and systems would have to be installed at $1.20 per watt of capacity. Current modules have 10 percent efficiency and an installed system cost of about $4 per watt. Progress is clearly needed, but the technology is advancing quickly; commercial efficiencies have risen from 9 to 10 percent in the past 12 months. It is worth noting, too, that as modules improve, rooftop photovoltaics will become more cost-competitive for homeowners, reducing daytime electricity demand.

In our plan, by 2050 photovoltaic technology would provide almost 3,000 gigawatts (GW), or billions of watts, of power. Some 30,000 square miles of photovoltaic arrays would have to be erected. Although this area may sound enormous, installations already in place indicate that the land required for each gigawatt-hour of solar energy produced in the Southwest is less than that needed for a coal-powered plant when factoring in land for coal mining. Studies by the National Renewable Energy Laboratory in Golden, Colo., show that more than enough land in the Southwest is available without requiring use of environmentally sensitive areas, population centers or difficult terrain. Jack Lavelle, a spokesperson for Arizona's Department of Water Conservation, has noted that more than 80 percent of his state's land is not privately owned and that Arizona is very interested in developing its solar potential. The benign nature of photovoltaic plants (including no water consumption) should keep environmental concerns to a minimum.

The main progress required, then, is to raise module efficiency to 14 percent. Although the efficiencies of commercial modules will never reach those of solar cells in the laboratory, cadmium telluride cells at the National Renewable Energy Laboratory are now up to 16.5 percent and rising. At least one manufacturer, First Solar in Perrysburg, Ohio, increased module efficiency from 6 to 10 percent from 2005 to 2007 and is reaching for 11.5 percent by 2010.

Photovoltaics

In the 2050 plan vast photovoltaic farms would cover 30,000 square miles of otherwise barren land in the Southwest. They would resemble Tucson Electric Power Company's 4.6-megawatt plant in Springerville, Ariz., which began in 2000. In such farms, many photovoltaic cells are interconnected on one module, and modules are wired together to form an array. The direct current from each array flows to a transformer that sends it along high-voltage lines to the power grid. In a thin-film cell, the energy of incoming photons knocks loose electrons in the cadmium telluride layer; they cross a junction, flow to the top conductive layer and then flow around to the back conductive layer, creating current.

Pressurized Caverns

The great limiting factor of solar power, of course, is that it generates little electricity when skies are cloudy and none at night. Excess power must therefore be produced during sunny hours and stored for use during dark hours. Most energy storage systems such as batteries are expensive or inefficient.

Compressed-air energy storage has emerged as a successful alternative. Electricity from photovoltaic plants compresses air and pumps it into vacant underground caverns, abandoned mines, aquifers and depleted natural gas wells. The pressurized air is released on demand to turn a turbine that generates electricity, aided by burning small amounts of natural gas. Compressed-air energy storage plants have been operating reliably in Huntorf, Germany, since 1978 and in McIntosh, Ala., since 1991. The turbines burn only 40 percent of the natural gas they would if they were fueled by natural gas alone, and better heat recovery technology would lower that figure to 30 percent.

Studies by the Electric Power Research Institute in Palo Alto, Calif., indicate that the cost of compressed-air energy storage today is about half that of lead-acid batteries. The research indicates that these facilities would add three or four cents per kWh to photovoltaic generation, bringing the total 2020 cost to eight or nine cents per kWh.

Electricity from photovoltaic farms in the Southwest would be sent over high-voltage DC transmission lines to compressed-air storage facilities throughout the country, where turbines would generate electricity year-round. The key is to find adequate sites. Mapping by the natural gas industry and the Electric Power Research Institute shows that suitable geologic formations exist in 75 percent of the country, often close to metropolitan areas. Indeed, a compressed-air energy storage system would look similar to the U.S. natural gas storage system. The industry stores eight trillion cubic feet of gas in 400 underground reservoirs. By 2050 our plan would require 535 billion cubic feet of storage, with air pressurized at 1,100 pounds per square inch. Although development will be a challenge, plenty of reservoirs are available, and it would be reasonable for the natural gas industry to invest in such a network.

Hot Salt

Another technology that would supply perhaps one fifth of the solar energy in our vision is known as concentrated solar power. In this design, long, metallic mirrors focus sunlight

Payoffs

- Foreign oil dependence cut from 60 to 0 percent
- Global tensions eased and military costs lowered
- Massive trade deficit reduced significantly
- Greenhouse gas emissions slashed
- Domestic jobs increased

Pinch Points

- Subsidies totaling $420 billion through 2050
- Political leadership needed to raise the subsidy, possibly with a carbon tax
- New high-voltage, direct-current electric transmission system built profitably by private carriers

onto a pipe filled with fluid, heating the fluid like a huge magnifying glass might. The hot fluid runs through a heat exchanger, producing steam that turns a turbine.

For energy storage, the pipes run into a large, insulated tank filled with molten salt, which retains heat efficiently. Heat is extracted at night, creating steam. The molten salt does slowly cool, however, so the energy stored must be tapped within a day.

Nine concentrated solar power plants with a total capacity of 354 megawatts (MW) have been generating electricity reliably for years in the U.S. A new 64-MW plant in Nevada came online in March 2007. These plants, however, do not have heat storage. The first commercial installation to incorporate it—a 50-MW plant with seven hours of molten salt storage—is being constructed in Spain, and others are being designed around the world. For our plan, 16 hours of storage would be needed so that electricity could be generated 24 hours a day.

Existing plants prove that concentrated solar power is practical, but costs must decrease. Economies of scale and continued research would help. In 2006 a report by the Solar Task Force of the Western Governors' Association concluded that concentrated solar power could provide electricity at 10 cents per kWh or less by 2015 if 4 GW of plants were constructed. Finding ways to boost the temperature of heat exchanger fluids would raise operating efficiency, too. Engineers are also investigating how to use molten salt itself as the heat-transfer fluid, reducing heat losses as well as capital costs. Salt is corrosive, however, so more resilient piping systems are needed.

Concentrated solar power and photovoltaics represent two different technology paths. Neither is fully developed, so our plan brings them both to large-scale deployment by 2020, giving them time to mature. Various combinations of solar technologies might also evolve to meet demand economically. As installations expand, engineers and accountants can evaluate the pros and cons, and investors may decide to support one technology more than another.

Direct Current, Too

The geography of solar power is obviously different from the nation's current supply scheme. Today coal, oil, natural gas and nuclear power plants dot the landscape, built relatively close to where power is needed. Most of the country's solar generation would stand in the Southwest. The existing system of alternating-current (AC) power lines is not robust enough to carry power from these centers to consumers everywhere and would lose too much energy over long hauls. A new high-voltage, direct-current (HVDC) power transmission backbone would have to be built.

Studies by Oak Ridge National Laboratory indicate that long-distance HVDC lines lose far less energy than AC lines do over equivalent spans. The backbone would radiate from the Southwest toward the nation's borders. The lines would terminate at converter stations where the power would be switched to AC and sent along existing regional transmission lines that supply customers.

The AC system is also simply out of capacity, leading to noted shortages in California and other regions; DC lines are cheaper to build and require less land area than equivalent AC lines. About 500 miles of HVDC lines operate in the U.S. today and have proved reliable and efficient. No major technical advances seem to be needed, but more experience would help refine operations. The Southwest Power Pool of Texas is designing an integrated system of DC and AC transmission to enable development of 10 GW of wind power in western Texas. And TransCanada, Inc., is proposing 2,200 miles of HVDC lines to carry wind energy from Montana and Wyoming south to Las Vegas and beyond.

Stage One: Present to 2020

We have given considerable thought to how the solar grand plan can be deployed. We foresee two distinct stages. The first, from now until 2020, must make solar competitive at the mass-production level. This stage will require the government to guarantee 30-year loans, agree to purchase power and provide price-support subsidies. The annual aid package would rise steadily from 2011 to 2020. At that time, the solar technologies would compete on their own merits. The cumulative subsidy would total $420 billion (we will explain later how to pay this bill).

About 84 GW of photovoltaics and concentrated solar power plants would be built by 2020. In parallel, the DC transmission system would be laid. It would expand via existing rights-of-way along interstate highway corridors, minimizing land-acquisition and regulatory hurdles. This backbone would reach major markets in Phoenix, Las Vegas, Los Angeles and San Diego to the west and San Antonio,

Dallas, Houston, New Orleans, Birmingham, Ala., Tampa, Fla., and Atlanta to the east.

Building 1.5 GW of photovoltaics and 1.5 GW of concentrated solar power annually in the first five years would stimulate many manufacturers to scale up. In the next five years, annual construction would rise to 5 GW apiece, helping firms optimize production lines. As a result, solar electricity would fall toward six cents per kWh. This implementation schedule is realistic; more than 5 GW of nuclear power plants were built in the U.S. each year from 1972 to 1987. What is more, solar systems can be manufactured and installed at much faster rates than conventional power plants because of their straightforward design and relative lack of environmental and safety complications.

Stage Two: 2020 to 2050

It is paramount that major market incentives remain in effect through 2020, to set the stage for self-sustained growth thereafter. In extending our model to 2050, we have been conservative. We do not include any technological or cost improvements beyond 2020. We also assume that energy demand will grow nationally by 1 percent a year. In this scenario, by 2050 solar power plants will supply 69 percent of U.S. electricity and 35 percent of total U.S. energy. This quantity includes enough to supply all the electricity consumed by 344 million plug-in hybrid vehicles, which would displace their gasoline counterparts, key to reducing dependence on foreign oil and to mitigating greenhouse gas emissions. Some three million new domestic jobs—notably in manufacturing solar components—would be created, which is several times the number of U.S. jobs that would be lost in the then dwindling fossil-fuel industries.

The huge reduction in imported oil would lower trade balance payments by $300 billion a year, assuming a crude oil price of $60 a barrel (average prices were higher in 2007). Once solar power plants are installed, they must be maintained and repaired, but the price of sunlight is forever free, duplicating those fuel savings year after year. Moreover, the solar investment would enhance national energy security, reduce financial burdens on the military, and greatly decrease the societal costs of pollution and global warming, from human health problems to the ruining of coastlines and farmlands.

Ironically, the solar grand plan would lower energy consumption. Even with 1 percent annual growth in demand, the 100 quadrillion Btu consumed in 2006 would *fall* to 93 quadrillion Btu by 2050. This unusual offset arises because a good deal of energy is consumed to extract and process fossil fuels, and more is wasted in burning them and controlling their emissions.

To meet the 2050 projection, 46,000 square miles of land would be needed for photovoltaic and concentrated solar power installations. That area is large, and yet it covers just 19 percent of the suitable Southwest land. Most of that land is barren; there is no competing use value. And the land will not be polluted. We have assumed that only 10 percent of the solar capacity in 2050 will come from distributed photovoltaic installations—those on rooftops or commercial lots throughout the country. But as prices drop, these applications could play a bigger role.

Underground Storage

Excess electricity produced during the day by photovoltaic farms would be sent over power lines to compressed-air energy storage sites close to cities. At night the sites would generate power for consumers. Such technology is already available; the PowerSouth Energy Cooperative's plant in Mc-Intosh, Ala. has operated since 1991. In these designs, incoming electricity runs motors and compressors that pressurize air and send it into vacant caverns, mines or aquifers. When the air is released, it is heated by burning small amounts of natural gas; the hot, expanding gases turn turbines that generate electricity.

2050 and Beyond

Although it is not possible to project with any exactitude 50 or more years into the future, as an exercise to demonstrate the full potential of solar energy we constructed a scenario for 2100. By that time, based on our plan, total energy demand (including transportation) is projected to be 140 quadrillion Btu, with seven times today's electric generating capacity.

To be conservative, again, we estimated how much solar plant capacity would be needed under the historical worst-case solar radiation conditions for the Southwest, which occurred during the winter of 1982–1983 and in 1992 and 1993 following the Mount Pinatubo eruption, according to National Solar Radiation Data Base records from 1961 to 2005. And again, we did not assume any further technological and cost improvements beyond 2020, even though it is nearly certain that in 80 years ongoing research would improve solar efficiency, cost and storage.

Under these assumptions, U.S. energy demand could be fulfilled with the following capacities: 2.9 terawatts (TW) of photovoltaic power going directly to the grid and another 7.5 TW dedicated to compressed-air storage; 2.3 TW of concentrated solar power plants; and 1.3 TW of distributed photovoltaic installations. Supply would be rounded out with 1 TW of wind farms, 0.2 TW of geothermal power plants and 0.25 TW of biomass-based production for fuels. The model includes 0.5 TW of geothermal heat pumps for direct building heating and cooling. The solar systems would require 165,000 square miles of land, still less than the suitable available area in the Southwest.

Concentrated Solar

Large concentrated solar power plants would complement photo-voltaic farms in the Southwest. The Kramer Junction plant in California's Mojave Desert, using technology from Solel in Beit Shemesh, Israel, has been operating since 1989. Metallic parabolic mirrors focus sunlight on a pipe, heating fluid such as ethylene glycol inside. The mirrors rotate to track the sun. The hot pipes run alongside a second loop inside a heat exchanger that contains water, turning it to steam that drives a turbine. Future plants could also send the hot fluid through a holding tank, heating molten salt; that reservoir would retain heat that could be tapped at night for the heat exchanger.

More to Explore

The Terawatt Challenge for Thin Film Photovoltaic. Ken Zweibel in *Thin Film Solar Cells: Fabrication, Characterization and Applications.* Edited by Jef Poortmans and Vladimir Arkhipov. John Wiley & Sons, 2006.

Energy Autonomy: The Economic, Social and Technological Case for Renewable Energy. Hermann Scheer. Earthscan Publications, 2007.

Center for Life Cycle Analysis, Columbia University: www.clca.columbia.edu

The National Solar Radiation Data Base. National Renewable Energy Laboratory, 2007. http://rredc.nrel.gov/solar/old_ data/nsrdb

The U.S. Department of Energy Solar America Initiative: www1.eere.energy.gov/solar/ solar_america

By 2100 renewable energy could generate 100 percent of the U.S.'s electricity and more than 90 percent of its energy.

In 2100 this renewable portfolio could generate 100 percent of all U.S. electricity and more than 90 percent of total U.S. energy. In the spring and summer, the solar infrastructure would produce enough hydrogen to meet more than 90 percent of all transportation fuel demand and would replace the small natural gas supply used to aid compressed-air turbines. Adding 48 billion gallons of biofuel would cover the rest of transportation energy. Energy-related carbon dioxide emissions would be reduced 92 percent below 2005 levels.

Who Pays?

Our model is not an austerity plan, because it includes a 1 percent annual increase in demand, which would sustain lifestyles similar to those today with expected efficiency improvements in energy generation and use. Perhaps the biggest question is how to pay for a $420-billion overhaul of the nation's energy infrastructure. One of the most common ideas is a carbon tax. The International Energy Agency suggests that a carbon tax of $40 to $90 per ton of coal will be required to induce electricity generators to adopt carbon capture and storage systems to reduce carbon dioxide emissions. This tax is equivalent to raising the price of electricity by one to two cents per kWh. But our plan is less expensive. The $420 billion could be generated with a carbon tax of 0.5 cent per kWh. Given that electricity today generally sells for six to 10 cents per kWh, adding 0.5 cent per kWh seems reasonable.

Congress could establish the financial incentives by adopting a national renewable energy plan. Consider the U.S. Farm Price Support program, which has been justified in terms of national security. A solar price support program would secure the nation's energy future, vital to the country's long-term health. Subsidies would be gradually deployed from 2011 to 2020. With a standard 30-year payoff interval, the subsidies would end from 2041 to 2050. The HVDC transmission companies would not have to be subsidized, because they would finance construction of lines and converter stations just as they now finance AC lines, earning revenues by delivering electricity.

Although $420 billion is substantial, it is less than the U.S. Farm Price Support program.

Although $420 billion is substantial, the annual expense would be less than the current U.S. Farm Price Support program. It is also less than the tax subsidies that have been levied to build the country's high-speed telecommunications infrastructure over the past 35 years. And it frees the U.S. from policy and budget issues driven by international energy conflicts.

Without subsidies, the solar grand plan is impossible. Other countries have reached similar conclusions: Japan is already building a large, subsidized solar infrastructure, and Germany has embarked on a nationwide program. Although the investment is high, it is important to remember that the energy source, sunlight, is free. There are no annual fuel or pollution-control costs like those for coal, oil or nuclear power, and only a slight cost for natural gas in compressed-air systems, although hydrogen or biofuels could displace that, too. When fuel savings are factored in, the cost of solar would be a bargain in coming decades. But we cannot wait until then to begin scaling up.

Critics have raised other concerns, such as whether material constraints could stifle large-scale installation. With rapid deployment, temporary shortages are possible. But several types of cells exist that use different material combinations. Better processing and recycling are also reducing the amount of materials that cells require. And in the long term, old solar cells can largely be recycled into new solar cells, changing our energy supply picture from depletable fuels to recyclable materials.

The greatest obstacle to implementing a renewable U.S. energy system is not technology or money, however. It is the lack of public awareness that solar power is a practical alternative—and one that can fuel transportation as well. Forward-looking thinkers should try to inspire U.S. citizens, and their political and scientific leaders, about solar power's incredible potential. Once Americans realize that potential, we believe the desire for energy self-sufficiency and the need to reduce carbon dioxide emissions will prompt them to adopt a national solar plan.

Ken Zweibel, James Mason and **Vasilis Fthenakis** met a decade ago while working on life-cycle studies of photovoltaics. Zweibel is president of PrimeStar Solar in Golden, Colo., and for 15 years was manager of the National Renewable Energy Laboratory's Thin-Film PV Partnership. Mason is director of the Solar Energy Campaign and the Hydrogen Research Institute in Farmingdale, N.Y. Fthenakis is head of the Photovoltaic Environmental Research Center at Brookhaven National Laboratory and is a professor in and director of Columbia University's Center for Life Cycle Analysis.

The Rise of Renewable Energy

Solar cells, wind turbines and biofuels are poised to become major energy sources. New policies could dramatically accelerate that evolution.

DANIEL M. KAMMEN

No plan to substantially reduce greenhouse gas emissions can succeed through increases in energy efficiency alone. Because economic growth continues to boost the demand for energy—more coal for powering new factories, more oil for fueling new cars, more natural gas for heating new homes—carbon emissions will keep climbing despite the introduction of more energy-efficient vehicles, buildings and appliances. To counter the alarming trend of global warming, the U.S. and other countries must make a major commitment to developing renewable energy sources that generate little or no carbon.

Renewable energy technologies were suddenly and briefly fashionable three decades ago in response to the oil embargoes of the 1970s, but the interest and support were not sustained. In recent years, however, dramatic improvements in the performance and affordability of solar cells, wind turbines and biofuels—ethanol and other fuels derived from plants—have paved the way for mass commercialization. In addition to their environmental benefits, renewable sources promise to enhance America's energy security by reducing the country's reliance on fossil fuels from other nations. What is more, high and wildly fluctuating prices for oil and natural gas have made renewable alternatives more appealing.

Overview

- Thanks to advances in technology, renewable sources could soon become large contributors to global energy.
- To hasten the transition, the U.S. must significantly boost its R&D spending on energy.
- The U.S. should also levy a fee on carbon to reward clean energy sources over those that harm the environment.

We are now in an era where the opportunities for renewable energy are unprecedented, making this the ideal time to advance clean power for decades to come. But the endeavor will require a long-term investment of scientific, economic and political resources. Policymakers and ordinary citizens must demand action and challenge one another to hasten the transition.

Let the Sun Shine

Solar Cells, also known as photovoltaics, use semiconductor materials to convert sunlight into electric current. They now provide just a tiny slice of the world's electricity: their global generating capacity of 5,000 megawatts (MW) is only 0.15 percent of the total generating capacity from all sources. Yet sunlight could potentially supply 5,000 times as much energy as the world currently consumes. And thanks to technology improvements, cost declines and favorable policies in many states and nations, the annual production of photovoltaics has increased by more than 25 percent a year for the past decade and by a remarkable 45 percent in 2005. The cells manufactured last year added 1,727 MW to worldwide generating capacity, with 833 MW made in Japan, 353 MW in Germany and 153 MW in the U.S.

Solar cells can now be made from a range of materials, from the traditional multicrystalline silicon wafers that still dominate the market to thin-film silicon cells and devices composed of plastic or organic semiconductors. Thin-film photovoltaics are cheaper to produce than crystalline silicon cells but are also less efficient at turning light into power. In laboratory tests, crystalline cells have achieved efficiencies of 30 percent or more; current commercial cells of this type range from 15 to 20 percent. Both laboratory and commercial efficiencies for all kinds of solar cells have risen steadily in recent years, indicating that an expansion of research efforts would further enhance the performance of solar cells on the market.

Solar photovoltaics are particularly easy to use because they can be installed in so many places—on the roofs or walls of homes and office buildings, in vast arrays in the desert, even sewn into clothing to power portable electronic devices. The state of California has joined Japan and Germany in leading a global push for solar installations; the "Million Solar Roof" commitment is intended to create 3,000 MW of new generating capacity in the state by 2018. Studies done by my research group, the Renewable and Appropriate Energy Laboratory at the University of California, Berkeley, show that annual production of solar photovoltaics in the U.S. alone could grow to 10,000 MW in just 20 years if current trends continue.

The biggest challenge will be lowering the price of the photovoltaics, which are now relatively expensive to manufacture. Electricity produced by crystalline cells has a total cost of 20 to 25 cents per kilowatt-hour, compared with four to six cents for coal-fired electricity, five to seven cents for power produced by burning natural gas, and six to nine cents for biomass power plants. (The cost of nuclear power is harder to pin down because experts disagree on which expenses to include in the analysis; the estimated range is two to 12 cents per kilowatt-hour.) Fortunately, the prices of solar cells have fallen consistently over the past decade, largely because of improvements in manufacturing processes. In Japan, where 290 MW of solar generating capacity were added in 2005 and an even larger amount was exported, the cost of photovoltaics has declined 8 percent a year; in California, where 50 MW of solar power were installed in 2005, costs have dropped 5 percent annually.

Surprisingly, Kenya is the global leader in the number of solar power systems installed per capita (but not the number of watts added). More than 30,000 very small solar panels, each producing only 12 to 30 watts, are sold in that country annually. For an investment of as little as $100 for the panel and wiring, the system can be used to charge a car battery, which can then provide enough power to run a fluorescent lamp or a small black-and-white television for a few hours a day. More Kenyans adopt solar power every year than make connections to the country's electric grid. The panels typically use solar cells made of amorphous silicon; although these photovoltaics are only half as efficient as crystalline cells, their cost is so much lower (by a factor of at least four) that they are more affordable and useful for the two billion

5,000 megawatts

Global generating capacity of solar power

37 percent

Top efficiency of experimental solar cells

20 to 25 cents

Cost per kilowatt-hour of solar power

Growing Fast, but Still a Sliver

Solar cells, wind power and biofuels are rapidly gaining traction in the energy markets, but they remain marginal providers compared with fossil-fuel sources such as coal, natural gas and oil.

The Renewable Boom

Since 2000 the commercialization of renewable energy sources has accelerated dramatically. The annual global production of solar cells, also known as photovoltaics, jumped 45 percent in 2005. The construction of new wind farms, particularly in Europe, has boosted the worldwide generating capacity of wind power 10-fold over the past decade. And the production of ethanol, the most common biofuel, soared to 36.5 billion liters last year, with the lion's share distilled from American-grown corn.

Jen Christiansen; Sources; PV News, BTM Consult, AWEA, EWEA, F.O. Light and *BP Statistical Review of World Energy* 2006.

Hot Power from Mirrors

Solar-thermal systems, long used to provide hot water for homes or factories, can also generate electricity. Because these systems produce power from solar heat rather than light, they avoid the need for expensive photovoltaics.

Solar Concentrator

A solar-thermal array consists of thousands of dish-shaped solar concentrators, each attached to a Stirling engine that converts heat to electricity. The mirrors in the concentrator are positioned to focus reflected sunlight on the Stirling engine's receiver.

Stirling Engine

A high-performance Stirling engine shuttles a working fluid, such as hydrogen gas, between two chambers. The cold chamber is separated from the hot chamber by a regenerator that maintains the temperature difference between them. Solar energy from the receiver heats the gas in the hot chamber, causing it to expand and move the hot piston. This piston then reverses direction, pushing the heated gas into the cold chamber. As the gas cools, the cold piston can easily compress it, allowing the cycle to start a new. The movement of the pistons drives a turbine that generates electricity in an alternator.

Don Foley, Source: U.S. Department of Energy

people worldwide who currently have no access to electricity. Sales of small solar power systems are booming in other African nations as well, and advances in low-cost photovoltaic manufacturing could accelerate this trend.

Furthermore, photovoltaics are not the only fast-growing form of solar power. Solar-thermal systems, which collect sunlight to generate heat, are also undergoing a resurgence. These systems have long been used to provide hot water for homes or factories, but they can also produce electricity without the need for expensive solar cells. In one design, for example, mirrors focus light on a Stirling engine, a high-efficiency device containing a working fluid that circulates between hot and cold chambers. The fluid expands as the sunlight heats it, pushing a piston that, in turn, drives a turbine.

In the fall of 2005 a Phoenix company called Stirling Energy Systems announced that it was planning to build two large solar-thermal power plants in southern California. The company signed a 20-year power purchase agreement with Southern California Edison, which will buy the electricity from a 500-MW solar plant to be constructed in the Mojave Desert. Stretching across 4,500 acres, the facility will include 20,000 curved dish mirrors, each concentrating light on a Stirling engine about the size of an oil barrel. The plant is expected to begin operating in 2009 and could later be expanded to 850 MW. Stirling Energy Systems also signed a 20-year contract with San Diego Gas & Electric to build a 300-MW, 12,000-dish plant in the Imperial Valley. This facility could eventually be upgraded to 900 MW.

The financial details of the two California projects have not been made public, but electricity produced by present solar-thermal technologies costs between five and 13 cents per kilowatt-hour, with dish-mirror systems at the upper end of that range. Because the projects involve highly reliable technologies and mass production, however, the generation expenses are expected to ultimately drop closer to four to six cents per kilowatt-hour—that is, competitive with the current price of coal-fired power.

Blowing in the Wind

Wind power has been growing at a pace rivaling that of the solar industry. The worldwide generating capacity of wind turbines has increased more than 25 percent a year, on average, for the past decade, reaching nearly 60,000 MW in 2005. The growth has been nothing short of explosive in Europe—between 1994 and 2005, the installed wind power capacity in European Union nations jumped from 1,700 to 40,000 MW. Germany alone has more than 18,000 MW of capacity thanks to an aggressive construction program. The northern German state of Schleswig-Holstein currently meets one quarter of its annual electricity demand with more than 2,400 wind turbines, and in certain months wind power provides more than half the state's electricity. In addition, Spain has 10,000 MW of wind capacity, Denmark has 3,000 MW, and Great Britain, the Netherlands, Italy and Portugal each have more than 1,000 MW.

60,000 megawatts
Global generating capacity of wind power

0.5 percent
Fraction of U.S. electricity produced by wind turbines

1.9 cents
Tax credit for wind power, per kilowatt-hour of electricity

In the U.S. the wind power industry has accelerated dramatically in the past five years, with total generating capacity leaping 36 percent to 9,100 MW in 2005. Although wind turbines now produce only 0.5 percent of the nation's electricity, the potential for expansion is enormous, especially in the windy Great Plains states. (North Dakota, for example, has greater wind energy resources than Germany, but only 98 MW of generating capacity is installed there.) If the U.S. constructed enough wind farms to fully tap these resources, the turbines could generate as much as 11 trillion kilowatt-hours of electricity, or nearly three times the total amount produced from all energy sources in the nation last year. The wind industry has developed increasingly large and efficient turbines, each capable of yielding 4 to 6 MW. And in many locations, wind power is the cheapest form of new electricity, with costs ranging from four to seven cents per kilowatt-hour.

The growth of new wind farms in the U.S. has been spurred by a production tax credit that provides a modest subsidy equivalent to 1.9 cents per kilowatt-hour, enabling wind turbines to compete with coal-fired plants. Unfortunately, Congress has repeatedly threatened to eliminate the tax credit. Instead of instituting a long-term subsidy for wind power, the lawmakers have extended the tax credit on a year-to-year basis, and the continual uncertainty has slowed investment in wind farms. Congress is also threatening to derail a proposed 130-turbine farm off the coast of Massachusetts that would provide 468 MW of generating capacity, enough to power most of Cape Cod, Martha's Vineyard and Nantucket.

The reservations about wind power come partly from utility companies that are reluctant to embrace the new technology and partly from so-called NIMBY-ism. ("NIMBY" is an acronym for Not in My Backyard.) Although local concerns over how wind turbines will affect landscape views may have some merit, they must be balanced against the social costs of the alternatives. Because society's energy needs are growing relentlessly, rejecting wind farms often means requiring the construction or expansion of fossil fuel-burning power plants that will have far more devastating environmental effects.

16.2 billion
Liters of ethanol produced in the U.S. in 2005

2.8 percent
Ethanol's share of all automobile fuel by volume

$2 billion
Annual subsidy for corn-based ethanol

Green Fuels

Researchers are also pressing ahead with the development of biofuels that could replace at least a portion of the oil currently consumed by motor vehicles. The most common biofuel by far in the U.S. is ethanol, which is typically made from corn and blended with gasoline. The manufacturers of ethanol benefit from a substantial tax credit: with the help of the $2-billion annual subsidy, they sold more than 16 billion liters of ethanol in 2005 (almost 3 percent of all automobile fuel by volume), and production is expected to rise 50 percent by 2007. Some policymakers have questioned the wisdom of the subsidy, pointing to studies showing that it takes more energy to harvest the corn and refine the ethanol than the fuel can deliver to combustion engines. In a recent analysis, though, my colleagues and I discovered that some of these studies did not properly account for the energy content of the by-products manufactured along with the ethanol. When all the inputs and outputs were correctly factored in, we found that ethanol has a positive net energy of almost five megajoules per liter.

We also found, however, that ethanol's impact on greenhouse gas emissions is more ambiguous. Our best estimates indicate that substituting corn-based ethanol for gasoline reduces greenhouse gas emissions by 18 percent, but the analysis is hampered by large uncertainties regarding certain agricultural practices, particularly the environmental costs of fertilizers. If we use different assumptions about

Plugging Hybrids

The environmental benefits of renewable biofuels would be even greater if they were used to fuel plug-in hybrid electric vehicles (PHEVs). Like more conventional gasoline-electric hybrids, these cars and trucks combine internal-combustion engines with electric motors to maximize fuel efficiency, but PHEVs have larger batteries that can be recharged by plugging them into an electrical outlet. These vehicles can run on electricity alone for relatively short trips; on longer trips, the combustion engine kicks in when the batteries no longer have sufficient juice. The combination can drastically reduce gasoline consumption: whereas conventional sedans today have a fuel economy of about 30 miles per gallon (mpg) and nonplug-in hybrids such as the Toyota Prius average about 50 mpg, PHEVs could get an equivalent or 80 to 160 mpg. Oil use drops still further if the combustion engines in PHEVs run on biofuel blends such as E85, which is a mixture of 15 percent gasoline and 85 percent ethanol.

If the entire U.S. vehicle fleet were replaced overnight with PHEVs, the nation's oil consumption would decrease by 70 percent or more, completely eliminating the need for petroleum imports. The switch would have equally profound implications for protecting the earth's fragile climate, not to mention the elimination of smog. Because most of the energy for cars would come from the electric grid instead of from fuel tanks, the environmental impacts would be concentrated in a few thousand power plants instead of in hundreds of millions of vehicles. This shift would focus the challenge of climate protection squarely on the task of reducing the greenhouse gas emissions from electricity generation.

PHEVs could also be the salvation of the ailing American auto industry. Instead of continuing to lose market share to foreign companies, U.S. automakers could become competitive again by retooling their factories to produce PHEVs that are significantly more fuel-efficient than the nonplug-in hybrids now sold by Japanese companies. Utilities would also benefit from the transition because most owners of PHEVs would recharge their cars at night, when power is cheapest, thus helping to smooth the sharp peaks and valleys in demand for electricity. In California, for example, the replacement of 20 million conventional cars with PHEVs would increase nighttime electricity demand to nearly the same level as daytime demand, making far better use of the grid and the many power plants that remain idle at night. In addition, electric vehicles not in use during the day could supply electricity to local distribution networks at times when the grid was under strain. The potential benefits to the electricity industry are so compelling that utilities may wish to encourage PHEV sales by offering lower electricity rates for recharging vehicle batteries.

Most important, PHEVs are not exotic vehicles of the distant future. DaimlerChrysler has already introduced a PHEV prototype, a plug-in hybrid version of the Mercedes-Benz Sprinter Van that has 40 percent lower gasoline consumption than the conventionally powered model. And PHEVs promise to become even more efficient as new technologies improve the energy density of batteries, allowing the vehicles to travel farther on electricity alone.

—D.M.K.

these practices, the results of switching to ethanol range from a 36 percent drop in emissions to a 29 percent increase. Although corn-based ethanol may help the U.S. reduce its reliance on foreign oil, it will probably not do much to slow global warming unless the production of the biofuel becomes cleaner.

But the calculations change substantially when the ethanol is made from cellulosic sources: woody plants such as switchgrass or poplar. Whereas most makers of corn-based ethanol burn fossil fuels to provide the heat for fermentation, the producers of cellulosic ethanol burn lignin—an unfermentable part of the organic material—to heat the plant sugars. Burning lignin does not add any greenhouse gases to the atmosphere, because the emissions are offset by the carbon dioxide absorbed during the growth of the plants used to make the ethanol. As a result, substituting cellulosic ethanol for gasoline can slash greenhouse gas emissions by 90 percent or more.

Another promising biofuel is so-called green diesel. Researchers have produced this fuel by first gasifying biomass—heating organic materials enough that they release hydrogen and carbon monoxide—and then converting these compounds into long-chain hydrocarbons using the Fischer-Tropsch process. (During World War II, German engineers employed these chemical reactions to make synthetic motor fuels out of coal.) The result would be an economically competitive liquid fuel for motor vehicles that would add virtually no greenhouse gases to the atmosphere. Oil giant Royal Dutch/Shell is currently investigating the technology.

The Need for R&D

Each of these renewable sources is now at or near a tipping point, the crucial stage when investment and innovation, as well as market access, could enable these attractive but generally marginal providers to become major contributors to regional and global energy supplies. At the same time, aggressive policies designed to open markets for renewables are taking hold at city, state and federal levels around the world. Governments have adopted these policies for a wide variety of reasons: to promote market diversity or energy security, to bolster industries and jobs, and to protect the environment on both the local and global scales. In the U.S. more than 20 states have adopted standards setting a minimum for the fraction of electricity that must be supplied with renewable sources. Germany plans to generate 20 percent of its electricity from renewables by 2020, and Sweden intends to give up fossil fuels entirely.

Even President George W. Bush said, in his now famous State of the Union address this past January, that the U.S. is

The Least Bad Fossil Fuel

Although renewable energy sources offer the best way to radically cut greenhouse gas emissions, generating electricity from natural gas instead of coal can significantly reduce the amount of carbon added to the atmosphere. Conventional coal-fired power plants emit 0.25 kilogram of carbon for every kilowatt-hour generated. (More advanced coal-fired plants produce about 20 percent less carbon.) But natural gas (CH_4) has a higher proportion of hydrogen and a lower proportion of carbon than coal does. A combined-cycle power plant that burns natural gas emits only about 0.1 kilogram of carbon per kilowatt-hour.

Unfortunately, dramatic increases in natural gas use in the U.S. and other countries have driven up the cost of the fuel. For the past decade, natural gas has been the fastest-growing source of fossil-fuel energy, and it now supplies almost 20 percent of America's electricity. At the same time, the price of $2.50 to $3 per million Btu in 1997 to more than $7 per million Btu today.

The price increases have been so alarming that in 2003, then Federal Reserve Board Chair Alan Greenspan warned that the U.S. faced a natural gas crisis. The primary solution proposed by the White House and some in Congress was to increase gas production. The 2005 Energy Policy Act included large subsidies to support gas producers, increase exploration and expand imports of liquefied natural gas (LNG). These measures, however, may not enhance energy security, because most of the imported LNG would come from some of the same OPEC countries that supply petroleum to the U.S. Furthermore, generating electricity from even the cleanest natural gas power plants would still emit too much carbon to achieve the goal of keeping carbon dioxide in the atmosphere below 450 to 550 parts per million by volume. (Higher levels could have disastrous consequences for the global climate.)

Improving energy efficiency and developing renewable sources can be faster, cheaper and cleaner and provide more security than developing new gas supplies. Electricity from a wind farm costs less than that produced by a natural gas power plant if the comparison factors in the full cost of plant construction and forecasted gas prices. Also, wind farms and solar arrays can be built more rapidly than large-scale natural gas plants. Most critically, diversity of supply is America's greatest ally in maintaining a competitive and innovative energy sector. Promoting renewable sources makes sense strictly on economic grounds, even before the environmental benefits are considered.

Jen Christiansen; Source: President's Committee of Advisors on Science and Technology: —D.M.K.

R&D is Key

Spending on research and development in the U.S. energy sector has fallen steadily since its peak in 1980. Studies of patent activity suggest that the drop in funding has slowed the development of renewable energy technologies. For example, the number of successful patent applications in photovoltaics and wind power has plummeted as R&D spending in these fields has declined.

Jen Christiansen; Source: Reversing the incredible shrinking energy R&D Budget;

"addicted to oil." And although Bush did not make the link to global warming, nearly all scientists agree that humanity's addiction to fossil fuels is disrupting the earth's climate. The time for action is now, and at last the tools exist to alter energy production and consumption in ways that simultaneously benefit the economy and the environment. Over the past 25 years, however, the public and private funding of research and development in the energy sector has withered. Between 1980 and 2005 the fraction of all U.S. R&D spending devoted to energy declined from 10 to 2 percent. Annual public R&D funding for energy sank from $8 billion to $3 billion (in 2002 dollars); private R&D plummeted from $4 billion to $1 billion.

To put these declines in perspective, consider that in the early 1980s energy companies were investing more in R&D than were drug companies, whereas today investment by energy firms is an order of magnitude lower. Total private R&D funding for the entire energy sector is less than that of a single large biotech company. (Amgen, for example, had R&D expenses of $2.3 billion in 2005.) And as R&D spending dwindles, so does innovation. For instance, as R&D funding for photovoltaics and wind power has slipped over the past quarter of a century, the number of successful patent applications in these fields has fallen accordingly. The lack of attention to long-term research and planning has significantly weakened our nation's ability to respond to the challenges of climate change and disruptions in energy supplies.

Calls for major new commitments to energy R&D have become common. A 1997 study by the President's Committee of Advisors on Science and Technology and a 2004 report by the bipartisan National Commission on Energy Policy both recommended that the federal government double its R&D spending on energy. But would such an expansion be enough? Probably not. Based on assessments of the cost to stabilize the amount of carbon dioxide in the atmosphere and other studies that estimate the success of energy R&D programs and the resulting savings from the technologies that would emerge, my research group has calculated that public funding of $15 billion to $30 billion a year would be required—a fivefold to 10-fold increase over current levels.

Greg F. Nemet, a doctoral student in my laboratory, and I found that an increase of this magnitude would be roughly comparable to those that occurred during previous federal R&D initiatives such as the Manhattan Project and the Apollo program, each of which produced demonstrable economic benefits in addition to meeting its objectives. American energy companies could also boost their R&D spending by a factor of 10, and it would still be below the average for U.S. industry overall. Although government funding is essential to supporting early-stage technologies, private-sector R&D is the key to winnowing the best ideas and reducing the barriers to commercialization.

Raising R&D spending, though, is not the only way to make clean energy a national priority. Educators at all grade levels, from kindergarten to college, can stimulate public interest and activism by teaching how energy use and production affect the social and natural environment. Nonprofit organizations can establish a series of contests that would reward the first company or private group to achieve a challenging and worthwhile energy goal, such as constructing a building or appliance that can generate its own power or developing a commercial vehicle that can go 200 miles on a single gallon of fuel. The contests could be modeled after the Ashoka awards for pioneers in public policy and the Ansari X Prize for the developers of space vehicles. Scientists and entrepreneurs should also focus on finding clean, affordable ways to meet the energy needs of people in the developing world. My colleagues and I, for instance, recently detailed the environmental benefits of improving cooking stoves in Africa.

But perhaps the most important step toward creating a sustainable energy economy is to institute market-based schemes to make the prices of carbon fuels reflect their social cost. The use of coal, oil and natural gas imposes a huge collective toll on society, in the form of health care expenditures for ailments caused by air pollution, military spending to secure oil supplies, environmental damage from mining operations, and the potentially devastating economic impacts of global warming. A fee on carbon emissions would provide a simple, logical and transparent method to reward renewable, clean energy sources over those that harm the economy and the environment. The tax revenues could pay for some of the social costs of carbon emissions, and a portion could be designated to compensate low-income families who spend a larger share of their income on energy. Furthermore, the carbon fee could be combined with a cap-and-trade program that would set limits on carbon emissions but also allow the cleanest energy suppliers to sell permits to their dirtier competitors. The federal government has used such programs with great success to curb other pollutants, and several northeastern states are already experimenting with greenhouse gas emissions trading.

Best of all, these steps would give energy companies an enormous financial incentive to advance the development and commercialization of renewable energy sources. In essence, the U.S. has the opportunity to foster an entirely new industry. The threat of climate change can be a rallying cry for a clean-technology revolution that would strengthen the country's manufacturing base, create thousands of jobs and alleviate our international trade deficits—instead of importing foreign oil, we can export high-efficiency vehicles, appliances, wind turbines and photovoltaics. This transformation can turn the nation's energy sector into something that was once deemed impossible: a vibrant, environmentally sustainable engine of growth.

Daniel M. Kammien is Class of 1935 Distinguished Professor of Energy at the University of California, Berkeley, where he holds appointments in the Energy and Resources Group, the Goldman School of Public Policy and the department of nuclear engineering. He is founding director of the Renewable and Appropriate Energy Laboratory and co-director of the Berkeley Institute of the Environment.

Article 13

What Nuclear Renaissance?

The nuclear industry isn't going anywhere. It's too costly and won't save us from global warming.

CHRISTIAN PARENTI

If you listen to the rhetoric, nuclear power is back. Smashing atoms will replace burning carbon-based coal, gas and oil. In the face of a disaster movie–like future of runaway climate change—bringing drought, floods, famine and social breakdown—carbon-free nukes are cast as the *deus ex machina* to save us at the last minute.

Even a few greens support nuclear power—most famously James Lovelock, father of the Gaia theory. In the popular press, discussion of nuclear energy is dominated by its boosters, thanks in part to sophisticated industry PR.

In an effort to jump-start a "nuclear renaissance," the Bush Administration has pushed one package of subsidies after another. For the past two years a program of federal loan guarantees has sat waiting for utilities to build nukes. Last year's appropriations bill set the total amount on offer at $18.5 billion. And now the Lieberman-Warner climate change bill is gaining momentum and will likely accrue amendments that will offer yet more money.

The Nuclear Regulatory Commission (NRC) expects up to thirty applications to be filed to build atomic plants; five or six of those proposals are moving through the complicated multi-stage process. But no new atomic power stations have been fully licensed or have broken ground. And two newly proposed projects have just been shelved.

The fact is, nuclear power has not recovered from the crisis that hit it three decades ago with the reactor fire at Browns Ferry, Alabama, in 1975 and the meltdown at Three Mile Island in 1979. Then came what seemed to be the coup de grâce: Chernobyl in 1986. The last nuclear power plant ordered by a US utility, the TVA's Watts Bar 1, began construction in 1973 and took twenty-three years to complete. Nuclear power has been in steady decline worldwide since 1984, with almost as many plants canceled as completed since then.

All of which raises the question: why is the much-storied "nuclear renaissance" so slow to get rolling? Who is holding up the show? In a nutshell, blame Warren Buffett and the banks—they won't put up the cash.

"Wall street doesn't like nuclear power," says Arjun Makhijani of the Institute for Energy and Environmental Research. The fundamental fact is that nuclear power is too expensive and risky to attract the necessary commercial investors. Even with vast government subsidies, it is difficult or almost impossible to get proper financing and insurance. The massive federal subsidies on offer will cover up to 80 percent of construction costs of several nuclear power plants in addition to generous production tax credits, as well as risk insurance. But consider this: the average two-reactor nuclear power plant is estimated to cost $10 billion to $18 billion to build. That's before cost overruns, and no US nuclear power plant has ever been delivered on time or on budget.

As Dieter Helm, an Oxford professor and leading economic expert on energy markets, has found, there never has been and never will be a nuclear power program totally dependent on the market.

Sixty years ago, the technology was swathed in manic space-age optimism—its electricity was going to be "too cheap to meter." While that wasn't true, nuclear power did serve a key role in the cold war: spent nuclear fuel rods are refined for weapons- grade plutonium and enriched uranium. That fact aside, rarely has so much money, scientific knowhow and raw state power been marshaled to achieve so little. By some estimates, an investment of several hundred billion dollars has led to a US nuke industry of 104 operating plants—about a quarter of the global total—that produces a mere 19 percent of our electricity.

In fact, the sputtering decline of nuclear power has been one of the greatest industrial failures of modern times. In 1985 *Forbes* called the nuke industry "the largest managerial disaster in history."

Atomic optimism run amok caused the largest municipal bond default in US history. In 1983 Washington Public Power Supply System abandoned three nuke plants in mid-construction. The projects were plagued by massive cost overruns—one infamous section of piping was reinstalled seventeen times, safety inspections were blatantly ignored, incompetent contractors were allowed to continue work and on and on. When the project finally died, unfinished costs had ballooned to $24 billion, and the utility walked away from $2.25 billion worth of bonds.

That project, like many others, drowned in the financial riptides of rising interest rates that were the central feature of the "Volcker recession" of the early '80s. (That was when Federal Reserve chairman Paul Volcker smashed inflation by jacking the Fed's interest rate from 8 percent in 1979 to more than 16 percent in 1982.) But nukes were also killed by the corruption and incompetence that so often plague large state projects, like Boston's Big Dig, the New Orleans levees, space-based weapons systems and Iraq's reconstruction.

Another reason atomic energy is so expensive is that its accidents are potentially catastrophic, and activists have forced utilities to build in costly double and triple safety systems. Right-wing champions of atom-smashing blame prohibitive costs on neurotic fears and unnecessary safety measures. They have a point in that safety is expensive, but safety is hardly excessive—details on that in a moment.

More important is the fact that nuclear fission is a mind-bogglingly complex process, a sublime, truly Promethean technology. Let's recall: it involves smashing a subatomic particle, a neutron, into an atom of uranium-235 to release energy and more neutrons, which then smash other atoms that release more energy and so on infinitely, except the whole process is controlled and used to boil water, which spins a turbine that generates electricity.

In this nether realm, where industry and science seek to reproduce the process that occurs inside the sun, even basic tasks—like moving the fuel rods, changing spare parts—become complicated, mechanized and expensive. Atom-smashing is to coal power, or a windmill, as a Formula One race-car engine is to the mechanics of a bicycle. Thus, it costs an enormous amount of money.

Worldwide, about twenty nuclear power plants are being built, but most are in Asia and Russia and are closely linked to nuclear weapons programs. Japan and France have large nuke programs, but both countries heavily subsidize their plants, use a single design and built their fleets not to make profits but to ensure some minimum strategic energy independence and, for France, to build an atomic arsenal.

Even if a society were ready to absorb the high costs of nuclear power, it hardly makes the most sense as a tool to quickly combat climate change. These plants take too long to build. A 2004 analysis in *Science* by Stephen Pacala and Robert Socolow, of Princeton University's Carbon Mitigation Initiative, estimates that achieving just one-seventh of the carbon reductions necessary to stabilize atmospheric CO_2 at 500 parts per billion would require "building about 700 new 1,000-megawatt nuclear plants around the world." That represents a huge wave of investment that few seem willing to undertake, and it would require decades to accomplish.

None of this has stopped the Bush Administration and Congress from channeling more money toward nukes. The current push to build nukes began in 2002, when the Administration launched its Nuclear Power 2010 program, which sought to spur construction of at least three major nuclear power plants. Then came the US Energy Policy Act of 2005, which offered three major forms of subsidy. New nuclear power plants could get production tax credits, federal loan guarantees and construction insurance against cost overruns and delays— together worth $18.5 billion.

The notion that nukes make sense and are the version of green preferred by grown-ups is being conjured by a slick PR campaign.

The notion that nukes make sense and are the version of green preferred by grown-ups is being conjured by a slick PR campaign. The Nuclear Energy Institute—the industry's main trade group—has retained Hill and Knowlton to run a green-washing campaign.

Part of their strategy involves an advocacy group with the grassroots-sounding name the Clean and Safe Energy Coalition. At the center of the effort are former EPA chief Christine Todd Whitman and former Greenpeace co-founder turned corporate shill Patrick Moore. (Moore is also a huge champion of GMO crops, which are notorious for impoverishing farmers in developing economies and using massive amounts of pesticides.) The industry also places ghostwritten op-eds under the bylines of scientists for hire.

All the major environmental groups oppose nuclear power. But the campaign is having some impact at the grassroots: the online environmental journal *Grist* found that 54 percent of its readers are ready to give atomic energy a second look; 59 percent of Treehugger.com readers feel the same way. In other words, people who understand climate change are feeling down right desperate.

But even the Oz-like magic of corporate spin, public subsidies and presidential speechifying have their limits. In late December the man whose name is synonymous with sound money turned his back on nuclear power.

Warren Buffett's MidAmerican Nuclear Energy Company scrapped plans to build a plant in Payette, Idaho, because no matter how many times its managers ran the numbers (and they spent $13 million researching it), they found that it simply made no sense from an economic standpoint.

South Carolina Electric and Gas has also suspended its two planned reactors, citing costs as the key factor. But the company says, "We remain very upbeat about the future of nuclear power."

If a nuke plant breaks ground soon, it will likely be NRG Energy's double-reactor plant, set to be erected in South Texas. But that one has also been delayed.

The fact that new nukes make little economic sense does not mean that old nukes are not profitable. In fact, these nightmarishly complex radioactive boondoggles have recently been turned into cash cows. Utilities achieved this remarkable transformation the old-fashioned way—they used socialism.

Old nuke plants have been turned into cash cows. Utilities achieved this remarkable transformation the old-fashioned way: they used socialism.

Beginning in the 1990s, most American energy markets were deregulated one state, one region at a time. In the process many old utilities were broken up into different firms: some generated power, others sold it, still others handled transmission. One of the crucial details of deregulation was allowing utilities to pass on to rate payers the "stranded costs"—the outstanding mortgage payments of their nuclear power plants.

Perhaps the most egregious example of this occurred in California. In 1996 the State Assembly passed legislation—written by utility lobbyists—that allowed Southern California Edison and Pacific Gas & Electric to hold rates high as prices dropped nationally. The two utilities were on target to receive $28 billion over four years. This money would pay off the stranded costs of the Diablo Canyon and San Onofre atomic plants. Halfway through the deal the California power crisis hit and deregulation was put on hold—utilities were forced to stop selling off their assets, and third-party speculation in energy markets was halted. But the state floated bonds to mop up the remaining stranded costs.

Similar deals were struck across the country. Once unburdened of old debts, the nuke plants—now having relatively low overhead costs—became valuable assets. A new generation of firms began buying them up. By 2002 ten companies owned seventy of the nation's 104 reactors. Among the big players in this game are Exelon, Entergy and Dominion Resources.

Many of the old plants went for a song. A particularly disturbing example of this is Vermont Yankee, a thirty-five-year-old reactor purchased by Entergy seven years ago for a mere $180 million. That's about half the price it would cost to build an equal-sized coal plant or wind farm.

Now Entergy is trying to run the power station as hard and as long as possible. In 2006 it received approval to increase power output at the plant by 20 percent. This "uprate" means the plant operates with 20 percent more pressure, heat and flow. And in just one year it earned Entergy $100 million in profits. Over the last decade, almost all US nuclear power plants have received uprates, but few match Vermont Yankee's full-throttle, 120 percent capacity.

Just after the uprate, one of Vermont Yankee's twenty-two cooling towers collapsed. That's right—it crumbled and fell over. Entergy officials said the collapse "baffled" them. The plant's spokesman, Rob Williams, admitted that "our inspections were not effective enough." Reached by phone, Gregory Jaczko, a commissioner at the NRC, admitted that the collapse "didn't look good." But he went on to reassure the public that the plant is essentially safe.

Now Entergy is petitioning the NRC to extend its operating license so that it can run the old plant for twenty years longer than was intended. Nationally, forty-eight facilities have had their licenses extended. In fact, despite critics' arguments that aging plants pose serious dangers, no license renewal requests have ever been denied.

"The NRC falls all over itself to facilitate the industry," says Ray Shadis, a consultant who has worked for both environmental groups and on NRC panels and research projects. The Project on Government Oversight and other watchdog groups point to a revolving door between the commission's staff and the nuclear industry. To take just one example, in 2007 former commissioner Jeffrey Merrifield joined the Shaw Group after spending his last months on the commission pushing to ease restrictions for precisely the type of construction activities that were the Shaw Group's specialty.

Diana Sidebotham, an antinuclear activist in Putney, Vermont, twenty miles north of the Vermont Yankee plant, thinks Entergy and the NRC are courting disaster. In 1971 Sidebotham helped found the New England Coalition on Nuclear Pollution, and she has been trying to shut down nuclear plants ever since. Her hillside farm looks out over the ridge lines of the Connecticut River Valley.

"One of these days a plant will blow," says Sidebotham, with just a touch of a genteel but steely New England accent. "And when it does, it will cause a great many deaths and widespread suffering, not to mention extraordinary economic damage."

Accidents do happen. In 2002 the Davis-Besse Nuclear Plant in Ohio was forced to close for two years after inspectors found a football-sized corrosion hole in the reactor's six-inch-thick steel cap. The plant was very close to a major accident. Repairs cost $600 million.

Democratic presidential candidate Barack Obama says he opposes any more relicensing of old nuclear plants. His rival Hillary Clinton has stopped just short of saying that. However, as was reported by the *New York Times,* Obama has close ties to the nuclear industry, particularly the Illinois-based Exelon, which has contributed at least $227,000 to his campaigns. Two of his top advisers have links to the firm, including his chief strategist, David Axelrod, who was a consultant for Exelon. Obama voted yes on the 2005 Energy bill, which lavished subsidies on oil, coal, ethanol and nukes; Senator Clinton, like almost half the Senate Democrats, voted against it. The Obama campaign says that as President he would not cut nuclear subsidies, only that he would boost subsidies for green power.

Activists like Sidebotham say the real issue is not how to build more nukes but how to handle the old, decrepit plants and their huge stockpiles of radioactive waste. Most of the atomic plants in this country are reaching the end of their life span; seventeen have been decommissioned. And increasingly the question is what to do with the accumulated waste—the extremely radioactive spent fuel rods. This is dangerous stuff. If exposed to air for more than six hours, spent fuel rods spontaneously combust, spewing highly poisonous radioactive isotopes far and wide. This spent fuel will be hot for 10,000 years.

Since 1978 the Energy Department has been studying Yucca Mountain in Nevada as a possible permanent repository for atomic waste. But intense opposition has held up those efforts. In the meantime, the partially burned uranium is stored at the old power plants, in pools of water called "spent fuel pools." Lying near great cities, on crucial river systems, in small rural towns, these pools are potentially a far greater risk than a reactor meltdown. Scenarios for how terrorists might attack and drain them range from driving a truck bomb to crashing an explosive-laden plane into them.

Just after 9/11, when security at nuke plants was supposed to be high, lead pellets started raining down on the containment structure and guard shack at Maine Yankee, in Wiscasset. (The plant has since been decommissioned.) A group of four men in camouflage, armed and intent on killing, had infiltrated into a swamp and were firing weapons from somewhere in the reeds. This "cell" turned out to be four local duck hunters who had no idea they were hitting the power plant.

Their foray against innocent mallards proved just how easy an attack could be. Activists demanded, and got, a safety review, which led to a shockingly blunt NRC document called "Report on Spent Fuel Pool Accident Risk," or NUREG-1738. The report found that containment structures, such as that at Vermont Yankee, "present no substantial obstacle to aircraft penetration." According to the NRC, a fire in the spent fuel pool at a reactor like Vermont Yankee (which stores 488 metric tons of spent fuel) would cause 25,000 fatalities over a distance of 500 miles if evacuation was 95 percent effective. But that evacuation rate would be almost impossible to achieve. The NRC claims to have the threat of terrorism under control, but for reasons of national security it can't explain how. And after 9/11 it admitted, "At this time, we could not exclude the possibility that a jetliner flying into a containment structure could damage the facility and cause a release of radiation that could impact public health."

Humanity's Faustian bargain with atomic power is a story still in its early stages. No one knows how long nuclear facilities will last or what will happen to them during future social upheavals—and there are bound to be a few of those during the next 10,000 years.

This much seems clear: a handful of firms might soak up huge federal subsidies and build one or two overpriced plants. While a new administration might tighten regulations, public safety will continue to be menaced by problems at new as well as older plants. But there will be no massive nuclear renaissance. Talk of such a renaissance, however, helps keep people distracted, their minds off the real project of developing wind, solar, geothermal and tidal kinetics to build a green power grid.

Christian Parenti is a *Nation* contributing editor.

The Biofuel Future: Scientists Seek Ways to Make Green Energy Pay Off

RACHEL EHRENBERG

Biofuels are liquid energy Version 2.0. Unlike their fossil fuel counterparts—the cadaverous remains of plants that died hundreds of millions of years ago—biofuels come from vegetation grown in the here and now. So they should offer a carbon-neutral energy source: Plants that become biofuels ideally consume more carbon dioxide during photosynthesis than they emit when processed and burned for power. Biofuels make fossil fuels seem so last century, so quaintly carboniferous.

And these new liquid fuels promise more than just carbon correctness. They offer a renewable home-grown energy source, reducing the need for foreign oil. They present ways to heal an agricultural landscape hobbled by intensive fertilizer use. Biofuels could even help clean waterways, reduce air pollution, enhance wildlife habitats and increase biodiversity.

Yet in many respects, biofuels are in their beta version. For any of a number of promising feedstocks—the raw materials from which biofuels are made—there are logistics to be worked out, such as how to best shred the original material and ship the finished product. There is also lab work—for example, refining the processes for busting apart plant cell walls to release the useful sugars inside. And there is math. A lot of math.

The only way that biofuels will add up is if they produce more energy than it takes to make them. Yet, depending on the crops and the logistics of production, some analysis suggest that it may take more energy to make these fuels than they will provide. And if growing biofuels creates the same environmental problems that plague much of large-scale agriculture, then air and water quality might not really improve. Prized ecosystems such as rain forests, wetlands and savannas could be destroyed to grow crops. Biofuels done badly, scientists say, could go very, very wrong.

"Business as usual writ larger is not an environmentally welcome outcome," states a biofuels policy paper authored by more than 20 scientists and published in *Science* last October.

Many scientists have expressed concern that political support for the biofuels industry has outpaced rigorous analysis of the fuels' potential impacts. Others see this notion as manure. Research needed to resolve that disagreement is now underway, as scientists in industry, national labs and universities across the country are assessing every aspect of these fuels, from field to tailpipe.

Researchers are growing crops, evaluating yields and comparing harvesting techniques. Computer models are providing stats on each crop's effect on environmental factors such as soil nutrients and erosion. The plant cell wall is under attack from several angles. And chemists and microbiologists are cajoling an expanding menagerie of microorganisms into producing higher fuel yields.

Green Goals

Ideally, high biofuel yields come with minimal environmental baggage and maximum efficiency at every step. The raw materials for these fuels run the gamut from corn to municipal waste to algae, and each has its own benefits and headaches. To make fuels, researchers must first process the raw material to create fermentable sugars or a crude oil-like liquid. Further refinement yields fuels such as ethanol, butanol, jet fuel or biodiesel.

In some cases, such as algae-based biodiesel, the technologies are far from mature. Squeezing ethanol from crops such as corn, on the other hand, uses a technology as old as whiskey. An infrastructure already exists for growing and moving grain, and distillation and fermentation techniques work at large scales.

But grain-based fuels raise several environmental issues, such as emissions of the potent greenhouse gas nitrous oxide from heavy fertilizer use. So, many scientists see corn ethanol as a bridging technology for use until the next-generation feedstocks fulfill biofuels' real promise. Nonfood plants rich in cellulose or even residual waste diverted from landfills may define the biofuel future.

Several studies attest to the benefits of fuels made from such feedstocks, although the degree of benefit varies depending on what factors are included in the analysis. Overall, dedicated energy crops such as switchgrass and waste residues from sources like commercial logging fare better than corn-based ethanol, concludes a recent modeling analysis and literature review citing more than 100 papers. Published online May 27 in Environmental Science & Technology, the analysis reports that municipal waste-based ethanol production emits an estimated

60 to 80 percent less greenhouse gas than corn-based ethanol production. Dedicated energy crops, especially when grown on marginal land, also fare better than corn in terms of greenhouse gas emissions, and require less water and generate less air pollution, report researchers from the National Renewable Energy Laboratory in Golden, Colo., and E Risk Sciences in Boulder, Colo.

Research also suggests that these new fuels will be priced competitively with gasoline from petroleum. A new assessment coauthored by Lee Lynd, head scientist and cofounder of the Boston-based ethanol start-up Mascoma Corp., found that the production costs of cellulose-based ethanol, when made on a commercial scale, could be competitive with gasoline at oil prices of $30 or more per barrel.

Both of these recent big-picture studies, while optimistic, call for continued research to improve existing production processes and better define each fuel's associated trade-offs.

Such research is in progress at the Idaho National Laboratory in Idaho Falls, where scientists David Muth Jr. and Thomas Ulrich take part in a coordinated, national effort to watch grass grow. In partnership with scientists at Oak Ridge National Laboratory in Tennessee and at several universities, Muth and Ulrich are keeping track of more than 50 field trials of various feedstocks across the country. The researchers are growing switchgrass and Miscanthus, an 11-foot tall perennial grass. Energy cane, an fiber-biomass relative of sugar cane, is also under study.

The research suggests that there is not one silver bullet source for biofuels. While there are some generally desirable plant characteristics—such as needing few nutrients and flourishing on degraded land—the future biofuels landscape will likely be a patchwork of different sources that work best in different regions.

"What's emerging pretty quickly is how site-specific both the production systems and problems are," says Muth.

Energy cane, for example, has "huge yields, but it is a water sink," he says. So it may be best for water-rich Gulf Coast states. Miscanthus, which has been tested in Europe for several years, produces very high yields and has the genes to withstand cold climates.

Part of biofuels' allure lies in the variety of ingredients from which the fuels may be spun. The Idaho National Lab is also investigating strains of algae that pump out oils as a raw material for biodiesel. At other sites agricultural and municipal waste, such as straw stalks, corn cobs and tree cuttings, are under investigation. Some researchers are focused on crops dedicated to energy, such as prairie grasses, and fast-growing softwoods, such as willow, poplar and eucalyptus. A pilot-scale system for growing the diminutive pond plant duckweed on wastewater is underway at North Carolina State University.

In Idaho, Muth is also using several computer models to calculate the effect that growing and removing the feedstocks has on factors such as soil's nutrients, carbon and water content. This information, along with yields and quality of plant material, is all being entered into a database to help predict which plants will grow best where.

Biomass Breakdown

Bioenergy is not just about growing crops up, though. It's even more about tearing them down. Biomass must be harvested from the field or forest, perhaps stored, and then shipped to a refinery for processing. Harvesting equipment, travel distances and processing methods must all be considered to determine whether biofuels make economic and energy sense.

"What is becoming a bigger and bigger issue to people is the logistics of it all—that's becoming a barrier to the whole thing," says J. Richard Hess, the technology manager of the Idaho National Lab program.

An essential part of biofuel logistics is the preprocessing of plants—cutting, baling and hauling the bales somewhere for storage before transporting them to a refinery. Those preprocessing steps pose problems with a material that isn't very dense or evenly shaped. "It's like moving air or feathers," Hess says.

Ideally, preprocessing would provide an end product that is uniform and easy to handle, like grain—the biomass equivalent of crude oil. "We're not aiming for a certain size, but a certain density that's easy to ship, is flowable," says INL's Christopher Wright.

Wright and Neal Yancey, also of INL, are trying to achieve the optimal density by finding the right balance of shredding and compacting, ultimately producing something like the alfalfa pellets fed to pet rabbits, or perhaps Matchbox car-sized blocks. This crude can then be shipped to a refinery to be heated into an oil-like liquid or broken down by enzymes into the desired fuel.

Breaking biomass down into fuel is no small task. The dominant method is known as biochemical conversion: Processes that use heat, chemicals or enzymes to turn the biomass into sugars that can be fermented by microbes such as yeast into ethanol. This ethanol is the same whether its origins are corn or other biomass. But it is currently a lot easier to get the fermentable sugars out of a starchy corn kernel than from something like wood chips or a weedy grass.

Plant cell walls are about 75 percent complex sugars, but getting at these sugars is a bit like trying to get the mortar and minerals out of a castle's rampart. Cell walls, one of the defining features of plants as a life-form, were made to resist degradation. Even termites and cows need special microbes in their guts to get the job done.

That's because those sugars are embedded in a complex architectural structure called lignocellulose—cellulose (long, unbranched chains of glucose) embedded in a matrix of more sugars (hemi-cellulose) embedded in the tough, gluelike lignin. (Biofuels researchers refer to the "recalcitrance" of the cell wall, as if it were an obstinate child.) Not only did cell walls evolve for strength, they also are a primary defense against microbial attack, and critters that are up to the task aren't common.

"Lignin is a highly problematic polymer from the point of view of processing, but an exemplary evolutionary achievement," researchers at the University of York in England commented in May 2008 in *New Philologist.*

To prep for the cell wall attack, plant matter is usually pretreated: the shredded, chopped or pelletized biomass is typically mixed with dilute acids or ammonia. At a biofuels symposium held in May in San Francisco, scientists presented work describing pretreatment with proton beam irradiation, steam explosion and microwave reactors. Ionic liquids—basically liquid salts—are also under investigation.

"Cellulose doesn't liquefy in minutes to hours—it's hours to days," says Jim McMillan of the national lab in Golden. This step is the main bottleneck in cellulosic fuel production, Lynd and several other researchers conclude in a February 2008 commentary in *Nature Biotechnology.*

Lignin is typically removed after pre-treatment and then burned in the refinery's boiler, replacing some fossil fuel use. The remaining plant matter is then broken into simple sugars, typically by a cocktail of microbial enzymes known as cellulases. Other microbes are then called in to ferment the sugars into ethanol.

Breaking down cellulose with enzymes is usually a separate step from fermentation—and a very costly one. But recent attempts to combine the conversion of cellulose to sugars with the conversion of sugars to fuel—called consolidated bioprocessing—have been successful. A strain of the soil-dwelling bacterium *Clostridium phytofermentans* will happily munch biomass such as wood pulp waste and will ferment it into ethanol. That discovery, by microbiologist Susan Leschine of the University of Massachusetts Amherst, led to the development of Qteros, a cellulosic-ethanol start-up in Marlborough, Mass. And in May, Mascoma researchers reported the engineering of a yeast and the bacterium *Clostridium thermocellum* to produce cellulases and ethanol in a single step.

At the San Francisco conference, posters reported on investigations of even more enzymes from various sources: bacteria that live in the deep sea, penicillin, diseased sea squirts, the bread mold Neurospora, a yeast that grows on wood-boring beetles and soil microbes from a Puerto Rican rainforest. Scientists are also fighting recalcitrance from the inside out by breeding lines of low-lignin plants.

Of course, getting a lot of ethanol in a benchtop flask is one thing. Scaling up to a silo-sized bioreactor is another. Industrial models exist—such as wringing pulp from trees for the paper industry or mass-producing cornstarch. "But we haven't done it with cellulose yet," says McMillan.

More than a dozen pilot plants for producing cellulosic ethanol are under construction and a handful are operating, with 2011 seen as the year for cellulosic technologies to walk the walk. The group at Idaho National Lab hopes to be able to demonstrate a system from field to refinery by autumn of 2010.

Environmental Cost

Yet concerns remain that the environmental side of the biofuels equation is still not worked out. Some argue that the numbers are too fuzzy to proceed with confidence that environmental burdens and benefits have been fully considered.

"There are people who say we don't have enough knowledge to move forward—to some extent that is true," says Michigan State University's Philip Robertson, coauthor of the *Science* policy paper. "But we do know a lot about sustainability—enough to implement logical science-based standards." This includes things like the strategic use of cover crops, fertilizer and tilling.

There is also the consideration of land-use changes—if forests are cleared for biofuels production, far more carbon will be released than is saved by the nonpetroleum fuels, several studies suggest. Such findings have led to scrutiny that has stung many in the industry who argue that biofuels are being held to a much higher standard than fossil fuels. If the petroleum isn't "charged" for the greenhouse gas emissions of twhe U.S. military keeping supply lanes open in the Persian Gulf, why should emissions from cleared forests be included in the biofuels ledger? asks Bruce Dale of Michigan State University in a recent editorial in the journal *Biofuels, Bioproducts & Biorefining.*

Congress is now considering legislation that may determine whether indirect land use can or cannot be a mark on the ruler used by the U.S. Environmental Protection Agency to measure biofuels' impacts. Eventually, many researchers hope, a more detailed picture will emerge of the benefits and costs across all stages of the life cycles of fossil and next-generation fuels.

"Some really interesting services are going to emerge from these crops," says Muth, of the Idaho National Lab. Some biofuel plants help sequester carbon in the soil, for example. A 2002 analysis reported that by the second or third planting year, switchgrass plots experience far less soil erosion than annual crops such as corn. Species that do well near wetlands can act as filters, preventing nitrates and phosphates from getting into the water, Muth says. "If there is a value on carbon sequestration . . . a value on clean water, there may be economic benefits for a lot of these crops."

Robertson adds, "If certain practices were being promoted with incentives, it would ensure that we have a biofuels industry that is sustainable with a net benefit, not a cost. We don't have that yet—I say 'yet' hopefully."

With appropriate carrots and sticks, biofuels could play a big role in the energy portfolio of the future. There may even be a day when, Back to the Future style, garbage can be thrown into a personal-sized bioreactor that yields fuel. (Trash biomass in the form of sugar beet pulp, tomato pomace, cashew apple, grape pomace, sweet gum and coffee pulp are all being investigated.) Several lines of research are investigating biofuel "coproducts," high-value molecules that can be extracted during processing, such as proteins for animal feed or aromatics for perfumes and drugs. These products will also bring the net costs of these fuels down, one of several variables that can help the biofuels math add up to success as a fossil fuel substitute.

"It's difficult to compare the costs of not changing with the costs of changing," Lynd said at the May meeting in San Francisco. "Asking is this or that realistic is well-intentioned, but all solutions involve changes—we don't have an option. Business as usual? Well, we think of it as a baseline, but it is a fantasy—even if you don't care about carbon—just as a supply issue. Fossil fuels will all be gone. They'll all be gone."

Putting Your Home on an Energy Diet

Simple steps with fast payback can cut family power bills.

Marianne Lavelle

Putting your house on an energy diet is simple: airtight construction, smart heating and cooling design, and high-efficiency appliances. But simple doesn't mean easy. You might as well tell Americans they ought to lay off nacho chips and sign up for a daily Zumba class. The nation's power demands, like our waistlines, are growing ever more bloated.

Look at just one of the new energy guzzlers: the digital photo frame. This always-on gadget burns a barely noticeable $9 extra a year into the average household electric bill, says the nonprofit Electric Power Research Institute. But the impact could be staggering. EPRI estimates that if every household in America owned one, it would take five medium-sized power plants just to keep those family photo slide shows rolling in the nation's living rooms. "I call these electronics the sleeping giants in our homes," says Thomas Reddoch, EPRI's director of energy utilization.

But there's a rising call for Americans to use less energy, either out of self—interested concern over escalating costs or genuine concern over the risk to the planet from global warming if the world's leading fossil fuel users continue on their current course. "The cleanest and cheapest kilowatt-hour is the one we do not have to produce," says Jim Rogers, chief executive of Duke Energy.

Goals

When the consulting firm McKinsey recently mapped out a possible pathway for U.S. carbon dioxide cuts at a cost that would not break the economy, almost 40 percent of the potential savings came from energy-efficiency steps that also would save people money. "It's a staggering amount of potential that could be an important step for achieving the carbon-abatement goals we have as a nation," says Ken Ostrowski, a McKinsey director.

Among the world's major economies, the United States is second only to Canada in energy use per person, but the nation's efficiency picture isn't all bad. Natural gas use per household is down significantly, thanks to vastly more efficient furnaces, better-insulated homes, and the population shift to the warmer South. As a result, overall energy use per U.S. household declined 26 percent between 1978 and 2001. But residential electricity use is surging, up 11 percent per household from 1993 to 2006 and 42 percent overall, as the number of gadget-filled households grows.

But research shows that with conservative measures that have fast payback, U.S. homes could become a third more energy efficient. "Green" builders everywhere know how to do it. Use 6-inch studs instead of two-by-fours for more wall cavity space to fill with insulation. In varying climates, use different kinds of high-performance windows to maximize sunlight or shield its intensity. And one simple, nontraditional step—designing ductwork so it's inside the home living space although still cleverly hidden—can cut family energy bills by a quarter to a third. Ductwork is so leaky that much of the heat or air conditioning in a home is lost. "Every time we build four new power plants to meet summer peak load, one of them is not necessary because it's generating nothing but cold air that's going into attics or crawl spaces," says Jeffrey Harris of the Alliance to Save Energy, a nonprofit coalition of business and environmental groups. The alliance is among groups pushing for a national minimum standard in building codes.

The National Association of Home Builders supports voluntary efforts but not mandates. "If you look at the places with more stringent energy requirements, you're looking at places with high housing costs," says Carlos Martín, assistant staff vice president of NAHB. "Especially in the market we have now, with foreclosures and people not able to afford even a slight price increase, that's a concern."

But a low-priced home is no bargain in the long run if it wastes energy. "You may pay a few dollars more on your mortgage payment, but you'll pay many dollars less on your utility bill," Harris says. "It's the sum of those, in cash flow to consumers, that really matters." Certainly, the payback on such investments is quicker in states with high electricity rates. But if the nation takes steps to cut carbon emissions, as seems inevitable, the price paid for fossil-fueled electricity will get higher even in states where efficiency makes less economic sense today. "These are half-century assets," Harris argues. "Why should we build infrastructure that basically is going to become another kind of dinosaur?" NAHB counters that the far bigger problem is the dinosaurs already among us: existing homes.

Reducing energy use in older homes is not easy. People are beginning to switch to twisty compact fluorescent light bulbs, which burn 75 percent less electricity than old-fashioned incandescents, with upfront costs recouped in less than a year. But the payback is neither so quick nor so clear on other items. Christine Rovner, director of Student Pugwash USA, a nonprofit that focuses on science and social responsibility, says she and her husband hired a heating and cooling system expert to advise them on steps they could take in their 60-year-old house in Washington, D.C. The windows, she knows, leak and need replacing, but she hasn't priced new ones. "We're too scared to," she says. She and her husband are weighing that possible outlay against other pressing costs, like day care for their young daughter and college savings. She thinks investing in appliances with the government's Energy Star label will be their next budget item. "We're watching our refrigerator and waiting for it to go," Rovner says.

Hogs 2.0

A swap to a new, efficient fridge would save enough energy to light the average household for nearly four months, Energy Star estimates, with an overall payback of about three years. But today, families can easily lose the gains they make in energy upgrades. At EPRI's Living Laboratory for Energy Efficiency in Knoxville, Tenn., researchers are tracking why the home electricity load is growing. Take, for instance, the set-top box—the converter needed to receive cable or satellite signals. These boxes are always in a ready state and draw as much power when they are turned off as when they are turned on. EPRI estimates each set-top box consumes about half the electricity of a new Energy Star refrigerator.

Of course, those set-top boxes are only one element of the home entertainment center. EPRI's testing shows that in energy use, those must-have 42-inch flat-screen TVs tower over the old 27-inch cathode-ray tube sets they're replacing. Reddoch notes that the traditional home centers of electricity use—heating, lighting, refrigeration, hot water—all have grown more efficient. "But the bad news is we've got this miscellaneous-electronics category—what we call the plug-in loads," he says. "They all are increasing."

Not far from EPRI's test center, researchers at Oak Ridge National Laboratory set out to prove how to drive down home energy use. Working with the nonprofit Habitat for Humanity and the Tennessee Valley Authority, they designed five homes with tight walls and windows, energy-sipping appliances, and ductwork inside the building envelope. The families, who didn't own homes before, helped build the modest three-bedroom dwellings. They pay $1 per day or less for electricity, while neighbors in similar homes fork over four to six times as much. Becky Clark says she has seen her monthly electric bill get down to $11 in the summer. And she likes that her 10-year-old son, Bradley, did a school science project on the home.

But electricity is still an essential fuel of our lifestyles. One December evening soon after the homes were built in 2003, Jeff Christian, Oak Ridge's director of building technology,

Heating Is Still No. 1, but New Devices Add to Electric Bills

Energy Costs Around Your House

Home Heating

The big-ticket item on the electric bill, especially in cold climates, for the 30 percent of U.S. households with electric heat.

- Average cost per year: $817

Refrigerators

Today's most efficient models use half the power of units made before 1993.

So the cost could triple if you plug in the old icebox in the basement.

- Average cost per year: $113

Plasma-Screen TV

Energy use rises and falls with the intensity of screen images, but the popular 42-inch-screen TVs with plasma technology can burn three times the power of old cathode-ray tube sets.

- Average cost per year: $50–$60

LCD TV

The 42-inch LCD TVs use less juice than plasma but twice the power of an old tube TV. In November, new Energy Star standards should help consumers seek efficient models.

- Average cost per year: $38–$40

Set-Top Box

The cable or satellite converter box adds to the energy cost of TV viewing by sucking electricity at the same rate whether turned on or off.

- Average cost per year: $27

Digital Photo Frame

Like many new gadgets, this adds just a little to each electric bill, but policymakers worry about the cumulative impact once the frames saturate the market.

- Average cost per year: $9

Sources: Electric Power Research Institute, Energy Information Administration, Tacoma Power.

rushed to see if the monitors were out of whack but found the sky-high readings were due to a festoon of Christmas lights and inflatable snowmen. He didn't mind: "These were families who had trouble affording to eat, who could enjoy the holidays this way because their electricity costs were so low."

But he'd like to see every house equipped with a little bit of solar energy, as the Oak Ridge scientists put on the rooftops of these homes. When the sun is shining and home energy use is low, the meter outside actually spins backward to show electricity savings. Solar panels are expensive, but Christian says the instant data on energy use would be worth it. "Then everyone could see how precious it is to collect energy from the sun," Christian says, "and how easy it is to shut off a light."

UNIT 4

Biosphere: Endangered Species

Unit Selections

16. **Forest Invades Tundra . . . and the New Tenants Could Aggravate Global Warming,** Janet Raloff
17. **America's Coral Reefs: Awash with Problems,** Tundi Agardy
18. **Seabird Signals,** Doreen Cubie
19. **Taming the Blue Frontier,** Sarah Simpson
20. **Nature's Revenge,** Donovan Webster

Key Points to Consider

- What is Nature telling us about our climate? What are the consequences of future climate change on the diversity and health of our environment?
- Are there ways to assess the value or worth of living organisms other than those from whom we derive direct benefits (our domesticated plant and animals species)? What are the relationships between economic assessments of the biosphere and moral or value judgments on the preservation of species?
- What kinds of changes are taking place in the chemistry, physics, and biology of the ocean—the world's largest ecosystem?
- Explain some of the interconnections between such things as changing ocean temperatures and the population of phytoplankton and the marine species that feed upon them.

Student Website

www.mhhe.com/cls

Internet References

Endangered Species
http://www.endangeredspecie.com

Friends of the Earth
http://www.foe.co.uk/index.html

Natural Resources Defense Council
http://nrdc.org

Smithsonian Institution Website
http://www.si.edu

World Wildlife Federation (WWF)
http://www.wwf.org

There have been five great extinction events over the last 550 million years. In each, more than 50 percent of all species were lost. We are now in the middle of the sixth. Some of the past extinctions have been attributed to specific causes, such as a meteorite impact, while the explanations for others are less clear. Our sixth extinction event is unambiguously related to human intervention. Early hunters efficiently reduced the number of large species. Agrarian practices and human settlement further reduced species diversity by modifying the local environment with animal sanctuaries becoming increasingly isolated and fenced off. And now the future of many species is threatened by climate change associated with global warming (the polar bear, sadly, being the "poster child" of vulnerable animals). In the past, animals and plants could adapt to environmental warming or cooling by migrating to more habitable climates. The physical isolation of animal habitats that exist today make that natural transition difficult to impossible.

The first three articles of this section deal with changes in sensitive ecosystems in response to global climate change. In "Forest Invades Tundra," dramatic changes are seen in the boreal ecotone between the Arctic tundra and the northern limits of trees. Tree size is changing, the temperature of the ground is changing, and the degree and severity of fires is changing. It is not only a "thermometer" of our planet's health, it forebodes a potential release of large amounts of CO_2 presently tied up in the soil.

Coral reefs are the rainforests of the oceans. They contain the greatest diversity and, just as the Arctic is the canary in the coalmine for the land, the coral reefs are the barometer for the health of the seas. In "America's Coral Reefs," Tundi Agardy describes how a number of factors have combined to decimate these fragile ecosystems. Another harbinger of the ocean's ills is the Cassin's auklets, a small sea bird found in northern waters. As described in "Seabird Signals," the dropping auklet population is related to warming ocean temperatures and changing ocean currents.

The final two articles look at a different sort of interaction between Man and Nature. "Taming the Blue Frontier" is an upbeat investigation of advances in responsible fish farming. Creative techniques, such as allowing netted fish populations to "migrate" throughout the ocean, or the juxtaposition of fish farms with filter feeder organisms (to trap the farms' waste) are examples of what can be done to maintain a sustainable fishing community.

The last article takes us on a trip to a unique world. Chernobyl is the site of the world's worst nuclear accident. On April 26, 1986, the Chernobyl nuclear power plant exploded, spewing high-level radioactive waste the world over. Other than clean-up crews, the town was abandoned immediately after the disaster. Now, over 20 years later, the "original" inhabitants of the region, the animals, are returning to the abandoned city from the surrounding forests; all the while Geiger counters scream out the super high levels of radiation.

U.S. Fish & Wildlife Service/David Bowman

Forest Invades Tundra . . . and the New Tenants Could Aggravate Global Warming

For the Arctic, green is the new black.

People frequently say "green" to mean "environmentally friendly." But encroaching conifer forests—really big greens—threaten to further spike the far North's already low-grade fever.

Temperatures in the high Arctic already are climbing "at about twice the global average," notes F. Stuart Chapin of the University of Alaska Fairbanks.

The newest data on the advance of northern, or boreal, forests come from the eastern slopes of Siberia's northern Ural Mountains. Here, north of the Arctic Circle, relatively flat mats of compressed, frozen plant matter—tundra—are the norm. This ecosystem hosts a cover of reflective snow most of the year, a feature that helps maintain the region's chilly temperatures. Throughout the past century, however, leading edges of conifer forests began creeping some 20 to 60 meters up the mountains, and in some places these forests are now overrunning tundra, scientists report in the July *Global Change Biology*.

Conifers here now reside where no living tree has grown in some 1,000 years, points out one of the authors, ecologist Frank Hagedorn of the Swiss Federal Institute for Forest, Snow and Landscape Research in Birmensdorf.

Ecologists and climatologists are concerned because emerging forest data suggest that the albedo, or reflectivity, of large regions across the Arctic will change. Most sunlight hitting snow and ice bounces back into space instead of being absorbed and converted to heat. So if a white landscape becomes open sea or boreal forest, what was once a solar reflector becomes a heat collector.

Sea-surface ice already is melting in the Arctic, and polar ice sheets are thinning. Warming threatens to further degrade these solar reflectors. So does the advance of boreal forests, Chapin says.

"Effects of vegetative changes will be felt first and most strongly locally—in the Arctic," he says. However, he adds, if the Arctic's albedo drops broadly, this could aggravate warming underway elsewhere across the planet.

Posturing

Tree rings from the Arctic Urals show that since the 15th century, many Siberian larch (Larixsibirica Ledeb.)—the primary tree species—have grown in a stunted, shrubby form, sporting multiple spindly trunks. This adaptation to harsh conditions helps the trees weather wind and snow. But the trees invest so many calories in making multistemmed clusters, Hagedorn says, that they end up puny and unable to make seeds. This infertility has thwarted the stand's spread.

After about 1900, these larches began to switch from their creeping, multi-stemmed form to tall trees with a more upright posture, though sometimes with up to 20 stems, Hagedorn and his Russian and Swiss collaborators report. Over time, new trees emerged with a single, upright trunk, at the same time bulking up with more biomass than shrubby, same-age kin. Overall, 70 percent of upright larches have emerged in just the past 80 years. Since 1950, 90 percent of local upright larches have been single-stemmed.

This forest advance into former tundra coincided with a nearly 1 degree Celsius increase in summer temperature and a doubling of winter precipitation.

"That's a good cocktail for growth," says arctic plant ecologist Serge Payette of Laval University in Quebec. Whether a tree grows up versus out depends on survival of its uppermost, or apical, buds. Good snow cover will protect those buds from winter damage, he says. Only if they are destroyed will the surviving lateral buds push growth horizontally, he explains.

Spruce are North America's more common boreal species at polar tree lines, Payette says. Some of these also assume a shrubby form, creating what he calls "pygmy forests" perhaps a meter high. But he has witnessed some of these trees assuming new, upright postures as areas warm and get wetter.

This process can create the "mirage" of tree line advance, he says. In fact, the trees may not move at all; in-place populations may simply recover from chronic stress and resume growth until they reach their normal height and mass.

Ecologist Andrea Lloyd of Middlebury College in Vermont has been studying the health of boreal tree lines throughout the warming Arctic. As in the Urals, warmth seemed to spur American spruce to move into new terrain. "I've also seen spruce advancing upwards," climbing up mountains to form dense stands, she says.

But that's only part of the story, she finds. Even where stands are advancing, "if you look at individual trees, some are starting to decline." They're growing increasingly slowly. Sometimes, as growth slows, tree numbers within a stand may be increasing. "It's a paradox," she acknowledges.

Forest ecologist Glenn Juday of Alaska-Fairbanks and his student Martin Wilmking have recorded similarly perplexing data from tree rings in 2,600 trees along two mountain ranges in polar Alaska. As the environment warmed, 42 percent of the trees grew more slowly and 38 grew more quickly.

Too little water seems a bigger factor affecting tree growth than temperature, although warming can foster drought, Juday acknowledges. Indeed, as the Arctic warms, it will likely become drier, he says. "So we can expect that at least in the western North American Arctic, there are going to be sites that eventually will get too dry to grow trees."

But their loss isn't likely to compensate for the tundra lost to trees, at least in Arctic-warming potential. In fact, their loss could further perturb the global climate because boreal forests currently hold huge amounts of carbon that had been emitted as carbon dioxide, a greenhouse gas. Until they decompose, they darken the land and remain solar collectors. Once they rot, their carbon will enrich already high atmospheric CO_2 levels.

Shrubs and Microbes

The threat of tundra displacement by trees has largely escaped notice, Juday says. And indeed, boreal forest advances in Alaska have been modest, at best. One reason: Seeds don't normally travel far in the Arctic, and even when they land on tundra, its dense mats resist implantation.

Except when those mats have been disturbed. A dry summer and warm September last year allowed a fire to ignite 100,000 hectares (about 250,000 acres) of Alaskan tundra. The huge footprint of disturbed land is now ripe for growing seeds. Fortunately, Juday says, boreal forests are on the other side of a mountain range from this scarred landscape.

Throughout the past half-century, a far more pervasive disturbance—what ecologists have taken to calling shrubbification—has been subtly transforming the tundra landscape. It starts with the arrival of tiny shrubs, such as spreading willows perhaps only 7.5 centimeters (about 3 inches) high, explains ecologist Ken Tape, also at Alaska-Fairbanks. He compared repeat photographs of Arctic tundra scapes taken around 1950 and again a few years back.

His calculations indicated that for the sites he studied, "there's been something like a 39 percent increase in shrub cover." It's consistent with data from satellite monitoring of Alaska's high Arctic that have shown "increases in biomass of a similar magnitude—about 25 to 30 percent," he says.

As these willows and other shrubs start moving in, they trap snow, which begins to insulate—and warm—the soil at their feet, explains Andy Bunn, an environmental scientist at Western Washington University in Bellingham. The warming will rouse sleeping bacteria in the soft, which will then begin to feed. In the process, they'll begin to spew much of the carbon that had been locked up in the formerly frozen soil. This fertilizes the shrubs, fostering the whole warming-growth cycle.

"There's what people call a big Arctic carbon bomb" waiting to go off, Bunn says. Up to 200 petagrams—that's 200 trillion kilograms—are stored in the top meter of Arctic tundra. For comparison, the atmosphere already has 730 petagrams of carbon in it, he adds. If shrub-related warming releases much of this carbon, it could undermine much of the carbon-limiting measures people are contemplating to slow global warming, he notes.

Although trees soak up carbon, boreal trees grow so slowly they'll likely never keep up with what the soil warming will spew, Bunn says. But forests could exacerbate the problem by darkening the still fairly light-colored shrubby landscape.

Warming has so changed the climate of a huge and growing span of tundra that it now hosts a temperature and moisture level that would support forests, Juday notes. "Today, if you planted a tree—in some cases very far up from the current tree line—it would survive in many parts of the tundra." Just 40 years ago, he says, it wouldn't.

America's Coral Reefs: Awash with Problems

Government must acknowledge the magnitude of the crisis and fully engage the scientific and conservation communities in efforts to solve it.

TUNDI AGARDY

America's coral reefs are in trouble. From the disease-ridden dying reefs of the Florida Keys, to the over-fished and denuded reefs of Hawaii and the Virgin Islands, this country's richest and most valued marine environment continues to decline in size, health, and productivity.

How can this be happening to one of our greatest natural treasures? Reefs are important recreational areas for many and are loved even by large portions of the public who have never had the opportunity to see their splendor firsthand. Coral reefs are sometimes referred to as the "rainforests of the sea," because they teem with life and abound in diversity. But although only a small number of Americans have ever had rainforest experiences, many more have had the opportunity to dive and snorkel in nearshore reef areas. And in contrast to the obscured diversity of the forests, the gaudily colored fish and invertebrates of the reef are there for anyone to see. Once they have seen these treasures, the public becomes transformed from casual observers to strong advocates for their protection. This appeal explains why many zoos have rushed in recent years to display coral reef fishes and habitats, even in inland areas far from the coasts (such as Indianapolis, site of one of the largest of the country's public aquaria). Coral reefs have local, national, and even global significance.

Even when one looks below the surface (pun intended) of the aesthetic appeal of reefs, it is easy to see why these biological communities command such respect. Coral reefs house the bulk of known marine biological diversity on the planet, yet they occur in relatively nutrient-poor waters of the tropics. Nutrient cycling is very efficient on reefs, and complicated predator-prey interactions maintain diversity and productivity. But the fine-tuned and complex nature of reefs may spell their doom: Remove some elements of this interconnected ecosystem, and things begin to unravel. Coral reefs are one of the few marine habitats that undergo disturbance-induced phase shifts: An almost irreversible phenomenon in which diverse reef ecosystems dominated by stony corals dramatically turn into biologically impoverished wastelands overgrown with algae. Worldwide, some 30 percent of reefs have been destroyed in the past few decades, and another 30 to 50 percent are expected to be destroyed in 20 years' time if current trends continue. In the Caribbean region, where many of the reefs under U.S. jurisdiction can be found, coral cover has been reduced by 80 percent during the past three decades.

The U.S. government is fully aware of the value of these marine ecosystems and the fact that they are in trouble. In 1998, the Clinton administration established the U.S. Coral Reef Task Force (USCRTF), a high-level interagency group charged with examining reef problems and finding solutions. Executive Order 13089 stipulated that a task force be established to oversee that "all Federal agencies whose actions may affect U.S. coral reef ecosystems shall: (a) identify their actions that may affect U.S. coral reef ecosystems; (b) utilize their programs and authorities to protect and enhance the conditions of such ecosystems; and (c) to the extent permitted by law, ensure that any actions they authorize, fund, or carry out will not degrade the conditions of such ecosystems." The task force comprises 11 federal agencies, plus corresponding state, territorial, and tribal authorities.

The USCRTF has looked for ways to better monitor the condition of reefs, share information, and coordinate management. Among the key government players are the National Oceanic and Atmospheric Administration (NOAA), Department of the Interior, Environmental Protection Agency (EPA), Department of Defense, Department of Agriculture, Department of Justice, Department of State, National Science Foundation (NSF), and NASA. Yet however well-intentioned this move on the part of government, coral reef health has continued to decline, and the USCRTF, while elevating the profile of the issue, has not been able to stem the degradation. The reasons for this ineffectiveness are complex and go beyond the "too little, too late" offered as the standard criticism. Although the response of the government may have indeed come too late for many of America's reefs, the shortcomings of the task force have more to do with its reluctance to fully engage with the scientific community, take advantage of emerging technologies, and raise awareness about the consequences of reef degradation. If this is happening to our most treasured marine environments, what can the future be for our less-well-loved, less charismatic marine areas?

Threats to U.S. Reefs

Even as we are becoming more fully aware of their enormous ecological and economic value, coral reefs are being lost in the United States, just as they are being destroyed in other parts of the world. Some 37 percent of all corals in Florida have died since 1996, and the incidence of coral disease at sampling sites there went up by 446 percent in the same short period. The U.S. has jurisdiction over a surprisingly large proportion of extant coral reefs, including the world's third largest barrier reef in Florida; a vast tract of reef systems throughout the Hawaiian Islands; and extensive reefs in U.S. territories such as Puerto Rico, the U.S. Virgin Islands, Guam, American Samoa, and the Northern Mariana Islands. These reef resources contribute an estimated $375 billion to the U.S. economy annually, yet virtually all of these reef ecosystems are under threat, and many may be destroyed altogether in the coming decades.

Although in many parts of the world coral reefs are deliberately destroyed in the process of coastal development or to obtain construction materials, in the United States coral reefs suffer the classic death of a thousand cuts. They are strongly affected by eutrophication: The overfertilization of waters caused by the inflow of nutrients from fertilizer, sewage, and animal wastes. The overabundance of nutrients causes algae to overgrow and smother coral polyps; in extreme cases, leading to totally altered and biologically impoverished alternate ecosystems. Reefs are also sensitive to sediments that increase turbidity and reduce the sunlight reaching the coral colonies. (Though corals are animals, they have symbiotic dinoflagellates called zooxanthellae living within their tissues. The photosynthesis undertaken by these plant symbionts provides corals with the extra energy needed to create the calcium carbonate that forms their skeletons and thus the reef structure.) Sedimentation is a common threat to U.S. coral reefs, especially in areas where unregulated coastal development or deforestation causes soil runoff into nearshore waters.

Because energy flows in coral reef ecosystems are largely channeled into ecosystem maintenance and little surplus is available for harvest, reefs are highly sensitive to overfishing. The removal of grazing fishes, for instance, increases the likelihood that algae will dominate the reef, causing a subsequent decline in productivity and diversity. Reef communities denuded of even relatively small numbers of fishes are also less likely to recover from episodic bleaching events, because recruitment is inhibited by the lack of grazing fishes to create settlement space. Similarly, declines in sea turtle species such as hawksbill and green turtles negatively affect reef ecology. The removal of top predators such as reef sharks, jacks, and barracudas can also cause cascading effects resulting in reduced overall diversity and declines in productivity. Despite these impacts, very few coral reef areas of the United States have fishing regulations expressly designed to prevent these ecological cascading effects from occurring. In fact, most people would be surprised to find out that even in seemingly protected reefs, such as those that occur within the Virgin Islands Biosphere Reserve around St. John, U.S.V.I., almost all forms of recreational and commercial fishing are allowed.

Coral reefs are also extremely vulnerable to changes in their ambient environment, having narrow tolerance ranges in temperature and salinity. Warming affects both coral polyp physiology and the pH of seawater, which in turn affects the calcification rates of hard corals and their ability to create reef structure. For this reason, even a slight warming of sea temperatures has dramatic effects, especially when coupled with other negative impacts such as eutrophication and overfishing. There is some indication that warming sea temperatures may render coral colonies vulnerable to the spread of disease or to increased mortality in response to normally nonpathogenic viruses and bacteria. The spread of known coral diseases and the emergence of new, even more debilitating diseases are alarming phenomena in the Florida Keys reefs and underlie many of the die-back episodes there in the past decade.

The effects of warming are most clearly manifested in coral bleaching. Bleaching is an event in which the zooxanthellae of the corals, which give corals their beautiful colors, are expelled from the coral polyps, leaving the colonies white. Bleached corals cannot lay down calcium carbonate skeletons and thus enter a period of stasis. A bleached coral is not necessarily a dead coral, however, and corals have been known to recover from bleaching events (we also know from paleoarcheology that bleaching is a natural event that preceded greenhouse gas-related warming of the atmosphere). Because some reefs do fully recover after bleaching, it is difficult to predict what consequences warming events such as periodic El Niños will have on the long-term health of any reef. This uncertainty has been seized on by both doomsayers and naysayers in the debate about the future of reefs: The doomsayers declare that the majority of reefs face certain death from bleaching, while the naysayers claim that bleaching is not only natural but adaptive. However, one thing is absolutely clear: Stressed reefs have a heightened sensitivity to temperature changes and are far less likely to recover from bleaching events. And with a few exceptions (some parts of the northwest Hawaiian Islands and Palmyra Atoll, for instance), all of the coral reefs in the United States are highly stressed by a combination of land-based sources of pollution, overfishing, and the destruction of habitats that are ecologically critical to reef communities, such as seagrass beds and mangrove forests. This does not bode well for a future in which sea temperatures will undoubtedly continue to rise.

These losses affect more than our personal environmental sensibilities. Reefs support some of the most important industries in the United States and the rest of the world: 5 percent of world commercial fisheries are reef-based, and over 50 percent of U.S. federally managed fishery species depend on reefs during some part of their life cycle. Herman Cesar, Lauretta Burke, and Lida Pet-Soede argue in a recent monograph on the economics of coral reef degradation that the costs of better managing reefs are far outweighed by the net benefits provided by reefs. In the Florida Keys, for example, they claim that a proposed wastewater treatment plant that would mitigate many of the threats to the Florida Keys reef tract would cost $60 to $70 million in capital costs and about $4 million in annual maintenance costs. At the same time, the benefits to the local population (estimated to be greater than the net present value of $700 million) would far eclipse the outlays. In Hawaii and the reef-fringed territories, coastal tourism is tightly coupled to intact reefs. Reefs in these regions not only provide tourist destinations, they also play important roles in controlling beach erosion and buffering land from storms. In such places, it is easy to see how an investment in better reef protection would be a small cost in contrast to the great benefits provided by sustained tourism revenues.

Inadequate Responses

The failure to respond to the coral reef crisis in this country has to do with many factors:

Incomplete understanding of the problem and communication failures. Although there is an appreciation of the crisis worldwide, there is still reluctance on the part of some U.S. managers to consider the crisis "our problem." Everyone is quick to lament the destruction of Southeast Asian and Indian Ocean reefs by dynamite fishers, or the use of cyanide in collecting coral reef fish in the Philippines, but the reefs under U.S. jurisdiction have hardly fared better. In the past decade, we have seen a slow awakening to the problems facing U.S. reefs, but the response has been to collect more data, slowly and painstakingly. At first independently, and then in a more coordinated fashion with the establishment of the USCRTF, government agencies have made greater efforts to monitor reefs in certain regions, but the massive amounts of data collected often create problems in data interpretation and management. Too little emphasis has been placed on either synthesizing the data collected or collecting data in new ways to make it more relevant for conservation. Lacking a synthesis or periodic syntheses, we end up burying our heads in the sand about what is happening to our coral reefs.

The USCRTF and the government agencies it represents have not actively looked for ways to partner with academic, scientific, and nongovernmental organizations to take advantage of information being collected and disseminated by them. Instead, the government has relied almost solely on the efforts of its own scientists. Many of its scientists, such as Charles Birkeland of the U.S. Geological Survey, are indeed world leaders in coral reef ecology and management, but collectively the research being undertaken by government agencies is either substandard, too conservative, or both. Virtually every new advance in coral reef ecosystem understanding has been made not by government scientists but by academics or researchers in the private sector. U.S. government scientists have not explored the potential of new technologies such as biochemical markers that indicate reef stress (pioneered by the private sector), nor have they properly harnessed the remote sensing technologies they have deployed in order to improve reef surveillance.

Even the knowledge that has been gained is inadequately communicated to the public and to decisionmakers. Part of the problem has been the rush to oversimplify what is actually a very complex set of issues, in the hopes that decisionmakers higher up will take both notice and action. In the Florida Keys, for instance, advocates for improving the water quality of the nearshore environment have fought against the restoration of the Florida Everglades, arguing that the increased water flows into Florida Bay would bring higher concentrations of pollutants to the reef tract. In casting the reef problems in such a simplistic light, proponents of singular solutions actually impede responsible government agencies from tackling reef problems head-on and in the comprehensive manner that is required.

The U.S. government has serious shortcomings when it comes to communicating and raising awareness about complicated environmental issues. For this reason, it would behoove the USCRTF to partner with organizations that have good outreach mechanisms in place, such as environmental groups. Such public-private partnerships would also ease the financial burden of the cash-strapped government agencies, allowing them to spend funds in short supply on management and on measuring management efficacy.

Poor use of cutting-edge science and the at-large scientific community. Although the United States is one of the most technologically advanced countries in the world, it has not adequately harnessed science to address the coral reef crisis. In a 1999 article in the journal Marine and Freshwater Research, Michael Risk compares the response of the scientific community to the coral reef crisis with its response to two other crises affecting the United States: Acid rain in the Northern Hemisphere and eutrophication of the Great Lakes. Risk argues that whereas there was effective engagement of the scientific community in tackling the latter two issues, neither U.S. nor international scientists have helped craft an effective response to the large-scale death of reefs.

Risk is right to ask why science has failed coral reefs, but I take issue with his assessment of the nation's inadequate response to the crisis. It is not the fault of the scientific community that the government has been slow to act to save reefs, but rather the fault of government in not knowing how to use science and scientists effectively. Decisionmakers have not engaged the scientific community and have failed to heed what scientific advice has been put forward. For instance, the government did not fully mobilize nongovernmental academic institutions and conservation organizations to help draft its National Action Plan to Conserve Coral Reefs, and as a result the plan has been criticized as lacking in rigor and ambition. It is telling that a World Bank project to undertake global coral reef-targeted research, which assembles international teams of leading researchers to address critical issues of bleaching, disease, connectivity, remote sensing, modeling, and restoration, has a paucity of U.S.

government scientists in all six of the working groups. This targeted research project is crucial: It intends to identify the key questions that managers need to have answered in order to better protect reefs, and it aims to do intensive applied research to answer those questions.

The National Action Plan to Conserve Coral Reefs was produced by the USCRTF and published on March 2, 2000. It is a general document describing why coral reefs are important and what needs to be done to protect them. There are two main sections: Understanding coral reef ecosystems and reducing the adverse impacts of human activities. The first section discusses four action items: (1) create comprehensive maps, (2) conduct long-term monitoring and assessment, (3) support strategic research, and (4) incorporate the human dimension (undertake economic valuation, etc.). The second section is a bit more ambitious: (1) create and expand a network of marine protected areas (MPAs), (2) reduce impacts of extractive uses, (3) reduce habitat destruction, (4) reduce pollution, (5) restore damaged reefs, (6) reduce global threats to coral reefs, (7) reduce impacts from international trade in coral reef species, (8) improve federal accountability and coordination, and (9) create an informed public. All well and good, but despite its moniker the action plan provides almost no guidance on how to do these things. It called for each federal agency to develop implementation plans (required by Executive Order 13089) by June 2000. However, those plans were only to cover fiscal years 2001 and 2002, and the plans were never formalized or made public. The USCRTF recognized that a greater investment needed to be made to figure out how each agency was going to contribute to carrying out the action plan and pushed agencies to develop post-2002 strategies. To date, only the Department of Defense and NOAA have completed such strategies. NOAA's plan is embodied in its National Strategy for Conserving Coral Reefs document published in September 2002. Both the action plan and the NOAA strategy are available on the USCRTF Web site (www.coralreef.gov).

The plans put forward by the USCRTF, however, place far too much emphasis on monitoring and mapping and far too little emphasis on abating threats and effectively managing reefs. The focus of research has been to monitor existing conditions rather than to set up applied experiments that would tell us which threats are most critical to tackle. This is not to say that all government research has been worthless. Regular monitoring in the Florida Keys allowed NOAA to understand the alarming "blackwater" event in January 2003 (in which fishermen noticed black water, later found to be a combination of a plankton bloom and tannins, moving from the Everglades toward the reefs) and reassure the public that it was a natural event, because they had several years of monitoring information with which they could hindcast. Similarly, the mapping investment, although too high a priority, has led to some interesting revelations: There are newly discovered reefs in the northeastern portion of the Gulf of Mexico that are now on the public's radar screen, for instance.

Although the USCRTF has recognized the importance of MPAs in conserving reefs, it has not given the government agencies that have responsibility for implementation guidance on how to optimally design these protected areas. The action plan thus codifies a dangerous tendency to use simplistic formulae for designing protected areas. The plan states these as its goals: "establish additional no-take ecological reserves to provide needed protection to a balanced suite of representative U.S. coral reefs and associated habitats, with a goal to protect at least 5% of all coral reefs . . . by 2002; at least 10% by 2005, and at least 20% by 2010." By adopting a policy of conserving 20 percent of reef areas within no-take reserves, without requiring planners to fully understand the threats to a particular reef and without guiding planners to locate such protected areas in the most ecologically critical areas, the plan pushes decisionmakers to implement ineffective MPAs, thus squandering opportunities for real conservation. In some jurisdictions, these area targets have already been reached, with 20 percent of reef areas set aside as no-take zones, but because these areas were chosen more for their ease of establishment and less for their ecological importance, little conservation has been accomplished. In a true display of lack of ambition and creativity, the USCRTF and its agencies have not considered using ocean zoning outside of MPAs to conserve reefs, and the MPA directives remain an old-school, one-size-fits-all approach.

Poor governmental coordination and lots of infighting. Since its formation in June of 1998, the USCRTF has made some strides toward better monitoring, information sharing, and management coordination for reefs under U.S. jurisdiction. This is no minor feat, because until the task force was established, no effort had been made to promote communication and cooperation between the multitude of agencies and bureaus that each have a role to play in coral reef management. NOAA's National Ocean Service and National Marine Fisheries Service, the Department of the Interior's U.S. Fish and Wildlife Service, and the EPA are the key players in the USCRTF, but also important are the National Parks Service, DOD, Department of Agriculture, Department of Justice, Department of State, NSF, and NASA. Although the major players (in particular, NOAA and the Department of the Interior) are engaged in internecine warfare over territorial claims and access to funding, some of the more minor players have taken their charge very seriously. DOD, for example, has developed its own plan for conserving the reefs under its jurisdiction, which include some of the most pristine reefs in the nation, such as the reefs of Johnston Atoll in the central Pacific.

Unlike many terrestrial habitats, coral reefs suffer both from human activity that directly affects the marine environment (such as dredging, fishing, and marine tourism) and from activity on land that has an indirect but highly insidious effect on reef health and productivity. Thus, in order to better understand and manage reefs, it is imperative that the United States continues, and now strengthens, coordinating mechanisms between the various government entities that control the wide array of human activities that damage reefs.

The USCRTF now has a roadmap to increase understanding about coral reefs and better protect them from further destruction, embodied in the National Action Plan to Conserve Coral Reefs. A subsequent report prepared by NOAA, in cooperation with the USCRTF, was submitted to Congress in 2002. The 156-page National Coral Reef Action Strategy provides a nationwide status report on implementation of the National Action Plan to Conserve Coral Reefs and the Coral Reef Conservation Act of 2000.

Will the USCRTF now be able to do what it could not in the first five years of its existence: Stem the tide of degradation affecting U.S. coral reefs? Or is the U.S. government merely creating a façade of improved management, while government researchers and managers continue to work in isolation from cutting-edge researchers in U.S. academe, nongovernmental organizations, and international institutions? Will new policy developments, such as the administration's support for broad environmental exemptions for DOD's military training and anti-terrorism operations, act to wholly undermine any substantive progress made by the USCRTF and the government agencies it represents?

Only time will answer these questions with certainty, but the initial impressions are not promising. The National Action Plan to Conserve Coral Reefs is too heavily invested in relatively easily accomplished activities such as mapping the nation's coral reefs, and its formulaic and simplistic approach to creating MPAs will not likely result in meaningful protection. Already overburdened and underfunded agencies are not getting the political mentoring they need to ensure that appropriations will be sufficient to allow them to carry out their mandates under these plans. Without public-private partnerships and private-sector financial support, too many elements of the plan will fall by the wayside. Neither the Action Plan nor the NOAA strategy provide adequate information on the true choices and tradeoffs that decisionmakers will have to consider and act on in order to create a revolution in the way we manage coral reefs. And clearly a revolution is needed; business as usual will only continue to put U.S. coral reef ecosystems in harm's way. In the end, the United States may fall far short of its goal of demonstrating how to effectively manage coral reefs in a way that all the world can see. Instead, it may well win the race to destroy the inventory of one of the world's most diverse and precious environments.

Global forces are at play: The United States is not an island. Were the United States suddenly to act more effectively to protect reefs under its jurisdiction, our reef ecosystems would still be in some peril, for many reasons. First, many damaging activities occur out of sight, especially in remote reef areas with little or no surveillance. Second, the open nature of marine systems means that reefs are affected by the condition of the environment far from the reef tracts themselves. Sometimes larval propagules travel long distances, and the origin of recruits is tens or hundreds of kilometers away, in areas that could be entirely outside U.S. jurisdiction. Similarly, pollution from outside the U.S. can easily find its way to reefs within America's borders. Finally, some threats to reefs are global in nature, such as rising temperatures caused by global warming. These threats will not diminish unless meaningful international agreements succeed in tackling the root causes of the threats. For all these reasons, protection of U.S. reefs will require more than administering the reefs within our borders; it will also require international negotiation, cooperation, and capacity-building.

Promising Sign

Is there hope for U.S. coral reefs? Yes, as long as we can more fully engage the private sector and the scientific community in the struggle to save reefs, and at the same time convince decisionmakers of the need to take significant steps to protect these fragile ecosystems. It is a promising sign that in June 2003, NOAA, EPA, and the Department of the Interior convened a meeting in Hawaii to discuss coral bleaching and ways to gain better collaboration between the scientific community at large and the government agencies charged with managing reefs. The USCRTF is beginning to reach out to scientists involved in coral reef research and management outside the United States, such as the coral reef-targeted research working groups formed under the recent World Bank initiative. In this way, the U.S. government can begin to take advantage of the significant strides in scientific understanding that have been made by nongovernmental researchers, both in the United States and abroad.

New advances in technology may help coral reefs as well, and just in time. For instance, the Planetary Coral Reef Foundation has teamed up with the Massachusetts Institute of Technology and other academic institutions to attempt to launch a satellite that will provide real-time information about the condition of reefs worldwide. Such a satellite mission would make it possible to know the extent of coral bleaching and the presence of fishing operations anywhere in the world at any time. With such a system in place, traditional surveillance could be cut back, allowing money to be redirected toward conservation. At the same time, donors could get a better sense of where their investments are paying off in terms of real conservation of reefs and could identify trouble spots quickly enough to get funds flowing to places where emergency measures are needed.

With the full engagement of the scientific community and with partnering to remove some of the burden from beleaguered government agencies, managers will be able to tailor responses to the given threats at any reef location. Where fishing is deemed to be a major stressor, the United States will have to find the political will to manage reef-based fisheries more effectively. Where pollution (whether nutrients, toxics, debris, or alien species) is undermining reef health and resilience, coastal zone and agricultural agencies will have to work to find ways to reduce pollutant loading. Where visitor overuse and diver damage are issues, managers will have to look for ways to prevent people from loving reefs to death. And in all areas, managers will have to resist oversimplifying the situation and begin to better inform the public and decisionmakers about the hard choices to be made.

The coral reef crisis is indeed our problem. It affects our natural heritage and the livelihoods of a great number of our citizens. Only when the people in power recognize the magnitude of the problem will effective steps be taken to engage the wider scientific and conservation community in safeguarding reefs. When future generations look back at the dawn of the millennium and the environmental choices that were made, they will either curse us for letting one of nature's most wondrous ecosystems be extinguished or praise us for recognizing the great value of reefs and moving to protect them. I hope it is the latter.

Tundi Agardy (tundiagardy@earthlink.net) is the executive director of Sound Seas in Bethesda, Maryland.

Seabird Signals

Off the U.S. West Coast, several seabird species are suffering, and biologists suspect that global warming's impact on the Pacific is to blame.

DOREEN CUBIE

Cassin's auklets may look nothing like canaries, but in a way that's what they are. These plump, dusky gray seabirds, like the proverbial birds in the coal mine, are telling scientists there is something wrong in the Pacific Ocean. In the 1970s, about 100,000 pairs of auklets nested on California's remote Farallon Islands. Today only 20,000 pairs remain. In 2005 and 2006, not one of the birds' eggs hatched, and only about a third of the pairs fledged young in 2007.

"We see signals in birds," explains seabird biologist Bill Sydeman. "They're the best indicators of what's going on." As president of the Petaluma, California-based Farallon Institute for Advanced Ecosystem Research, Sydeman and his colleagues have been following the islands' seabirds for the past 35 years. "We know there are changes to the ecosystem," he says, and those changes are also hurting other species, namely tufted and horned puffins and rhinoceros auklets.

These birds, all members of the auk family, spend their lives at sea, coming ashore only to raise their young. More at home in the ocean than on land or even in the air, they use their wings to "fly" underwater to catch prey, often diving as deep as 100 to 300 feet. Over the past century, auk populations have been depleted by introduced predators on nesting islands, oil spills, pollution and fishing nets that entangle and drown them. But today there are new threats. The seabirds' food web appears to be unraveling, and scientists suspect that global warming is at the heart of the problem. Certainly Cassin's auklets seem to be running up a red flag. These diminutive birds, about the size and shape of quail, lay their eggs in shallow burrows or rock crevices on coastal islands from Baja California to Alaska. Once their chicks hatch, the adults leave the young alone during the day, flying back each night with food. During good times, mom and dad auklets return with their expandable throat pouches stuffed with krill. But these are not good times. Krill, which feed a variety of wildlife ranging from salmon to blue whales, have been in short supply the last few years when Farallon auklets are trying to raise their chicks.

Tufted puffins are another "barometer of biological change," says John Piatt, a research biologist with the U.S. Geological Survey's Alaska Science Center in Anchorage. "They reflect the ocean regimes they live in," says Piatt, who calls the birds "seabird tough guys" both for their appearance, which he likens to bikers in leather regalia, and their lifestyle, which includes winters spent far out to sea surviving on squid and lanternfish.

As colorful as Cassin's auklets are plain, most tufted puffins nest in North America, from the Aleutians to California. The birds' northernmost colonies are thriving but—tough guys or not—their southern colonies are declining dramatically. At Oregon's Three Arches Rocks, for example, puffin numbers have plummeted, and only a fraction remain. In Washington, nearly a third of all colonies have disappeared.

Rhinoceros auklets are also facing tough times. Closely related to puffins, these birds get their name from a small "horn" that grows on the base of their bill. Rhinos, as they're often called, range from California to Alaska and across the Pacific to Russia, Japan and Korea. Their largest breeding colony, some 600,000 birds, is found on the Sea of Japan's Teuri Island. In the United States, biologist Julie Thayer of Petaluma, California-based PRBO Conservation Science has been studying the birds on the Farallones and on central California's Año Nuevo Island.

Unlike the Cassin's auklets, the rhinos and puffins feed fish rather than krill to their chicks. Nonetheless, from California to British Columbia, they, too, are having trouble finding enough food for their youngsters. In times past, when the birds laid their eggs, the waters surrounding their nesting islands were roiling with rockfish, sand lance, sablefish and squid. Today the fish are often AWOL. "We are seeing an alarming decrease in appropriate-sized prey," says Thayer.

One explanation may be an ocean that's heating up. From 1937 to 2002, the sea surface temperature in the vicinity of British Columbia's Triangle Island has fluctuated from year to year but increased overall by nearly 1 degree C. Researchers from the Canadian Wildlife Service and several universities have monitored breeding seabirds on the island since 1975, including a colony of 50,000 tufted puffins. They've found the puffins' fledgling success is virtually zero when the sea surface exceeds 9.9 degrees C. When the water is that warm, biologists believe prey fish move elsewhere, forcing adult puffins to follow and leave their eggs and chicks behind.

"Breeding birds are tied to specific places," explains Thayer. Although some auk species can take their chicks to sea when they are only a few days old, the young of puffins and auklets must stay in or near their burrows until they are able to fly. As a result, parents are tethered to their nesting island for as long as three months, usually unable to forage more than about 30 miles away. "The timing of prey is very important to these birds," adds Thayer.

Changing climate seems to be throwing that timing out of kilter. In particular, changes to coastal upwelling—the critical movement of nutrient-rich water from the depths of the ocean to its surface—may be having disastrous consequences. Upwelling delivers food to phytoplankton, the single-celled plants that are the foundation of marine food chains. Because phytoplankton can live only in the top 100 feet or so of the ocean, where light permeates, their existence depends on upwelling.

Today it often takes longer for upwelling to occur, and sometimes it doesn't happen at all. When mixing does take place, the waters usually come up from a shallower level, which means they are poorer in nutrients. This sets off a domino effect in the food chain: disappearing phytoplankton, a drop in krill and other animal plankton, a scarcity of certain fish and starving seabirds.

Rhinoceros auklets, which time their breeding to coincide with coastal upwelling and its abundant prey, are especially hard hit. Because mixing is often delayed, the rhinos are now nesting later. But there is still often a mismatch, times when the fish simply are not there. "Some years, there's massive abandonment of eggs," says Thayer.

A newer theory—so far not widely accepted—is that changes in major ocean currents are disrupting Pacific food chains. One of these "rivers in the sea," as Sydeman calls them, is known as the North Pacific Drift. The current flows from Japan to the coast of North America, where it divides into the northern Alaska Current and the southern California Current. Exactly where that split occurs is very important to marine life, says Sydeman. "When it bifurcates more to the north, the water reaching British Columbia and California is colder and saltier, with more subarctic plankton," he says. "But when the current splits more to the south, the water is warmer and has less plankton."

"Colder is better for seabirds," says Sydeman. In 2002, when the North Pacific Drift divided at the Queen Charlotte Islands in northern British Columbia, Cassin's auklets on the Farallones fledged many chicks. But in 2005–2006, when it split off central Oregon, "the auklets suffered complete reproductive failure," he says. Oceanographers remain uncertain whether such fluctuations are part of a natural cycle or another indication that the Pacific is starting to go haywire.

Whatever their causes, changes to the Pacific food web may be extending far beyond the breeding season. According to Julia Parrish, an associate professor of zoology at the University of Washington, winter seabirds seem to be sending a message by washing up dead on beaches in ever-increasing numbers. As executive director of the Coastal Observation and Seabird Survey Team, she has been tracking the number of beached seabirds in California, Oregon, Washington and Alaska since 2000—a program that gives biologists a snapshot of how many birds are dying. "It mirrors what's going on in the colonies," says Parrish, adding that the team has noticed a spike in beached Cassin's and rhinoceros auklets every January. "We didn't see this before," she says. "It suggests there might be something new going on, perhaps a change in the amount of winter food resources."

The team also has seen an uptick in the number of dead horned puffins. This look-alike of the Atlantic puffin breeds primarily in Alaska, where it has begun nesting farther to the north as the summer pack ice shrinks. Horned puffins winter in the Pacific, usually far from land, and are rarely found washed up on the coast. But that has changed over the past few years, mystifying researchers who wonder if more birds are dying or if shifting conditions are forcing them to forage closer to shore.

Despite such changes, Parrish remains cautiously optimistic. "Sea-birds are pretty adaptable," she says. Piatt agrees. "These birds are very long-lived," he says, pointing out that it's not unusual for auks to survive 20 years or more. "A pair only has to raise two chicks to replace themselves," adds Piatt. If they can manage that, they will sustain their populations.

Because of the birds' long life spans, some biologists suspect it may be typical for auks to have more years with breeding failures than successes. Worldwide, the populations of most puffins and auklets remain high, often numbering in the millions. Nonetheless, when these "canaries" start dying, it's time to pay attention to what's happening in the coal mine. Parrish cautions that we must not ignore what the birds are saying. "They're shouting at us," she says. "We tend to have trouble listening."

South Carolina journalist **Doreen Cubie** was the winner of NWF's 2007 Trudy Farrand/John Strohm Magazine Writing Award.

Article 19

Taming the Blue Frontier

Ten thousand years ago, humans made the shift on land from hunting and gathering to farming. Now the same transformation is taking place at sea. This time, can we get it right?

SARAH SIMPSON

A shipment of 100,000 fresh, sushi grade cobia, each fish amounting to about five pounds of firm, white meat, arrives on schedule in the Port of Miami. In this case, "fresh" does not mean beheaded and ice-packed—these fish are very much alive and swimming. As fingerlings, they were set adrift in a 3-million-liter pen which latched onto a current traveling the Caribbean in a predictable, clockwise path. Nine months later, a frenzy of splashes erupt at the water's surface as the underwater corral emerges from the depths. After rounding the western tip of Cuba and skirting a storm near the Yucatán (via remotely operated thrusters), the floating farm has made port just as the fish reach harvestable size.

Aquatic engineer Clifford Goudey had this futuristic vision dancing in his head last July when he tested the world's first self-propelled, submersible fish pen. A geodesic sphere measuring 19 meters in diameter, the cage proved surprisingly maneuverable when outfitted with a pair of 2.5-meter propellers, says Goudey, who directs MIT Sea Grant's Offshore Aquaculture Engineering Center. In his Caribbean current scenario, Goudey imagines launching dozens of floating farms in a steady progression, each a week behind the other. His work marks a breakthrough in the quest to raise fish in parts of the oceans that are too deep for traditional, anchored cages. It also amounts to a key step toward what a few cutting-edge thinkers have been craving for years: the wholesale taming of the sea.

The oceans provide about 20 percent of the world's protein, and pressure to deliver this critical food stream has led to extreme overharvesting. As wild fish stocks decline, aquaculture is the logical candidate to pick up the slack, and some are looking to it as a way to rebuild commercial fish stocks. In a 2005 Nature commentary, oceanographer John Marra argued that widespread ocean farming is inevitable. "We have already accepted domestication of the land," Marra wrote. "Now is the time to accept the same for the seas."[1]

Such visions have long been anathema to many environmentalists who fear the spread of present-day aquaculture's myriad ills. Many of today's coastal fish farms have decimated habitat and spread disease into local fish populations. Making matters worse, fish farms represent a net drain on populations of wild fish, which are often caught just so they can be ground into feed for salmon and other species.

Despite these concerns, a shift is underway. Some members of the environmental community are concluding that widespread aquaculture must be pursued if we are to save the oceans and feed the planet. Aquaculture production must double by 2050 just to keep up with per capita demand. But merely scaling up current methods would only exacerbate the problems.

In other words, the world needs new, sustainable aquaculture practices, and it needs them fast. It took 10,000 years for domestic agriculture to transform the land, but viable ocean farming schemes must be developed in one one-hundredth of that time if they are to forestall the oceans' demise. This urgency is spurring some leading environmentalists and scientists to lend their knowledge and support, instead of their opposition. In a recent lecture, the renowned marine ecologist Jeremy Jackson discussed the threat of overfishing and announced, "the most important scientific challenge we now face is how to make aquaculture ecologically sustainable."

The mobile fish pens are just one example of the cutting-edge technologies emerging to surmount Jackson's challenge. Also in the works are intriguing methods of recycling fish sewage, new feed formulations that use dramatically smaller amounts of wild fish, and onshore farms where salt-water species are tricked into living in fresh water. As these developments solve some of aquaculture's seemingly intractable problems, they could also be the first steps toward widespread, sustainable domestication of the oceans.

From his post at eastern Canada's Bay of Fundy, Thierry Chopin has seen first-hand how salmon farms devastate bays and inlets. Typically located just a stone's throw from shore, the farms are relentless sources of excrement and pollution. Some estimates suggest the nutrient outfall from salmon farms in Scotland, for example, is comparable in volume to the untreated sewage from half its human population. What's more, salmon are sloppy eaters that typically consume only about 80 percent of the food that comes their way. The nutrient-laden effluent spills from cages, sometimes triggering harmful algal blooms and other pollution problems. But Chopin, a University of New Brunswick marine biologist who has spent eight years trying to mitigate these problems, says cleaning up salmon farms may be as simple as re-casting poop and uneaten food scraps as a resource.

In conjunction with Cooke Aquaculture, a Canadian company pursuing sustainable farming, Chopin is experimenting with so-called integrated multitrophic aquaculture, or integrated farms. This innovative approach positions salmon pens in close proximity to plants and animals that actually consume the pollution. The farms aim to absorb the vast majority of the salmon waste, sparing the surrounding waters while nurturing species that can be sold on the global seafood market.

Seen from above, Chopin's operation looks like a tray of soda cans, with circular salmon pens anchored in a square grid. It amounts to a carefully calibrated ecosystem, and Chopin, along with Shawn Robinson at Canada's Department of Fisheries and Oceans, has helped identify which species can thrive within it. "It's all about choosing species based on their function," Chopin explains.

Seaweeds, for instance, are amazingly efficient waste recyclers that can extract about 40 percent of the dissolved nutrients available during their growing season. To take advantage of this, Chopin's team positions seaweed on ropes dangling from rafts located downstream from the pens. The kelp thrive in this fertilizer bath, which is primarily ammonia released from salmon gills and decaying food pellets; some species grow as much as 46 percent faster than they do in salmon-free areas.

Sharing the same grisly appetite for salmon waste are filter-feeding mussels, which play a different role in the clean-up process by extracting particles of excrement and food scraps. Chopin's system places the mussels in cages alongside the fish pens. Thanks to their close quarters with salmon, the mussels grow as much as 50 percent faster as they absorb about half of the fine waste particles. About three years ago, though, Chopin's team realized some waste particles were too big for the mussels to manage. That's where sea cucumbers and urchins, which thrive on the heftier scraps and are placed in trays directly below the salmon pens, come into the picture. "One man's trash is another's treasure" takes on new meaning when you see how remarkably plump a culinary delicacy such as urchin roe grows in a cloud of salmon sewage.

There is, of course, a more obvious way to manage the effluent problem: move fish farms away from the coast, into deeper waters where pollution would be carried off and diluted by the sea. That turns out to be much harder than it sounds. For starters, offshore cages have to be built to withstand the pressure and currents that come with being located farther out at sea—keeping enormous pens steady in 60 meters of open water is a tricky proposition. They must also keep their plump inhabitants safe from sharks and other predators looking for an easy lunch.

A handful of companies are overcoming these challenges with innovative cage designs that allow fish to be farmed in deeper water than ever before. Anchored varieties of the spherical AquaPod, developed by Ocean Farm Technologies in Searsmont, Maine, range up to 27 meters in diameter, which translates to almost a billion liters in capacity. Another innovator is Bainbridge Island, Washington–based OceanSpar, which has developed its SeaStation pens in the shape of oversized toy tops. Built around galvanized steel frames, the pens are covered in Kevlar-like netting that prevents wily fish from chewing their way out—or in.

Half a mile off the Hawaiian coast, Kona Blue Water Farms is using eight 3-million-liter SeaStations to house some 480,000 Hawaiian yellowtail. The pens are tethered by a network of 22 anchors, each weighing 3.5 tons and anchored by a one-ton chain. All told, Kona Blue spent around $500,000 to set up the infrastructure 30 meters beneath the sea (except during maintenance and harvest), which allows excrement and uneaten food to be swept away in brisk subsurface currents. Water quality downstream from the pens is the same as at sites upstream. The innovations deliver a glimpse of the future, when industrial techniques may transform the continental shelf into a sprawling network of farms playing a vital role in global food production.

Neil Sims, cofounder of Kona Blue Water Farms, is taking aim at another critical hurdle to sustainability: the need to harvest vast amounts of wild fish just to feed the ones being farmed. Salmon and other carnivorous fish must be fed large quantities of fish oil and fish meal to gain the taste and texture that consumers crave. It typically takes 2.3 kilograms of so-called "forage fish" to produce half a kilogram of farmed fish. (And that's using carefully formulated feed pellets; many fish farms around the world still use raw fish, which pushes the necessary kilograms to nine or higher.) Fortunately, Sims is gaining ground in his crusade to rewrite this equation.

Sims oversees a booming operation that produces one of the world's most prized farmed fish: Hawaiian yellowtail sold under the name Kona Kampachi. Sold in swank restaurants across the U.S., a serving of Kona Kampachi sashimi can fetch a price upwards of $15, in part because sushi connoisseurs prize the fish's firm, yet tender, flesh. Sims worries that, as aquaculture grows, it will further harm the species that form the basis of fishmeal. In the past 25 years, farming of marine fish and shellfish has grown by 10 percent per year. That surge translates into ever-increasing pressure on populations of forage fish such as anchovies and sardines. So Sims has launched an ambitious effort to find replacement sources for the fatty acids and amino acids his fish need.

When Kona Blue anchored its first offshore pen in 2005, their feed was 80 percent Peruvian anchovy fishmeal and fish oil. By early 2008, the company had reduced that percentage to 30, thanks to careful experimentation that allowed Kona Blue to substitute soybean meal and chicken oil for the fish products. Sims is thrilled to say it now takes only 1.4 kilograms of Peruvian anchovies to produce one kilo of Kona Kampachi. Indeed, this breakthrough—in combination with Kona Blue's other conscientious practices—made U.S.–farmed yellowtail the first ocean-farmed fish to earn a "good alternative" rating from Seafood Watch, Monterey Bay Aquarium's popular sustainable seafood advisory list.

Sims acknowledges the battle is ongoing and Kona Blue is aggressively pursuing a 1:1 ratio. To achieve this, the company is looking at soy protein concentrates as well as canola and soy oils. Kona Blue is also keeping an eye on a particularly exciting biotech breakthrough in which scientists have coaxed the coveted omega-3 fatty acid DHA out of microscopic algae. One animal-nutrition firm is now testing fish feeds enhanced with the same algal-based DHA already marketed in infant formula, milk, and juice.

Sims isn't alone. Similar feed formulations have been developed for cobia and other fish, but the limiting factor is cost. Until the price of fish meal and fish oil rise to reflect the world's depleted stocks, Sims says, it will be hard for many aquaculturists to justify pricier, more sustainable feed.

Even those at aquaculture's leading edge have difficulty predicting just how quickly the new practices might translate into large-scale ocean domestication. But they do agree that difficult barriers remain and that the biggest challenges may be not technical but political.

Take Goudey's self-propelled AquaPod. Sending flotillas of corralled fish to fatten up on the high seas is already feasible from an engineering standpoint, he says. A first step might be free-floating farms riding in and out and back again with the tide, returning to the same spot every 12 hours. Eventually, Goudey envisions full transoceanic voyages: penned fingerlings launched from Miami hitch a ride on the Gulf Stream to Europe, where they are harvested and replaced with a new, young brood for the return voyage to America. It would require only the integration of the self-propelled cage (such as the one he tested last summer) with a surface buoy carrying an automatic feeder (imagine a giant version of what you leave for your cat when you go on vacation)—plus navigation, tracking, and communications networks such as those already well-honed for research submersibles and Mars rovers.

Even though the technologies are within reach, progress toward offshore farming has been sluggish. Of some 50 offshore installations worldwide, only five U.S. commercial marine fish and shellfish farms have ventured into open water. Goudey thinks more aquaculture entrepreneurs would jump into the fray if the U.S. put into place the appropriate legislation and permitting systems. This would not only give aquaculturists the green light but also help guide the industry toward a sustainable future. Introduced in 2007, the National Offshore Aquaculture Act, for instance, would have authorized the U.S. government to grant aquaculture permits throughout the U.S.–exclusive economic zone, which extends 200 nautical miles from each coast. Issuance of these permits could be tied to sustainable practices. But the legislation has languished in Congress since 2005 and has not been reintroduced this session.

That reality has forced at least two U.S. offshore fish farms, frustrated with the permitting chokehold, to investigate expanding their operations to Mexico and Panama—or move them there entirely. In other words, businesses that might be goaded into pursuing sustainable aims via legislation now have incentives to migrate to other waters where the aquaculture mentality might be more akin to "anything goes."

Halting such overseas moves would also give the U.S. opportunity to improve food security. Among natural resources, the country's $9 billion annual trade deficit in seafood is second only to its dependence on foreign oil. To help offset that food imbalance, the U.S. Department of Commerce has declared it would like to quintuple the value of annual domestic aquaculture production, currently just shy of $1 billion, by 2025.

If the world can muster the unprecedented political will and international cooperation necessary to domesticate the high seas, critics ask, why not put those energies toward restoring the oceans rather than risk degrading them further? As Julia K. Baum wrote in *Nature* in reply to Marra's 2005 call to tame the seas, offshore aquaculture "is not 'inevitable.' It is a course of action that can be chosen—or not."[2]

Given the world's food needs, such a wholesale rejection of aquaculture might amount to accepting the status quo: a fishing industry that is devastating wild stocks, decimating the oceans, and generating enormous amounts of CO_2. As Scripps's Jeremy Jackson points out, "sustainable fishing is an oxymoron." Jackson draws a comparison to hunting and gathering and argues that sending flotillas of fishing boats out to round up wild fish is on a par with hunting down bears and elk for food. "If we're going to get lots of protein from the ocean," he concludes, "the only solution is aquaculture."

That doesn't mean today's latest methods are the final solution. After all, farmed yellowtail and salmon, like cod and tuna, are luxury foods that most people in the world will never taste. Together with academics such as Stanford University's Rosamond Naylor, Jackson believes we won't be able to save the oceans until we abandon our taste for fish that live high on the food chain. In that case, the world's ocean based protein would have to come from anchovies, shellfish, and other species operating at lower trophic levels.

From this point of view, the latest aquaculture innovations are best seen as an incomplete step in an important direction. Perhaps offshore farming will ultimately provide all the high-trophic-level species the world demands and also mass-produce the species many environmentalists find more favorable. Free-floating farms, for instance, could also be used to grow sardines, Goudey says. The University of New Brunswick's Chopin adds his own twist to the idea: he would add shellfish and seaweed rafts trailing behind.

As we pursue these goals, Chopin reminds us to be patient and to understand that a rapid transition to aquaculture means there will be missteps along the way. "Even after centuries of agriculture, we don't have all the best practices," he says. "In aquaculture, we want to solve everything in a few decades."

Notes

1. Marra, J. 2005. When will we tame the oceans? *Nature* 436:175–176.
2. Baum, J., J. McPherson, and R. Myers. 2005. Farming need not replace fishing if stocks are rebuilt. *Nature* 437:26.

Further Readings

Halweil, B. 2008. Farming fish for the future. Worldwatch Report 176. Eagle, J., R. Naylor, and W. Smith. 2004. Why farm salmon outcompete fishery salmon. *Marine Policy.* 28(3):259–270.

Michler-Cieluch, T., G. Krause, and B.H. Buck. 2009. Reflections on integrating operation and maintenance activities of offshore wind farms and mariculture. *Ocean & Coastal Management* 52:57–68.

Nature's Revenge

DONOVAN WEBSTER

Inside Chernobyl's Zone of Alienation, one of the largest wildlife preserves in Europe teaches us a lesson about the durability of nature and the fragility of man.

The giant ferris wheel—rising an enormous 150 feet into the sky—is becoming entangled. Trees have sprouted through the asphalt around its base, and in the day's breeze, the vegetation's full, green, midsummer leaves flutter against the wheel's yellow passenger gondolas, while streaks of rust work slow, abstract-expressionist decay down its steel supports.

To the wheel's left are other carnival attractions: a rusted "flying umbrella" ride, its red coat of paint peeling. Farther left, the bumper-car arena stands shattered: its roof is gone, its overhead electrical grid a shambles, the cars rusted and crumbling. All of it is amok with weeds.

In any direction, a city of stolid 15-story apartment blocks—a metropolis once home to 45,000 people—stands empty. And as the day's gusts blow stronger, the only noise beyond wind in the trees is the slamming of unsecured doors as drafts weave down the empty apartment-building hallways, having slipped in through left-open balcony doors and shattered windows.

Standing in this abandoned amusement park, beneath a blue sky tufted with bright cumulus clouds, the off-kilter truth of this place smashes home. To really comprehend man's thumbprint on earth so far, not to mention see what nature is capable of when left untended by human beings, everyone should vacation in one of Europe's largest wilderness parks—though they'll probably want to do it only once. That's because this vast wildlife preserve is a place called Chernobyl.

A towering irony of the 1986 Chernobyl nuclear disaster—the worst environmental accident in history—is that it began as a safety check.

Near midnight on April 25, 1986, in the state of Ukraine along the Soviet Union's European boundary, the Chernobyl facility's Reactor 4 was to be shut down for routine maintenance. In advance of taking Reactor 4 off-line, though, supervisors decided to see if—during its powering down—enough electricity was still generated to run its emergency cooling system, which existed to drop core temperatures should a crisis arise.

At 1:23 A.M. on April 26, as Reactor 4 was proceeding off-line without incident, steam conduits between the reactor core and the turbines used to create electricity suddenly strangled shut. Within seconds—and for reasons not fully understood, though human error is prominently mentioned—reactor power spiked to 100 times normal, overheating and rupturing fuel rods inside the core. As the nuclear reaction ran out of control, pressures inside the sealed core container spiked, and Reactor 4's containment vessel exploded, literally blowing its 640-ton top. When air hit the nuclear core, there was a second explosion, throwing a sparking, spitting fireball nearly a mile into the sky and hurling fragments of burning nuclear fuel across the landscape.

On the placid pasturelands of Ukraine, an atomic nightmare blew in. At least 14 tons of radioactive uranium were released into the sky, 400 times the amount spewed at Hiroshima. Compounding the problem, the graphite moderating rods inside the containment vessel—used to slow the reaction—had caught fire and were blazing.

"The reactor fire was very hot; we measured at least 7,000 degrees," says Vladimir Verbitskiy, 47, an officer at the control center of the Chernobyl Exclusion Zone who worked there at the time, and who still works there today. "But the bigger problem was the radiation itself. Initially, Reactor 4's core was emitting 40,000 roentgens an hour, and deadly levels for humans are 600 an hour. And the containment vessel was shattered facing west, with its top blown off and lying on its side. There was no way to contain the radiation; it was directed straight up and also horizontally away in beams through the cracked vessel. Those beams were very powerful. If you got stuck in the way of one for a long period of time, your flesh would simply disintegrate."

With Reactor 4 aflame, an ashy plume of radioactive products (including uranium, xenon gas, iodine, strontium, and caesium) lifted into the atmosphere, carried northward on winds toward Ukraine's borders with Belarus and Russia. Flaming bits of radioactive debris were still falling back to earth, setting outlying fires.

While the secondary fires were extinguished within hours, the fire inside the containment vessel raged on, with no public statement from the Soviet government. But if the Kremlin was taciturn after the explosion, Chernobyl was in a frenzy. "The roads were washed 24 hours a day, just to keep them clear of radioactive ash," says Verbitskiy. "Everyone was ordered inside with the windows shut. Helicopters were brought in to drop sand, clay, boron, and lead onto the fire from above, trying to smother it and trap the radioactivity. Dropping lead was a big mistake, since it melted before it hit the fire and turned into a molten and now radioactive liquid that spread. Of course, they were trying anything . . . everything. After all, nothing like this had ever happened before."

Two days after the explosion, Chernobyl's radioactive smoke had drifted more than 1,000 miles northwest, far enough to be detected by workers at a Swedish nuclear plant, which forced the Russians into a public declaration. With their announcement, the Russians also ordered the evacuation of dozens of cities near the reactor, ultimately displacing three towns and 91 villages and 326,000 people. The model "atomopolis" of Pripet (population: 45,000), which housed reactor workers a few miles from the site, was emptied within 36 hours.

As the fire burned on, the nuclear cloud passed across some of Scandinavia, Western Europe, and the United Kingdom, with local governments advising their citizens to remain indoors with the windows shut. Then, pushed by an approaching weather system, the cloud drifted back southeast. Nine days after the explosion—on May 4, 1986—the Chernobyl fire was finally declared under control. By then, its nuclear smoke had blanketed most of Europe.

In the end, two people were killed in the explosion, and 28 firefighters and emergency cleanup workers died in the first three months from acute radiation sickness. Eventually, all four reactors were closed. To isolate the most radioactive areas from human contact, the Soviet government drew two circles around Reactor 4. The first, called the Inner Zone, had a 10-kilometer (6.2-mile) diameter. The second, at a 30-kilometer (18.6-mile) diameter from the reactor, was called the Exclusion Zone, or—with a nice, Soviet-style flourish—the Zone of Alienation.

All told, following the Reactor 4 explosion and fire, 21,000 square miles were contaminated for the foreseeable future, with the meat, milk, vegetables, and fruit there all showing higher than acceptable levels of radioactivity. At Chernobyl, the combined zones are Europe's largest and most aggressively unpeopled place. Which means, ironically, that wildlife has come back with a vengeance.

Then and now, all roads leading into Chernobyl's zones are guarded by sentries. There are penalties for going in without proper approval, which, for the worst offenders, results in five years in jail. That is, if you're not shot for perceived spying.

As the brick Exclusion Zone guardhouse falls behind us, Sergei Ivanchuk, 39, slides the clipboard with our visitor-approval paperwork between the seats of his Chevy SUV. He stares out across the Exclusion Zone's broad fields of tall golden grass, fires up his yellow handheld Geiger counter, and smiles.

"We're inside the 30-kilometer zone now," he says. "Since 1986, nothing comes out of here. No wood products, no agriculture, no hunting. No one under 18 years old is allowed inside either, since they're worried about the health of young people who may be still growing."

Sergei has the straight-ahead manner of a Red Army soldier, which he once was. In the early 1990s, he was a United Nations administrator in his hometown of Kiev—about 65 miles south of Chernobyl. After he left the U.N. in 1994, he heard that the Ukrainian government and the International Atomic Energy Agency would be opening the Exclusion Zone to limited visitation. Initially, those callers were largely curious scientists, but the group was eventually expanded to include journalists and environmentalists, and then finally to controlled visits by day tourists. Sergei's tourism agency, SoloEast Travel, in Kiev, has prospered since then.

He points out the window to his left, across enormous fields broken on the horizon by a green line of forest. "Look here for wild horses," he says. "Przewalski's horses, the Mongolian-style horses, with long tails and chin beards. They've been living here for several years."

Across the fields, at the moment, no horses are in evidence, though a few hundred yards into the fields a short-toed eagle—large, gold, and heavy winged—lifts out of the grasses, clutching a rodent in its talons. In the years after the Chernobyl disaster, wild animals from wolves to moose, elk to wild boar, have returned. The Ukrainian government even went so far as to release a few endangered Asian bison, hoping to prop up their population.

In the morning sunshine, our car roars up the otherwise empty pavement, past signs marking the abandoned villages of Cherevach (former population: 460) and Zalissya (former population: 2,849). Yellow triangular signs with red "radioactive" symbols line the roadsides. Trees have grown through the houses, lifting their roofs and leaving each house tilted and slowly disintegrating as it twists off its foundation after 22 years of neglect.

"You see?" says Sergei, smiling and shaking his head. "Now the forest is taking back the towns. It's amazing how strong the bushes and trees are. They grow up in the houses, and slowly tear the houses apart. You will see things in the next days that, later, when you think about them, will seem like a dream."

Just ahead is the town of Chernobyl. The highway enters it by ducking beneath an archlike silver pipe that passes above the pavement ("Those are new heating- and water-supply pipes," says Sergei. "Because it was easier and cheaper to install and maintain them, they don't bury them."), then we pass by a large white-brick marker, its blue writing in Cyrillic, a yellow cartoon atom rising from its top.

"It's official," says Sergei. "You are now at Chernobyl. Its population was 13,400 before the accident. Today, there are 3,000 workers who still maintain the place, but most don't live here, they go home at night. In some ways, it's like any town. People come and work, taking care of things, monitoring the reactor sites."

The roads are broad, smooth, deserted, and very clean: The pavement is still washed daily. The sun shines, dappling the speckless asphalt with shade from trees. We're to meet our mandatory tour officer, plus do a little paperwork, at a place called the Chernobyl InterInform Agency. A two-story blockhouse off the main road, the InterInform building is run by the Ukrainian Ministry of Emergencies and Affairs of Population Protection, a bureau created after the Chernobyl accident. When we pull into the parking lot, a sturdy, unsmiling man in a brown digi-camo uniform—trousers tucked into tall black lace-up boots—is standing out front. This is Vladimir Verbitskiy, the control center officer; he will be our guide. "Call me Volodya," he says as we shake hands.

After being led inside, there's a presentation, with photos and maps showing how the airborne radiation migrated over time. Then, before we climb into the car to start our visit, there's a little business to conduct. It's a contract, and photographer Christopher Sturman and I must sign it before we can proceed.

Contained in the deal are rules: *No weapons, no alcohol or drugs; you are not allowed to eat or smoke in the open air, to touch structures or vegetation, sit or place photo equipment on the ground, take items out of the zone, violate the dress code (no open-toe shoes, shorts, skirts . . .), or be in the Exclusion Zone without the officer.*

Then it gets stranger. *Accordingly, I, participant of the delegation coming to the Exclusion Zone . . . agree that the State Department Administration of the Exclusion Zone shall not be liable for possible further deterioration of my health as a result of the visit to the Exclusion Zone.*

This is not comforting. Still, Chris and I sign off, and, with the contract memorialized, Volodya lightens up. "Let's go have a look around," he says, with Sergei translating.

We pass through Chernobyl's largely deserted blockhouse town—heating ducts snaking everywhere above ground—before the car rolls out the far side into untended forests. Huge power lines on steel trestles etch their way from disparate compass points toward a common spot a few miles ahead. Clearly, we are approaching a large power source.

"This is the Red Forest," says Sergei. "The old forest here was killed by radiation and bulldozed, with all the trees buried. The trees killed by radiation had their trunks and bark burned to a reddish color. That's where this forest gets its name—and its radioactivity. It is one of the hottest places in the Exclusion Zone. These trees? They're new. They're 22 years old."

As we roll closer to the reactor site, the numbers on Sergei's Geiger counter begin to rise. From a normal background reading of about 14, they're now jumping: 23. . .25. . .27. . .37. . . .

When Sergei sees me tracking this ascent in my notebook, he waves a hand as an amused smile crosses his face. "Not to worry," he says. "These readings are nothing. You will soon experience thousands."

"Back in 1986," Sergei is saying, "the Chernobyl Nuclear Site was the Soviet Union's proudest model city. When foreign delegations came to visit, Chernobyl is what they were shown. It was a source of great national pride."

A mile ahead are two more enormous buildings. The first, low and painted white with four white stacks rising into the sky, are Reactors 1 and 2. Then, in the distance—and familiar in the same way the Chrysler Building and the pyramids at Giza are known to people who've never seen them firsthand—stand Reactors 3 and 4. The building, with its circular, scaffold-encased smokestack and Stalin-gothic gray steel and poured concrete, feels familiar and somewhat doomed. As if a cloud hangs over it, even on this blue-sky day.

We approach it in Sergei's SUV, and—as we pull into a parking area near a 15-foot-tall marble memorial of two hands holding a building—Sergei's yellow Geiger counter goes bonkers. A grating *beep. . .beep. . .beep* alarm is sounding. The monitor's readings soar past 300 micro-roentgens. As Sergei brakes the car to a stop near the memorial, he smiles. "I set the alarm for 300," he says, "just to advise us that we are still safe, but getting hotter."

We step from the car, and—in hot noon sunshine—walk closer to the memorial. As we do, the readings on Sergei's Geiger counter keep rising: 352. . .362. . .400. Strangely, Reactor 4 no longer looks like it did in postexplosion photos. Back then, with its containment vessel and walls and roof blown open—objects lying tossed around in the blackened blast zone—Reactor 4 in 1986 resembled a gigantic gutted animal, its viscera ripped open and trailing stringy guts. Today, with successive concrete and steel sarcophagi encasing the core, Reactor 4's structure is all smooth walls and squared corners. Reactor 4 looks like any large industrial building. They could be making airplane components in there, or maybe a satellite launch vehicle.

In the hot sun, Volodya—who has been quiet—suddenly springs to life. "I remember that night," he says. "People who were trapped in the control area of the reactors had to jump outside and run to the administration building, where they could wash. Everyone was covered in radioactive ash and they wanted to get it off. Unfortunately, as they ran, they ran past the ruptured reactor, which was facing where we are standing now, so they were getting 200 to 300 micro-roentgens an hour—a huge dose."

Volodya shakes his head. "Many people working near the reactor on the night of the explosion died within the next few weeks or months," he says. "By the end of 1986, we realized it would probably be impossible for anyone to live or work here ever again, due to the radiation. It had become a dead zone."

As I listen, Sergei's Geiger counter keeps beeping, its readings rising: 405. . .415. . .418. But Volodya and Sergei don't seem concerned. Instead, they are talking and pointing. Showing where the reactor's 640-ton top landed inside the industrial yard after the explosion. . . .

"Initially, 200,000 people were brought in to clean the place," says Sergei. "They called them liquidators. They worked in or close to the reactor, wore lead suits, and would expose themselves to the radiation for less than a minute at a time. The first task was to extinguish fires on the roof of Reactor 4 and not let it spread to the roof of Reactor 3. Then the task was to drain the contaminated water from the bubbler pool under the reactor to avoid a steam explosion. Later, when helicopters threw thousands of tons of sand inside and when the reactor fire was out, they began to build the sarcophagus." He points. "They constructed it on the far side of the reactor, assembling it where the radioactivity was least dangerous. Using bulldozers—some of them remote-control ones—they covered the pieces with six to eight layers of clay, then made huge cement steps, and set the sarcophagus into place, by November 1986."

The initial sarcophagus soon began to show signs of wear from the demanding Ukrainian weather—before long, birds were said to be nesting in the cracked concrete—and since then, regular repairs have been made to the structure and the roof. By 2012, a much-longer-term containment structure, valued at $1.2 billion, will be in place.

Still, there are big concerns about what the sarcophagi contain. Due to the chaotic nature of the fire extinguishing, the radioactivity thrown off by the unshielded core, and the haste to cover it all in a way that would leave it safer, no one can be sure about the configuration of the remaining core and the 180 tons of nuclear fuel still inside.

"One of the great worries goes unspoken," says Sergei. "They can't control the processes inside the reactor—lavalike fuel material turns from its present form into dust and mixes with rain, which gets through roof holes. It creates a potentially dangerous mess. Nobody can predict how dangerous."

Now—after 15 minutes of assessing Reactor 4 at close range, with Sergei's yellow Geiger counter yowling nonstop—Volodya and Sergei say it's probably time to move on. As we head back to the car, Sergei sums up: "So, officially, 3.5 million people had their lives affected by the explosion at Chernobyl, with 1.2 million suffering in some way from the explosion itself," he says. "According to the Ukrainian government, there were 6,400 casualties related to the Chernobyl explosion over 20 years: Elevated cancers, leukemia, that type of thing. And that is certainly too bad. But the hardest part is, we just don't know how long this will last, how dangerous it will stay, how expensive it will be. Once released, radiation is very hard to manage, and radiation security has expenses you cannot even begin to imagine."

Sergei fires up the SUV, and we race away from Reactor 4. As we pass in front of the cracked reactor core, the Geiger counter soars past 835. . .839. . .soon we're beyond 1,000. We roll quickly across this stretch of road, passing back into the new, 22-year-old trees of the highly radioactive Red Forest—the land once fixed in the irradiating beams of Reactor 4—and Volodya sighs. "We still wash these roads every day in summer," he says, "and still the meter readings are high."

He points out the window as Sergei's SUV rips down the highway, trying to get past the most radioactive stretch of the trip.

"See how we cut the grass?" asks Volodya.

We all look outside. Along the roadside, the grass is tended to perhaps 10 yards away from the pavement, with taller grass, something like hay fields, growing up beyond that.

"We cut the grass like this for a little better visibility," he says. "But even that is expensive. Wherever we cut grass here, we have to rake it up and, after compacting it, it goes into long-term nuclear containment. Even the grass in this part of the Exclusion Zone is a radioactive hazard. That gives you an idea of the cost, the expense, of a nuclear accident."

The freak-show centerpiece of the visit, however, isn't Reactor 4 or the tide of new wildlife said to be rolling across the zone, or even the five-foot catfish that now inhabit the cooling ponds. (The fish are said to grow that large not from mutation but from lack of competition and because workers treat them like pets and throw them bread.)

Instead, it's Pripet.

This city, where most of Chernobyl's nuclear engineers and workers once lived, was built in 1970 and evacuated forever 16 years later. As we roll toward it, a few miles from Reactor 4, its haunting grip begins taking hold. Formerly the pride of the Soviet Union, Pripet was a community that wanted for nothing. It had the best schools—12,000 children attended school there—plus theaters, stores, gyms, swimming pools, soft ice cream, restaurants, and the first true supermarket in the Soviet Union.

"First, we'll go to the amusement park," says Sergei.

As we drive through the city, it's instantly obvious that the place was evacuated so quickly that everything in it was simply dropped . . . and now seems poised on the edge of being picked up and used again. Liquidators came through, collecting much of the garbage and personal items left on the streets and inside buildings for burial, but other things remain. As the SUV makes its way through the abandoned streets toward the amusement park, we pass city squares shaggy with trees growing through marble paving stones and poured concrete. Metal shopping carts and children's toys sit abandoned on the sidewalks. The apartment-block buildings lining the streets stand empty; the interiors of some of their lobbies look like overgrown terrariums.

Earlier, as we drove toward the containment zone, talking about Pripet, photographer Christopher Sturman asked Sergei: "What's the recommended amount of time allowed to spend in Pripet?"

Sergei smiled. He waited a long moment. "None," he said.

We pass by Pripet's old train station, and a yard-long brown snake slithers to safety beneath a rusty, long-abandoned rail car. I ask about wildlife in the city, and Volodya and Sergei nod.

"I saw a wolf yesterday," says Volodya.

"I saw deer my last visit," adds Sergei. "The animals are here, but they're wild. They don't like being seen."

Then, up ahead, through the overgrown trees, the giant Ferris wheel at the city's center comes into view. The postapocalyptic stretch of our visit has begun.

As we get out and begin exploring the amusement park, Sergei pulls a cooler and some folding chairs from the back of his SUV. "Usually, we eat lunch at the InterInform building, which has very good cooks and good food—all of it brought in from Kiev, so there is no radioactivity," he says. "But today, a working lunch. A picnic."

As Sergei lays out the bread and cured meats, cheese, mustard, fruit, and pickles, Volodya is talking. "Nowadays here, we have all the usual animals of Ukraine: bears, foxes, wolves, puma, rabbits, wild boar. But the strangest thing about the Exclusion Zone isn't wild animals in this urban setting. It's that, after the accident, nature reacted crazily. For the first few years, some trees in the containment zone had abnormally huge leaves. Many times normal size. In one village, the strawberries grew two to three times as large as normal. No one could eat them, of course. And after the explosion, the plants were knocked off their clocks. Farther south, in Kiev for example, in the autumn the leaves would already be brown and fallen, but in the Exclusion Zone, with winter coming, the trees would still be a rich green. Nature was off its usual schedule. Slowly, it has gotten back to normal."

After a sandwich, we wander the amusement park, with Sergei following me past the Ferris wheel and swing sets and flying rides.

"You want to see something interesting?" he asks. He walks to a patch of moss in the shade of the bumper-car arena. "The moss and most plants that take nutrients from the ground, as they take the nutrients, they take in the radioactivity too. Moss is among the most radioactive plants in the zone."

He sets his yellow Geiger counter down on the moss's cool-looking brownish green, and, once again, the readings shoot up as it beeps its warnings: 352. . .500. . .1,320.

Sergei snatches up the counter, and, quickly, the number falls back to a slightly high background reading of 42. He smiles. "So now you see why you don't want to stand on the moss, right?"

After we put away our picnic, it's time to explore Pripet.

The word most often used to describe the experience—voyeuristic—is accurate. At Pripet's white-painted high-rise hotel on the central square, for example, we step through the carpet of broken picture-window glass at the entrance and find the guest logs still at the front desk, pages open to April 1986. There are also letters of introduction, civil orders, and cash receipts intended for accounting.

Sergei steps behind the counter. "You need a room for the night, sir?" he jokes.

We wander the hotel's empty hallways and rooms, climbing stairs and seeing old steel bed frames and bathroom showers, and, in the kitchen areas, huge steel stove vents hanging from the ceilings. In some rooms, newspapers from April 26, 1986, sit on the floor.

It's strange. Disorienting. Skeletal ficus trees stand in big, round wooden planters—the trees themselves now 22 years dead—each in the same spot on every floor. Sergei and I climb the stairs to the hotel's roof and stare across the empty city. The wind blows. The hotel's Cyrillic-lettered electric sign topping the roofline has largely fallen down. Everywhere, in every direction, are signs of human habitation, but no humans. Birds flit in the trees. Some sort of animals—a handful of deer, perhaps—move noisily along an overgrown sidewalk that winds from the amusement park behind this hotel and toward the broad, tea-colored Pripet River to the east. The fishing there is said to be terrific, though the river-bottom silt has its own radioactivity problems.

I've been so busy looking out across this dead, empty city, in fact, that I haven't noticed the patch of rooftop moss beneath my shoes. Fortunately, both Sergei's Geiger counter and the screening machine back at the InterInform building pick up only a little radioactivity, nothing to be concerned about. Still, when the trip is over, I'll throw away the shoes.

Over two days, the sights of Chernobyl and Pripet become no less strange or sobering (we spend the evening in a new, radiation-shielded, $95-a-night hotel in Chernobyl, a place intended for diplomatic visitors). Late one afternoon, we visit the sports complex in Pripet: a basketball court and gymnasium, with the gymnast's vaulting pad still sitting in the middle of the basketball floor. The basketballs—brown dust having sifted all over them—are now deflated and crumpled; they resemble enormous dried seeds.

At Pripet's River Station—a sort of ferry terminal on the river that connected the atomopolis with the capital of Kiev—poured-concrete docks and moorings stand waiting. The River Station feels like an abandoned airport or train stop. Facing the river is the passenger waiting area: a ticket desk, a refrigerator behind the bar, lots of benches for travelers. On the other side of the building, facing south, is a fancy restaurant. Its huge picture windows, with stained-glass imagery of idealized people, are now broken and patchy; they must have thrown amazing light across this room, once filled with diners.

We wander outside and into a new growth of woods. As I walk, I look down at my feet and realize I'm pushing a dozen brown, five-inch lizards through the low weeds ahead of me.

Though the place looks deserted, wildlife is everywhere. Life stepped right back in when people disappeared. In fact, without human management, nature sometimes wheels out of balance. At lunch, Volodya said, "We often have to cull animal herds that have grown too big. Their horns and hooves are radioactive. The wolves that eat them are radioactive. The fish in the cooling ponds are 100 times more radioactive than a normal fish . . . but wolves seem to be the animals that really succeed here. They have lots to eat and no predators. Each year, we shoot about 25 wolves. We don't know the exact population of any of the animals, but we watch the situation closely. Basically, when local dogs start to be chased by wolves, that's the signal."

There is a movie theater near the hospital: its facade is decorated with thousands of red and yellow and green tiles, creating a sort of wave pattern. The city's Palace of Culture sits next to the big hotel on Pripet's central square. The palace's enormous auditorium, once home to theatrical and concert programs, nowadays has rain dripping through its roof. The stage curtain hangs in shreds.

Still, backstage at the Palace of Culture is fascinating. There are huge paintings of Soviet-party bosses, clearly meant for regular display, as well as theater props: A brown wooden throne with a gold crown painted on the seat's back, painted boxes that served as desks, painted rolls of oilcloth backdrops. Torn papers and wooden planks stray across the floor. What was the last production? Who was the last person to leave? The backstage entrance to the outside world is eternally open: Twenty-foot doors are swung out like a barn's, and they now sit on rusted hinges. Rain or shine, in summer or snow, the Palace of Culture is in an eternal state of waiting.

"You see now?" asks Sergei. "These people were nuclear scientists and technicians. They were highly educated and highly paid, particularly by Soviet Union standards. They were rich. They were the future. Then, because of an accident, the future ended."

Probably the most disorienting spot in all of Pripet is the school. You enter the white, multistory Bauhaus-style building—lots of openable windows and airy classrooms—to find broad hallways, bright tile designs, and a sectioned-off area near the front door filled with empty cubbyholes and coat hooks. Children's drawings are on the walls and tables, smashed record albums litter the hallways, and drawing books and flyblown paperwork are scattered on counters and the floor.

Going classroom to classroom, it's easy to see what was being taught where back in 1986. The science rooms have periodic tables on the walls, deep plastic tubs for student experiments, and elaborate stands for plants. The history and language classrooms are full of books, blackboards, and posters. The mathematics rooms look like many all over the world: lots of blackboards. In each room, rows of desks await, dusty and empty.

Not too far from the school, just past an ice-cream shop so overgrown it is impossible to enter, sits the music store, a building fronted by more shattered picture windows. The glass cases that once displayed instruments and tuning equipment are now empty. Only the piano sales floor still has inventory. Beneath a water-bowed insulated ceiling, a sea of pianos stretches wall to wall. Sergei stops and begins tapping a keyboard. When nothing happens, he plucks the piano's strings, making a creepy, off-key noise.

Late one afternoon, we leave Pripet and Chernobyl behind for Paryshiv, an abandoned village inside the Exclusion Zone near the Ukrainian border with Belarus. As Sergei's SUV streaks down another empty ribbon of highway, a brownish wild boar—its body dense-looking and round; its head too small—stands at the edge of the roadside, watching us pass. "That's a young one," says Volodya. "The mature ones are black. They're everywhere."

As we cross the Pripet River on a deserted bridge, the floodplain stretches in all directions and a large painted turtle is crossing the empty road. "In the years after the explosion and evacuations, some people came back," says Sergei. "Across the Exclusion Zone, there are about 300 resettlers, mostly older people and those who couldn't fit in somewhere else. We don't usually bother them; some are not friendly. Still, though it was illegal to resettle, the authorities have finally said okay."

The SUV slows and turns right onto a side road. There are little houses; most are in ruins. We pass into untended fields and patches of trees. We take another right onto an overgrown dirt two-track. Just ahead, at a fenced dacha, an old woman in a red dress, a blue apron, and a bright-yellow head scarf is sitting peacefully on a bench in the shade of a tree. Just inside the fence, chickens wander the property. A large, tended vegetable garden sits behind the house, its furrows straight.

"This is Maria Shylan, 78 years old," says Sergei. As he shuts off the vehicle, she stands.

"Come, come," she says, clearly happy and excited to have visitors. She opens the wooden gate to her house and leads us inside. "Let me show you my garden, though I'm fighting the potato bugs at the moment. I'm the only one out here to fight them. In the end, the bugs may get the potatoes, and the wolves will get me."

She leads us back to the garden, which covers perhaps an acre. "I have tomatoes, potatoes, onions, carrots, green vegetables, anything you might need. I eat some and preserve some for the winter. Scientists come to check my crops, to make sure they're safe to eat. They check my well too, to make sure the drinking water is safe. It is. I have pigs and chickens for meat and eggs." She points to a muddy pen and a full chicken coop. "And about once a month, a grocery truck comes, and I buy the few things I need: sugar, mostly. Once a week, a bakery truck comes, but the bread gets hard in two days, so once a week isn't enough. Mostly, my bread is potato pancakes that I make myself."

Is she here alone?

"I used to have a little dog. But wolves got him. You see the wolves all the time; they come out at dawn and dusk. One night my little dog was barking, so I knew something was around. Then it grew quiet, and my dog was gone. I never saw him again."

Standing in the sun, fists on her hips, Maria is smiling a near-toothless grin. "Sorry about my teeth," she says, "but they fell out after the explosion, which we couldn't see from here. But the next day, there was much activity. They loaded us all onto buses and moved us away, to Kiev. Many people went into older-care homes. Not me. After a time, I grew homesick, so I came back. This house is where my family has always lived, my father and his father. Now I have all this to myself. I enjoy this simple life, though it is lonely."

Suddenly, tears come to her eyes. "My husband died a long time ago," she says, "but both my sons, 46 and 51, died because of Chernobyl. One was a crane operator there, the other an electrician. I said, 'You will grow sick because of that place,' and they both said, 'Mother, no. It's fine.' I have only one living relative. A grandson. He lives in Kiev. Has a wife and a baby. He visits every few months, but he is busy. Come, I have something for you."

Maria heads into the house, her walk an energetic waddle. The interior is spotlessly clean. The ceilings are low. The plaster walls are covered in bright-blue paint. Curtains partition off different areas, likely to better retain the huge earthen stove's heat in winter. Maria motions to sit at the kitchen table. Then, from a wooden cabinet, she pulls out a plate of potato pancakes ("I made these just today"), a big bowl of honey, and a jar with some clear liquid inside.

"Eat," she says. She grabs my hand, puts a pressed metal fork into it, stabs a potato cake with the fork, and then dunks it into the honey. "Eat. I love to feed guests."

Then she pours two small glasses of liquid and motions for us to lift and drink. We toast to her, simultaneously. It's moonshine. Vodka. It smells like water and burns like Satan; my eyes cloud up, my stomach recoils. To settle it, I take a bite of the potato pancake and honey. It's dense but tasty. Sort of like corn bread.

She refills the glasses. "Again," she says. "I make this myself. I put sugar, potatoes, and bread with yeast and water in a jar. It takes two weeks of brewing, then I cook it and strain it. In summer, I work outside to make this place better. In winter, I have the television. It's easy now, though I don't have my health."

Having spent the day gardening, and then knocking back shots of moonshine, Maria appears healthier than many 40-year-olds I know. But the day is growing late—sunlight is now slanting inside the house's windows at a low angle—and Maria is winding down from her efforts. She wanders to a glass-fronted chest on the kitchen's far side, opens it, and pulls out a picture frame. It holds black-and-white portraits of her sons in their formal Soviet uniforms, big brimmed hats on their heads. Her eyes mist over as she looks at the photos, stroking them with her hand.

"You have a life, a family, a place where your family has always lived . . . and then . . . an accident," she says. "All I have now is pictures. Pictures and a garden in a place where no one else lives because more than 20 years ago, someone made a mistake."

The Chernobyl Exclusion Zone closes each night promptly at five o'clock. And wherever you are when that hour tolls, that's where you'll spend the night. "We have to get back now," says Sergei, hustling us out of Maria's place and back to the car. "Really, we have to hurry."

So on this, our last night spent in and around Chernobyl, we leave Maria's house and floor it back to the InterInform building. There we'll have a tasty beef Stroganoff dinner before eventually bedding down at the hotel, watching Ukrainian variety shows on TV.

Sergei's Chevy roars up the road, passing another plate-size painted turtle crossing the asphalt near the empty, poured-concrete arch of the Pripet River bridge. It's strange. Spending a few days at Chernobyl, you can't help thinking that nature is exacting some sort of revenge on mankind, pointing out the limitations of what humans can do and build . . . and maintain. And yet there's no violence or anger in the reprisal. Instead, nature's reacquisition of what man has done and built is more like a slow flood or a relentless, steady grasping back.

The return of life in the Chernobyl Exclusion Zone is a reminder that, when humans talk about "saving the environment," they're overlooking the fact that the environment is going to be fine. The earth's force of life remains robust—and is always there . . . waiting. Instead, it's the future of humans that requires our work and protection. But the earth itself? It's gonna be okay.

Around us, the river's floodplain is lit in sunset gold: Its overgrown grasses and tufted green trees are fluttering in a thousand gilded shades. To our left, a "ship graveyard" of beached barges sits at the river's edge; each boat's rusted iron, peeling paint, and irradiated cargo lit bright in the evening light. In another century, the rust will have dissolved them completely. There will be nothing left of them but pictures.

"I saw a lynx there the other day," says Volodya, pointing. He smiles. "That's the thing about the Exclusion Zone that never changes. Every day, it surprises you. You never know when you're going to see something new—and then, there it is."

UNIT 5

Resources: Land and Water

Unit Selections

21. **Tracking U.S. Groundwater: Reserves for the Future?,** William M. Alley
22. **How Much Is Clean Water Worth?,** Jim Morrison
23. **Searching for Sustainability: Forest Policies, Smallholders, and the Trans-Amazon Highway,** Eirivelthon Lima et al.
24. **Diet, Energy, and Global Warming,** Gidon Eshel and Pamela A. Martin
25. **Landfill-on-Sea,** Daisy Dumas

Key Points to Consider

- How are groundwater reserves in the United States being monitored in terms of the differences between the amount of water extracted and the amount of water returned to groundwater reserves? Does current scientific analysis hold the idea that many of our groundwater supplies are being depleted to the point where they will no longer yield usable water?
- Is there a relationship between water pollution and the cost of water? How can conservation methods alter the price of water for commercial, domestic, industrial, and agricultural needs?
- Is it possible to reach a balance between the demands of commercial timber industries for tropical hardwoods and the utility of the forest environment for agriculture? Has the development of a transportation system in the Amazon improved or worsened the environmental prospects for the world's largest tropical forest area?
- What is the environmental impact of plastics to our environment? What is the Great Pacific Garbage Patch?
- Can changing your diet reduce carbon emissions?

Student Website

www.mhhe.com/cls

Internet References

Global Climate Change
http://www.puc.state.oh.us/consumer/gcc/index.html

NASA Global Climate Change site
http://climate.nasa.gov

National Oceanic and Atmospheric Administration (NOAA)
http://www.noaa.gov

National Operational Hydrologic Remote Sensing Center (NOHRSC)
http://www.nohrsc.nws.gov

Terrestrial Sciences
http://www.cgd.ucar.edu/tss

Over 20 percent of the world's population lives in extreme poverty. Increased population, loss of arable land, expansion of farming into marginal regions, drought, degradation in the quality and availability of drinking water, spread of disease, and civil/political conflict all contribute to the misery. Each of these factors is interrelated, each conspiring to make the others worse. As populations grow, there is an ever-greater effort to extract more from the land. But more intense farming practices cause significant loss of topsoil. As a result, people are forced to inhabit and cultivate more marginal farmlands. Increased populations mean higher water demands, more draining of aquifers and consuming (as well as polluting) available surface water sources. Coupled with overgrazing, removal of wood for fuels and higher temperatures due to global warming, much of the world is in the midst of a crisis, and there is no hope for relief anytime soon.

The tragic situation described above violates most of the fundamental tenets of sustainable development. We utilize numerous resources provided to us by the Earth. Some are free, such as the air we breathe, while others we must pay for. Some are renewable, such as trees, wind, water, and sunshine. Others, such as fossil fuels, are limited, and once gone, will remain gone, for all practical purposes, forever. For future generations to benefit from the same resources we have today, we must practice sustainable development. In other words, we cannot use a resource at such a rate that that it will be forever lost for future generations. We must utilize technologies that do not damage the environment and do not deplete the natural resource base. At the same time, people must have access to economic progress, and we must invest in human growth and well-being. And finally, we must, at some point, reach some stable population. The concept of sustainable development is scalable, so that it applies communities, nations, and continents. If a community, nation, or entire planet squanders its available resources, the next generation will lose the benefit enjoyed by the one preceding it.

No one knows for sure what the ultimate carrying capacity of the planet is. But what is clear is that many resources that we have taken for granted are limited, in some cases critically. In the first article "Tracking U.S. Groundwater: Reserves for the Future?", William M. Alley discusses the problems of depleted groundwater reserves. Groundwater contain more than 70 times the quantity of fresh water as all lakes and rivers combined. Groundwater is generally clean and available throughout the year. We have used groundwater to turn semi-arid lands into vast farms and to turn deserts into sprawling cosmopolitan regions. But the answer to the question—How much groundwater is available?—can, for many regions of the world, be quite unsettling. Aquifers are being drained, or in some places become saline due to influx of saltwater. Fortunately, some communities, especially desert communities, are beginning to come to terms with the limitations of groundwater. Alley holds out hope that with proper monitoring, we will be able to make the smart choices to insure that groundwater supplies are not depleted. The economic sense of conserving and respecting our water supplies is illustrated in the next short article, "How Much is Clean Water Worth" by Jim Morrison. Here, we see that a simple cost-benefit analysis makes it clear to New York City planners that repairing the health of the watershed in the Catskill Mountains was far cheaper than the alternative, building a multi-billion-dollar water treatment facility.

Few things are as depressing as reading about the rapid rate of destruction of the equatorial rain forests. Because of some unique properties of these ecosystems, they are very susceptible to irreparable damage. Once a forest is clear cut, it is pretty much gone. Eirivelthon Lima and others describe how tropical forests can be turned into sustainable ecosystems in "Searching for Sustainability." Through flexibility, concessions and decentralization, both sustainability and economic growth are possible. Along the same lines, Eshel and Martin write about the high environmental costs to a non-vegetarian (or vegan) diet in "Diet, Energy, and Global Warming." If you want to conserve the rain forests, stop eating at McDonalds!

A very different ecosystem is mentioned in the last article of this section, "Landfill-on-Sea." The Central Pacific Gyre (CPG) is the largest uniform ocean realm on the planet. And it is filled with garbage. Wind and water currents naturally concentrate flotsam in the CPG, but the huge amount of plastic (twice the size of France) is a sobering reminder of just how much plastic we use and how bad we are at recycling it.

It is easy for us to look around and see that things are bad and getting worse. But as these essays point out, environmentalists, economists, and policymakers can make a difference. There is much cause for hope. The easy way forward is the "hard path," the business-as-usual model where practices of the past are continued into the future. The "soft path" is one where we look at the world in a new way, recognizing the worth of our environment and work to preserve it. All it takes is imagination, creativity, and courage.

Article 21

Tracking U.S. Groundwater: Reserves for the Future?

WILLIAM M. ALLEY

During the past 50 years, groundwater depletion has spread from isolated pockets to large areas in many countries throughout the world. Groundwater occurs almost everywhere beneath the land surface. Its widespread occurrence is a major reason it is used as a source of water supply worldwide. Moreover, groundwater plays a crucial role in sustaining streamflow between precipitation events and especially during protracted dry periods. In addition to human uses, many plants and aquatic animals are dependent upon groundwater discharge to streams, lakes, and wetlands.

A growing awareness of groundwater as a critical natural resource leads to some basic questions. How much groundwater do we have left? Are we running out? Where are groundwater resources most stressed? Where are they most available for future supply? To address these basic and seemingly simple questions requires consideration of several complexities of defining groundwater availability and a review of how one monitors groundwater reserves.

The term "groundwater reserves" is used to emphasize the fact that groundwater, like other limited natural resources, can be depleted. This potential for depletion is a key concept, despite the fact that unlike nonrenewable resources such as mineral deposits, most groundwater resources are replenished. On the other hand, some "fossil" groundwaters in arid and semiarid areas have accumulated over tens of thousand of years (often under cooler, wetter climatic conditions) and are effectively nonrenewable except by artificial recharge of surface water or treated wastewater.

Groundwater management decisions in the United States are made at a local level, which may be a state, municipality, or special district formed for groundwater management. Thus, monitoring of groundwater reserves should be designed to provide the information needed by these entities as a primary consideration. The issues to be addressed are varied and occur at many scales from preservation of a small spring fed by a nearby water source to the management of groundwater development throughout a large aquifer system or river basin.

The nation's groundwater reserve is not a single vast pool of underground water, but rather is contained within a variety of aquifer systems.[1] In general, the locations of the nation's aquifers are known, so much of the current research focuses on characterizing aquifer systems and how they respond to human activities.[2]

Many aquifers cross political divides, including county, state, and international boundaries. This characteristic (as well as the specialized nature of the science of groundwater hydrology) drives the need for a federal role and multijurisdictional collaboration in groundwater monitoring. Concerns about groundwater reserves have become more regional, national, and even global in scale in recent years, as exemplified by interstate and international conflicts over the salinization, contamination, or overexploitation of groundwater.[3] The effects of groundwater development may require many years to become evident. Thus, there is an unfortunate tendency to forgo the data collection and analysis that is needed to support informed decision making until after problems materialize.

How Much Groundwater Is Available?

The volume of water stored as groundwater is often compared to other major global pools of water within the Earth's hydrological cycle. For example, if one ignores water frozen in glaciers and polar ice, groundwater comprises more than 95 percent of the world's freshwater resources. This statistic illustrates the considerable value of groundwater, but it is also misleading, as it misses the variation in quantity, quality, and availability from location to location. The volume of groundwater in storage, its quality, and the yield to wells vary greatly across the planet. Typically, groundwater is used locally, so the effects of localized pumping on a given region are the primary concern of hydrogeologists.

Estimates of the volume of groundwater are poorly known relative to other pools of water. For example, the volume of the Earth's oceans has been well known for many years, whereas global estimates for groundwater storage vary by orders of magnitude (see Table 1). In part, this variability is due to different considerations of depth and salinity in defining the global groundwater pool. In addition, the variability reflects less knowledge about groundwater than other global pools of water. Early

Table 1 Volume of Water Attributed to Oceans and Groundwater Over Time

	Cubic Kilometers of Water (in thousands)	
Date	**Oceans**	**Groundwater**
1945	1,372,000	250
1967	1,320,000	8,350
1978	1,338,000	10,530–23,400
1979	1,370,000	4,000–60,000
1997	1,350,000	15,300

Note. These data come from different studies of the world water balance. Significant figures are largely retained from original sources.

Source: W. M. Alley, J. W. LaBaugh, and T. E. Reilly, "Groundwater as an Element in the Hydrological Cycle," in M. Anderson, ed., *The Encyclopedia of Hydrological Sciences* (Chichester, UK: John Wiley and Sons Ltd., 2005), 2215–28.

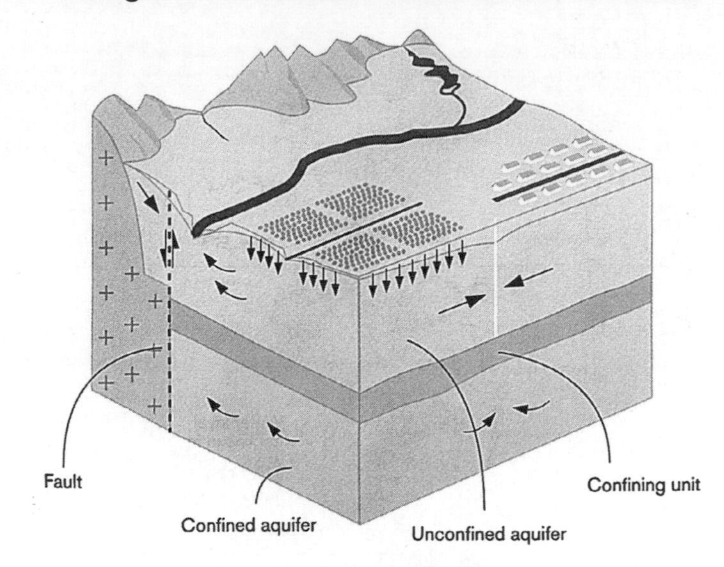

Figure 1 Hypothetical basin-fill aquifer system.

Note. The arrows show the direction of groundwater flow. Among the features shown are an unconfined aquifer overlying a confining unit and confined aquifer, a gaining stream, recharge from irrigated agriculture, and mountain-front recharge.

Source: Modified from S. A. Leake, *Modeling Ground-Water Flow with MODFLOW and Related programs,* U.S. Geological Survey Fact Sheet, 121–97 (Washington, DC. 1997).

estimates of the global groundwater pool greatly underestimated its volume. It was not until after development began in earnest in the mid-twentieth century that an appreciation of the large storage volume of groundwater emerged universally. More recently, scientists have viewed this resource as an important component of the world water cycle and have expressed increasing interest in quantifying its role.[4]

As a practical matter, it is virtually impossible to remove all water from storage with pumping wells. However, the volume of recoverable groundwater in storage for a particular area or aquifer can be estimated as the product of the area, saturated thickness, and specific yield (accounting as appropriate for differences in the estimates of saturated thickness and specific yield among multiple layers or zones).[5] To assess the value and limitations of estimates of groundwater in storage, it is helpful to first consider how aquifers are drained and then look at their dynamic links to the surface environment.

Aquifer Drainage

The mechanism of aquifer drainage depends on whether an aquifer is unconfined or confined (see Figure 1). In an unconfined aquifer, the upper surface of the saturated zone (water table) is free to rise and decline. The principal source of water from pumping an unconfined aquifer is the dewatering of the aquifer material by gravity drainage. The volume of water that is usable in practice is limited by the aquifer's permeability (how easily water moves through a rock unit), water quality, cost of drilling wells, and design of the well and pump.

Consider as an example the unconfined High Plains aquifer, which underlies an area stretching from southern South Dakota to the end of the panhandle of Texas and is the most heavily pumped aquifer in the United States. Depletion of aquifer storage from pumping has had substantial effects on irrigated agriculture in the High Plains, particularly in the southern half, where more than 50 percent of the saturated thickness has been dewatered in some areas. In Kansas, scientists have estimated the lifespan of the aquifer by projecting past trends into the future until the saturated thickness of the aquifer reaches a level at which groundwater pumping for irrigation becomes impractical.[6] The results suggest that many areas in western Kansas have less than 50 years of usable groundwater remaining. Thirty feet of saturated thickness was the critical level in this study, although the researchers noted additional studies that suggest that 30 feet is not enough saturated thickness to provide sufficient well yields for irrigation.

Changes in groundwater levels throughout the High Plains aquifer are tracked annually through the cooperative effort of the U.S. Geological Survey and state and local agencies in the High Plains region (see Figure 2a). Despite the considerable effects of storage depletion in much of the High Plains, only 6 percent of the volume of water in the High Plains aquifer has been depleted since pumping began (see Figure 2b), illustrating how aggregated information about storage depletion over large areas can mask significant local effects.

Confined aquifers, which underlie low permeability confining systems, are filled by water under pressure and respond to pumping differently. The water for pumping is derived not from pore drainage but rather from aquifer compression and water expansion as the hydraulic pressure is reduced. Pumping from confined aquifers results in more rapid water-level declines covering much larger areas when compared to pumping the same quantity of water from unconfined aquifers. If water levels in an area are reduced to the point where an aquifer changes from a confined to an unconfined condition (becomes dewatered), the source of water becomes gravity drainage as in an unconfined aquifer. A major complication arises, however, because the drawdowns in the confined aquifer will induce leakage from adjacent confining units. Slow leakage over large areas can result in the confining unit supplying much, if not most, of the water derived from pumping.[7] Therefore, it is particularly difficult to relate

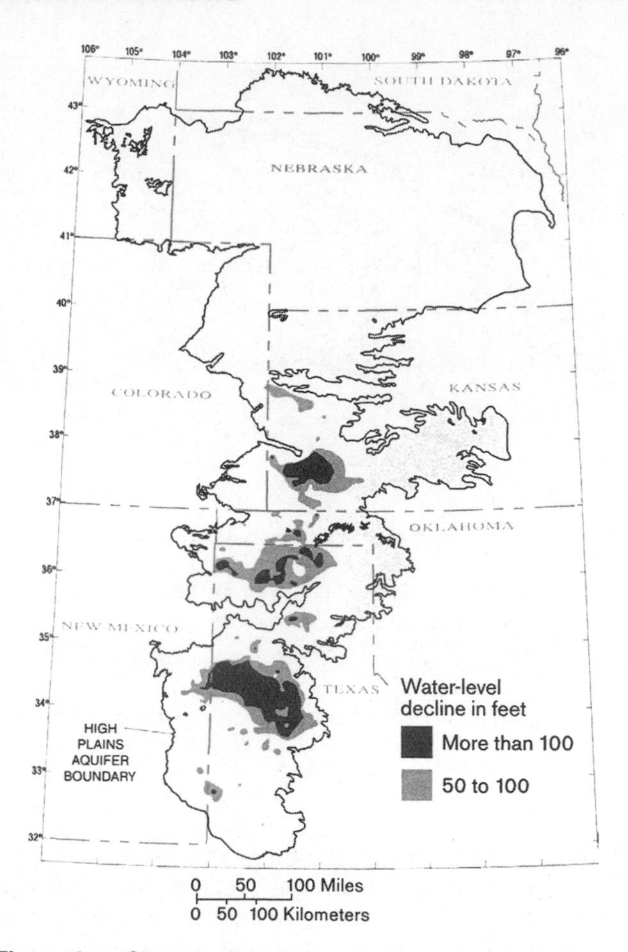

Figure 2a Changes in groundwater levels in the high plains aquifer from predevelopment to 2000.

Source: V. L. McGuire et al., *Water in Storage and Approaches to Ground-Water Management, High Plains Aquifer, 2000,* U.S. Geological Survey Circular 1243 (Reston, VA, 2003).

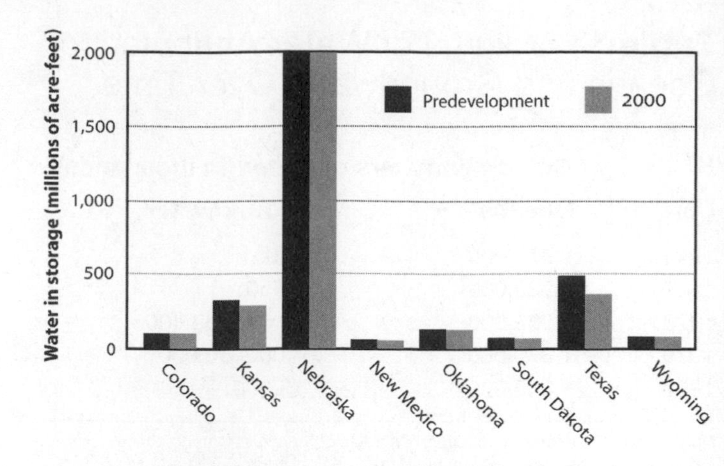

Figure 2b Comparison of predevelopment and 2000 groundwater in storage by state.

Source: V. L. McGuire et al., *Water in Storage and Approaches to Ground-Water Management, High Plains Aquifer, 2000,* U.S. Geological Survey Circular 1243 (Reston, VA, 2003).

estimates of the volume of groundwater in storage to the usable volume of groundwater in confined aquifers.

Further complications may arise for those aquifers with silt and clay layers that can permanently compact as a result of pumping. Consider, for example, the Central Valley aquifer in California, the nation's second-most-pumped aquifer. By 1977, about 28 percent of the decrease in aquifer storage of 60 million acre-feet was the result of permanent reduction of pore space by compaction, resulting in land subsidence throughout much of the area.[8] Farmers in the Central Valley have drawn on imported surface water as a major source of irrigation water to reduce groundwater depletion and associated subsidence. The decrease in aquifer storage of 60 million acre-feet, although very large, represented only a small part of the more than 800 million acre-feet of freshwater stored in the upper 1,000 feet of sediments in the Central Valley.[9]

Similarly, subsidence caused by groundwater pumping in the low-lying coastal environment of Houston, Texas, has increased its vulnerability to flooding and tidal surges. As a result, Houston has undertaken an expensive shift from sole reliance on its vast groundwater resource to partial reliance on surface water for its water supply.

Several key points arise from the examples and discussion thus far. First, measurement of storage depletion should be placed in the context of individual aquifer systems. For example, in evaluating water-level declines, one has to distinguish carefully between confined and unconfined aquifers, as the two respond very differently to pumping. Second, aquifer-wide estimates of recoverable water in storage have limited utility without considering the distribution of water-level changes and their effects. Finally, depletion of a small part of the total volume of water in storage can have substantial effects that become the limiting factors to development of the groundwater resource. These issues are further reinforced when one considers the response of surface-water bodies to groundwater pumping.

Interactions with Surface Water

Groundwater flows from areas of recharge to areas of discharge. Recharge includes water that naturally enters a groundwater system and water that enters the system at artificial recharge facilities or as a consequence of human activities such as irrigation and waste disposal. Discharge may occur to the atmosphere by transpiration; to streams, lakes, and other surface-water bodies; or through a pumping well. The balance between groundwater recharge and discharge controls groundwater levels and storage in a manner analogous to how deposits and withdrawals control savings in a bank account. If recharge exceeds discharge for some period, groundwater levels and storage will increase. Conversely, groundwater levels and storage will decline during periods when discharge exceeds recharge.

A common misperception is that the development of a groundwater system is "safe" if the average rate of groundwater withdrawal does not exceed the average annual rate of natural recharge. People sometimes make the erroneous assumption that natural recharge is equivalent to the basin sustainable yield.[10] Even further misinterpretations suggest that pumping at less than the recharge rate will not cause water levels and groundwater storage to decline.

To understand the fallacy inherent in these conclusions, one needs to consider how groundwater systems respond to pumping. Under natural conditions, a groundwater system is in long-term equilibrium. That is, averaged over some period (and in the absence of climate change), the amount of water recharging the system is approximately equal to the amount of water leaving (discharging from) the system. Withdrawal of groundwater by pumping changes the natural flow system, and the water is supplied by some combination of increased recharge, decreased discharge, and removal of water that was stored in the system.

Initially, water levels in pumping wells will decline to induce the flow of water to these wells, and water is removed from storage. Subsequently, the groundwater system readjusts to the pumping stress by "capturing" recharge or discharge. Also, the storage contribution to the water budget decreases with time for any given withdrawal. If the system can come to a new equilibrium, the changes in storage will cease (at a new reduced level of groundwater storage), and inflows will again balance outflows. Thus, the long-term source of water to discharging wells becomes a change in the inflow and outflow from the groundwater system.

The amount of groundwater available for use depends upon how the changes in inflow and outflow affect the surrounding environment and upon the extent to which society is willing to accept the resultant environmental changes. Consequences include reduced availability of water to riparian and aquatic ecosystems and reduced availability of surface water for use by humans. Further complicating matters, the effects of pumping on surface-water resources can be spread out over a long period of time, as illustrated by the alluvial aquifer example in Figure 3.

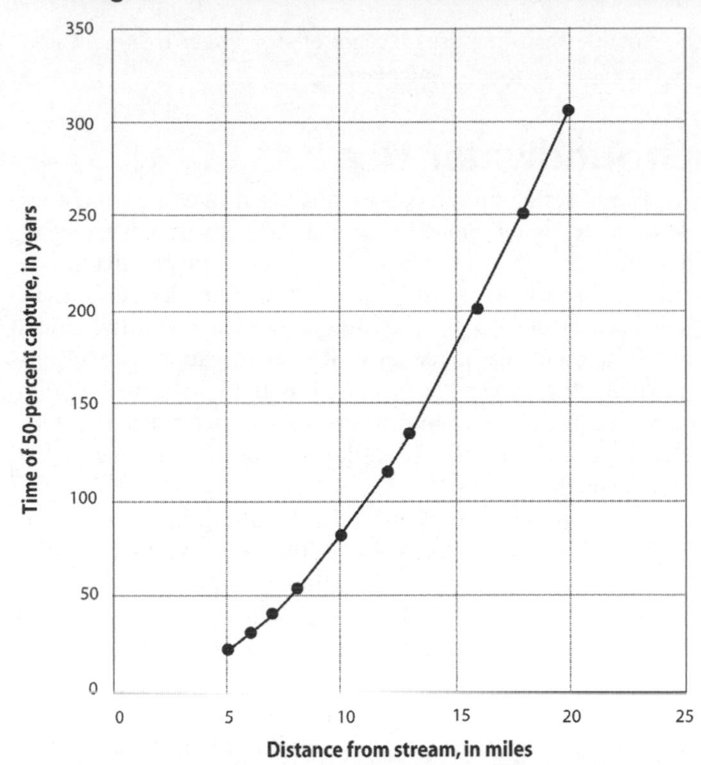

Figure 3 Effect of pumping on surface-water resources.

Note. Time of 50-percent capture is the number of years until 50 percent of the pumping rate is accounted for as reduced groundwater discharge to the stream. The relation is for a fully penetrating stream in an aquifer having a transmissivity to storage ratio of 110,000 square feet per day.

Source: C. Fillippone and S. A. Leake, "Time Scales in the Sustainable Management of Water Resources," *Southwest Hydrology* 4, no. 1 (2005): 17. © SAHRA–University of Arizona.

In many areas, the effects of groundwater pumping on surface-water resources, and importantly, the large uncertainties associated with these effects, become the limiting factors to groundwater development. For example, University of Arizona water law and policy expert Robert Glennon in his popular book *Water Follies* describes controversial situations from throughout the United States where groundwater pumping affects streams and lakes.[11] The effects on surface water can occur with relatively little depletion of the total amount of groundwater in storage. One of the areas Glennon discusses is the Upper San Pedro River Basin in southeastern Arizona, where concerns about streamflow depletion have caused conflicts between development and environment interests in this ecologically diverse riparian system. The health of the riparian system is dependent on the groundwater level and hydraulic gradient near the stream. A key question is how pumping in the basin affects these components of riparian system health. Congressionally mandated efforts are under way to reduce the annual storage depletion (overdraft) in the Sierra Vista area—a subwatershed of the Upper San Pedro Basin.[12] Current overdraft in the Sierra Vista subwatershed is about 10,000 acre-feet per year, which is small, relative to estimates ranging from 20 to 26 million acre-feet of total groundwater storage.[13] A monitoring plan is an important element for verifying the effectiveness of management measures in reducing overdraft in the Sierra Vista subwatershed, with the ultimate goal of mitigating impacts on the riparian system.

Water-Quality Limitations

Groundwater contamination from human activities clearly places constraints on groundwater availability. Likewise, water-quality constraints on groundwater availability can result from pumping. Perhaps best known are the many cases of saltwater intrusion from pumping groundwater along coastal areas. Groundwater pumping also can induce movement of saline water from underlying aquifers in inland areas. Likewise, shallow polluted groundwater may be induced or accelerated downward and throughout an aquifer by prolonged pumping, such that contaminated groundwater penetrates further and more quickly than otherwise anticipated. The removal of water from storage also changes the quality of the remaining groundwater because good quality water commonly is withdrawn first, and the residual often includes poorer quality groundwater from elsewhere in the aquifer or groundwater that has leaked into the aquifer from adjacent units in response to declining water levels. All these and other possible changes in water quality need to be considered in conjunction with information about changes in water levels and water in storage in evaluating the availability of groundwater. In some cases, the quality of groundwater will be suitable for some uses but not others. Water treatment may be necessary to meet some needs.

Groundwater Use

An average of 85 billion gallons of groundwater are withdrawn daily in the United States. More than 90 percent of these withdrawals are used for irrigation, public supply (deliveries to homes, businesses, industry), and self-supplied industrial uses. Irrigation is the largest use, accounting for about two-thirds of the amount. The percentage of total irrigation withdrawals provided by groundwater increased from 23 percent in 1950 to 42 percent in 2000. Groundwater provides about half the nation's drinking water with nearly all those in rural areas reliant upon groundwater.[14]

The importance of groundwater withdrawals in the United States is similar to that in the rest of the world, with some variations from country to country. Rapid expansion in groundwater use occurred between 1950 and 1975 in many industrial nations and subsequently in much of the developing world. The intensive use of groundwater for irrigation in arid and semi-arid countries has been called a "silent revolution" as millions of independent farmers worldwide have chosen to become increasingly dependent on the reliability of groundwater resources, reaping abundant social and economic benefits but with limited management controls by government water agencies.[15] Perhaps as many as two billion people worldwide depend directly upon groundwater for drinking water. The dependence on groundwater for drinking water is particularly high in Europe, where about 75 percent of the drinking-water supply is obtained from groundwater.[16]

Water-use data, when coupled with a scientific understanding of how aquifers respond to withdrawals, are crucial for water planning. Yet information on groundwater use is spotty and often inaccurate within the United States and worldwide. In the United States, practices for collecting water-use data vary significantly from state to state and from one water-use category to another, in response to laws regulating water use and interest in water-use data as an input for water management. Programs to collect water-use data in each state are summarized in a review by the National Research Council.[17]

Some water-use data, such as withdrawals for drinking water and other household uses and withdrawals by some industrial users are obtained by direct measurement, and some may be estimated as the amount reported or allowed by permit. Many uses, such as for self-supplied domestic use, agriculture, and some industries, are often estimated using coefficients relating water use to another characteristic, such as number of employees, number of units manufactured, irrigated acreage, or number of livestock. For example, self-supplied domestic water withdrawals are typically determined by multiplying an estimate of the self-supplied population by a per-capita use coefficient. Likewise, water use for a particular type of industry might be estimated using information on employment or production and estimates of gallons per day per employee or per unit of product. Ideally, coefficients used for water-use estimation are grounded in representative data records. In practice, they are often derived empirically or developed using data that are sparsely sampled in time and space and perhaps extrapolated beyond the climatic, technological, and economic conditions for which they were originally developed. Other complications arise in these calculations because it may be difficult to separate surface-water and groundwater withdrawals without site-specific data and because small-scale use may be excluded from official statistics.

In determining the effects of pumping, it is important to recognize that not all the water pumped is necessarily consumed. For example, some of the water pumped for irrigation is lost to evapotranspiration, and some of the water returns to the groundwater system by infiltration, canal leakage, and other paths of irrigation return flow. Of course, water that is not used for consumption can undergo substantial changes in quality between withdrawal and recharge. Ideally, information on groundwater use includes estimates of consumptive use and return flow as well as withdrawals.

Groundwater Sustainability and Management

Achieving an acceptable tradeoff between groundwater use and the long-term effects of that use is a central theme in the evolving concept of groundwater sustainability.[18] Initially, people viewed groundwater as a convenient resource for general use, and they focused their attention on the economic aspects of groundwater development. Sustainability concerns, emerging in the early 1980s, have brought environmental viewpoints and an intergenerational perspective to the forefront in discussions about groundwater availability.

Groundwater sustainability is commonly defined in a broad context as the development and use of groundwater resources in a manner that can be maintained for an indefinite amount of time without causing unacceptable environmental, economic, or social consequences. The amount of time it takes for the effects of pumping to be manifested elsewhere in the environment reinforces the importance of sustainability as a concept for groundwater management but also makes sustainable solutions difficult to apply in practice. Application of sustainability concepts to water resources requires that the effects of many different human activities on water resources and the overall environment be understood and quantified to the greatest extent possible over the long term. Thus, sustainability likely requires an iterative process of continued monitoring, analysis, application of management practices, and revision. For some cases, particularly in arid areas, the groundwater resource is treated as nonsustainable.[19]

The tradeoff between the water used for consumption and the effects of groundwater withdrawals—on maintenance of instream-flow requirements for fish and other aquatic species, the health of riparian and wetland areas, and other environmental needs—is the driving force behind discussions about the sustainability of many groundwater systems. Considerable scientific uncertainty is associated with disputes over whether pumping will have a specific impact on a particular river or spring. Further complicating matters is the fact that although they are linked through the hydrologic cycle, groundwater and surface water are typically managed separately under different laws and administrative bodies.

Groundwater management strategies are composed of a small number of general approaches:[20]

- use of sources of water other than local groundwater, by shifting the local source of water (either completely or in part) from groundwater to surface water or importing water from outside the local water-system boundaries (the California Central Valley and Houston have implemented these approaches);
- changing rates or spatial patterns of groundwater pumping to minimize existing or potential unwanted effects (examples include moving well fields inland to avoid saltwater intrusion, shifting from deep to shallow groundwater or vice versa, and maintaining sufficient distances between wells to avoid excessive drawdown);
- control or regulation of groundwater pumping through implementation of guidelines, policies, taxes, or regulations by water management authorities (these imposed actions may include restrictions on some types of water use, limits on withdrawal volumes, or establishment of critical levels for aquifer hydraulic heads);
- artificial recharge through the deliberate introduction of local or imported surface water—whether potable, reclaimed, or waste-stream discharge—into the subsurface for purposes of augmenting or restoring the quantity of water stored in developed aquifers (options include infiltration from engineered impoundments, direct-well injection, and pumping designed to induce inflow of freshwater from surface waterways);
- use of groundwater and surface water through the coordinated and integrated use of the two sources to ensure optimum long-term economic and social benefits;
- conservation practices, techniques, and technologies that improve the efficiency of water use, often accompanied by public education programs on water conservation;
- reuse of wastewater (gray water) and treated wastewater (reclaimed water) for non-potable purposes such as irrigation of crops, lawns, and golf courses;
- desalination of brackish groundwater or treatment of otherwise impaired groundwater to reduce dependency on fresh groundwater sources.

These general approaches are not mutually exclusive; that is, the various approaches overlap, or the implementation of one approach will inevitably involve or cause the implementation of another. For example, many approaches involve combinations of surface water, groundwater, and artificial recharge. During periods of excess surface-water runoff, and when surface-water impoundments are at or near capacity, surplus surface water can be stored in aquifer systems through artificial recharge. Conversely, during droughts, increased groundwater pumping can be used to offset shortfalls in surface-water supplies. Depleted aquifer systems can be seen as potential subsurface reservoirs for storing surplus imported or local surface water.

It is important to frame the hydrologic implications of various alternative development strategies in such a way that their long-term implications can be properly evaluated, including effects on the water budget. For example, changing the rates or patterns of groundwater pumping will lead to changes in the spatial patterns of recharge to or discharge from groundwater systems. As another example, in some areas of extensive use of artificial recharge, such as parts of southern California, water from artificial recharge may have replaced much of the native groundwater.

Monitoring Groundwater Reserves

Water-level measurements in observation wells provide the primary source of information about groundwater reserves. Water-level data collected over periods of days to months are useful for determining an aquifer's hydraulic properties; however, data collected over years to decades are required to monitor the long-term effects of aquifer development and management.

The amount of effort in collecting long-term water-level data varies greatly from state to state, and many long-term monitoring wells are clustered in certain areas.[21] Although they are difficult to track, the number of long-term observation wells appears to be declining because of limitations in funding and human resources. For example, the number of long-term observation wells monitored by the U.S. Geological Survey (USGS) declined by about half from the 1980s to 2000.

For many decades, hydrogeologists and others have been making periodic calls for a nationwide program to obtain more systematic and comprehensive records of water levels in observation wells. O. E. Meinzer, a longtime chief of the USGS Ground Water Division and considered by many to be the father of the science of hydrogeology, described the characteristics of such a program about 70 years ago:

> The program should cover the water-bearing formations in all sections of the country; it should include beds with water-table conditions, deep artesian aquifers, and intermediate sources; moreover, it should include areas of heavy withdrawal by pumping or artesian flow, areas which are not affected by heavy withdrawal but in which the natural conditions of intake and discharge have been affected by deforestation or breaking up of prairie land, and, so far as possible, areas that still have primeval conditions. This nation-wide program should furnish a reliable basis for periodic inventories of the ground-water resources, in order that adequate provision may be made for our future water supplies.[22]

More recently, the Heinz Center report *The State of the Nation's Ecosystems* indicated that data on groundwater levels and rates of change are "not adequate for national reporting."[23] This report advocated supplementing existing networks to develop a national indicator of trends in groundwater levels.

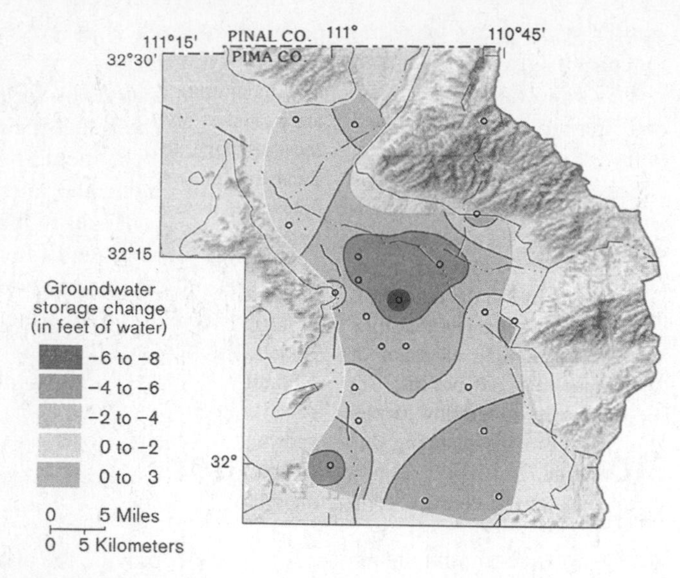

Figure 4 Change in groundwater storage in the Tucson Basin, 1989–1998.

Note. Change in storage was estimated using microgravity measurements.

Source: D. R. Pool, D. Winster, and K. C. Cole, *Land Subsidence and Ground-Water Storage Monitoring in the Tucson Active Management Area, Arizona,* U.S. Geological Survey Fact Sheet 084–00 (Reston, VA, 2000).

The U.S. Government Accountability Office noted that no federal agencies are collecting groundwater data on a national scale and only the USGS and National Park Service are collecting water-level data on a regional scale.[24]

Historically, water-level measurements were simply tabulated, recorded in a paper file, and possibly published in reports. Today, many agencies use the Internet to enhance users' access to current and historical monitoring data. Furthermore, continuous collection, processing, and transmission of water-level data on the Internet in "real time" (typically updated every few hours) is becoming more of a standard procedure. Real-time groundwater data are useful in formulating drought warnings, as they suggest potential effects on water levels in shallow domestic wells. Real-time capability can lead to improved data quality (from continual review of the data) as well as to increased interest in groundwater conditions on the part of the general public.

In addition to water-level monitoring, certain geophysical techniques can enhance the delineation and interpretation of water-level changes over a region. For example, microgravity methods can be used to measure the small gravitational changes that result from changes in groundwater storage (including water stored in the unsaturated zone) over an area (see Figure 4). More recently, researchers have proposed satellite-based measurements of gravity to measure changes over areas the size of a large part of the High Plains aquifer.[25] Meeting the majority of needs for water-level information requires much finer detail than that which satellite measurements can provide, but future technologies may improve this technique. Land- and satellite-based gravity measurements provide area-wide information on changes in the volume of water in storage but do not provide information on vertical changes in heads (water levels) in aquifer systems.

A second geophysical technique, Interferometric Synthetic Aperture Radar (InSAR), uses repeated radar signals from space to measure land-surface uplifts or subsidence at high degrees of measurement resolution and spatial detail.[26] Like gravity methods, InSAR has the advantage of being able to make measurements over large areas and between monitoring wells. The InSAR information can provide additional insights into the areal extent of groundwater depletion where it is linked to subsidence, and can even detect uplift from artificial recharge. InSAR has been found to be particularly useful in identifying faults and geologic structures that may impede groundwater flow and affect the response of an aquifer system to pumping.

Integrated Monitoring and Assessment of Groundwater Reserves

As previously noted, the desire for a national network for monitoring groundwater levels has been discussed since the early 1900s but remains unfulfilled. Meanwhile, groundwater issues have evolved beyond early concerns focused on the hydraulics of individual wells and well field development to encompass many aspects of groundwater, including quantity, quality, and interactions with surface water. Technological advances have been made in sensors, communications, and electronic control systems to monitor groundwater, and computer modeling has become widely used to evaluate groundwater systems. From today's perspective, what might an ideal national program involve?

First and foremost, a national water-level monitoring program should be a collaborative process that involves discourse among local, state, and federal governmental agencies, nongovernmental organizations, and the public.[27] Ideally, data collected would serve double-duty by contributing to the larger regional and national picture while meeting local needs. There should be sufficient consistency in approach to describe the status of groundwater reserves across the country and to show how different constraints affect utilization of the nation's aquifers. A major early goal would be to identify critical gaps in existing coverage.[28]

Many of the primary issues affecting groundwater availability require analysis at the scale of aquifers to achieve a meaningful perspective. To that end, monitoring programs should be designed in the context of the specific characteristics of each aquifer system. A comprehensive national monitoring program should track major aquifers that are affected by groundwater pumping, areas of future groundwater development, and areas of groundwater recharge. Water levels should be measured in wells open to different depths and in the context of the three-dimensional groundwater-flow system.

A long-term record of water-level measurements should encompass the period between the natural and developed states of aquifer systems. Other approaches, such as gravity measurements to estimate subsurface-water storage changes, should be considered in conjunction with the water-level monitoring program. Establishing links between water-level and water-quality requires an understanding of groundwater-flow systems. Studies of natural mixing in aquifers suggest that existing damage to

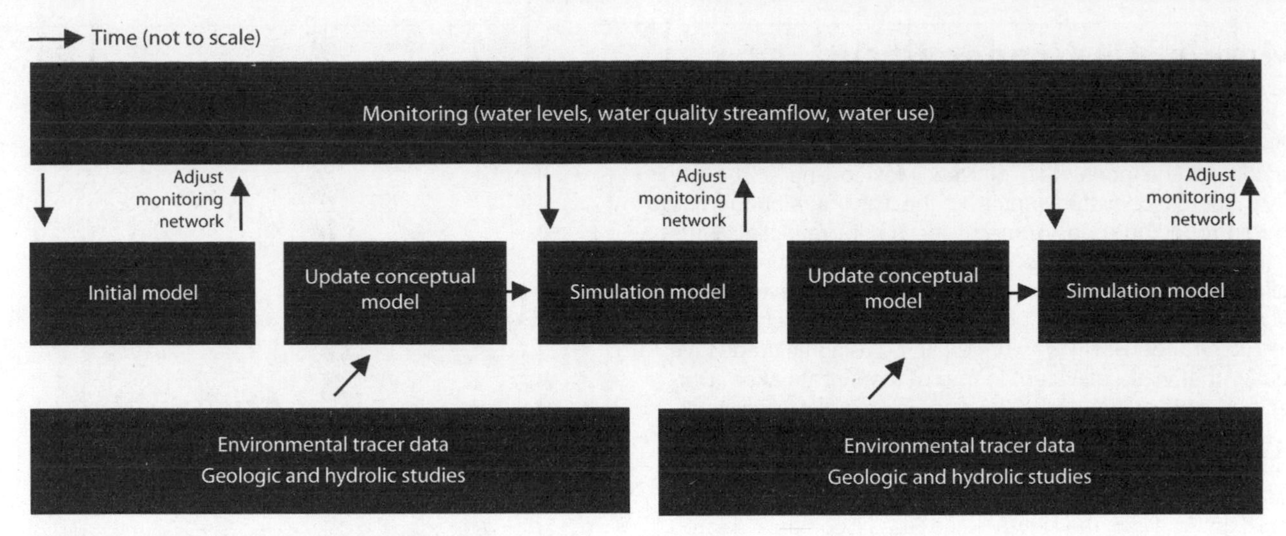

Figure 5 Tracking groundwater resources.

groundwater quality may be lasting and could gradually extend deeper into aquifer systems, thereby reducing further groundwater availability.[29]

Data on changes in groundwater levels provide essential information about changes in groundwater storage and provide the simplest way to convey the extent of groundwater depletion. However, changes in groundwater storage are only part of the story. As noted in previous examples, the status of groundwater reserves should be placed in the context of the complete water budget for that aquifer system. Thus, the monitoring of surface water and groundwater should be linked, particularly measurements of streamflow during low-flow periods when groundwater discharge is the primary component of streamflow. This might require an increase in the number of streamgaging stations in targeted basins to estimate the groundwater contribution to streamflow.

In addition to monitoring data on natural systems, estimation of water withdrawals and consumptive use is an essential part of computing a water budget for a developed aquifer system. Groundwater pumping is one component of the water budget that is physically possible to measure; yet it is commonly one of the most uncertain components of the water budget. Water-use information should be an integral part of evaluations of groundwater quantity and quality and other environmental conditions. Where multiple overlying aquifers are used, efforts should be made to estimate withdrawals from each.

Groundwater systems are dynamic and adjust over decades or more to pumping and other stresses. Many aquifer systems have undergone several decades of intensive development and may be far from equilibrium. Thus, it is challenging to place current conditions in the context of the dynamic but slow changes that may be taking place. A simple snapshot of current conditions may not indicate, for example, how future streamflow depletion will evolve from the pumping that has already occurred.

During the past several decades, computer models for simulating groundwater and surface-water systems have played an increasing role in the evaluation of groundwater development and management alternatives. Groundwater modeling serves as a quantitative means of evaluating the water balance of an aquifer, as it is affected by land use, climate, and groundwater withdrawals, and how these changes affect streamflow, lake levels, water quality, and other important variables. Generally, monitoring and computer modeling are treated as distinct activities, but to be most effective, the two should be linked. Such a framework is considered further below, and its essential elements are illustrated in Figure 5.

Monitoring groundwater reserves serves as primary information used in the development and calibration of computer models. Likewise, the process of model calibration and use provides insights into which components of the system are best known, which components are poorly known, and which components are more important than others. Thus, the experience gained from modeling should provide a basis for a periodic evaluation of the monitoring network.

As its basis, every simulation model has a conceptual model that represents the prevailing theory of how the groundwater system works. The appropriateness of this conceptual model is tested as a numerical model is built, and field observations are compared to the model simulations. Unfortunately, more often than not, data will fit more than one conceptual model, and good calibration of a model does not ensure a correct conceptual model. Thus, conceptual and numerical modeling should be viewed as an iterative process in which the conceptual model is continuously reformulated and updated as new information is acquired.[30] The importance of this approach and its link to monitoring data is recognized explicitly in Figure 5 as a key step prior to each new stage of groundwater modeling.

Additional scientific studies conducted at the time of modeling or during intervening periods can provide insights into the adequacy of the conceptual model that underlies the computer model as well as help in adjustment of model parameters. Such studies include use of environmental tracers, studies of the geologic framework, and geophysical studies. For example, an increasing number of chemical and isotopic substances are being measured in groundwater to identify water sources, trace directions of groundwater flow, and measure the age of the water (time since recharge). Comparison of the results from these environmental tracers with information from computer modeling can

Middle Rio Grande Basin

The Middle Rio Grande Basin encompasses about 38 percent of the population of New Mexico and is the primary source of water supply to the City of Albuquerque and surrounding area. Some of the most productive parts of the aquifer system are in eastern Albuquerque, where, coincidentally, most of the initial groundwater development occurred. This led to the popular belief that the entire Middle Rio Grande Basin was underlain by a highly productive aquifer that was equivalent to one of the Great Lakes. During the 1980s and early 1990s, a combination of large water-level declines measured in monitoring wells (greater than 150 feet in some areas) and new insights into the geologic framework of the basin led to serious questions about this paradigm. In 1995, the New Mexico State Engineer declared the Middle Rio Grande Basin a "critical basin" faced with rapid economic and population growth for which there is less than adequate technical information about the available groundwater supply.

To fill some of the gaps in information, an intensive 6-year effort was undertaken to improve understanding of the hydrogeology of the basin.[1] Geological, geophysical, and environmental tracer studies provided new insights into the source areas for recharge to different parts of the aquifer; indicated that mountain-front recharge is less than previously estimated; showed that the hydraulic connection between the Rio Grande and the aquifer is less than previously thought in some areas; identified new faults that may affect groundwater flow; and suggested that the aquifer is less productive in some areas than previously thought (see the figure at right). The new information was incorporated into a revised groundwater model for the region. In conjunction with the study, new monitoring wells were established in the Albuquerque area, generally as nests of several wells completed at different depths in the aquifer and located to minimize short-term fluctuations caused by nearby high-capacity production wells. The combined approach of monitoring, modeling, and scientific studies has been instrumental in helping the City of Albuquerque revise its water-use and future water-supply strategy.

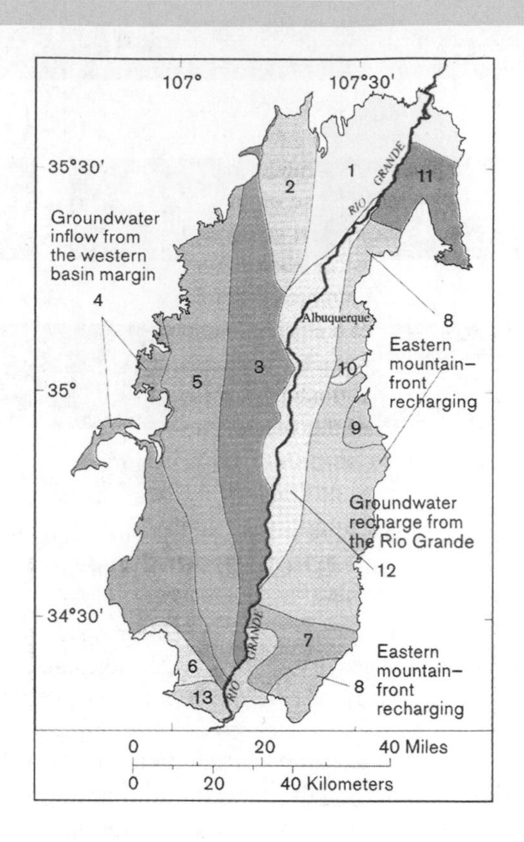

1. J. R. Bartolino and J. C. Cole, *Ground-Water Resources of the Middle Rio Grande Basin, New Mexico,* U.S. Geological Survey Circular 1222 (Reston, VA, 2002), http://pubs.water.usgs.gov/circ1222

lead to either increased confidence in the conceptual model of a groundwater system or recognition of the need for changes. The Middle Rio Grande Basin in central New Mexico (see box above) provides an example of how long-term water-level monitoring combined with environmental and geologic studies has contributed to an evolving series of conceptual and simulation models used to help manage the groundwater resources of the basin.

Not all aquifer systems lend themselves to the exact same approach. For example, consolidated geological formations with fractures, joints, or solution cavities can be difficult to model, given the discontinuous nature of their permeability. These rocks commonly are highly vulnerable to contamination, and the wide range in water-level fluctuations can cause shallow domestic wells to go dry during extended droughts. Interpretation of water-level monitoring from individual wells is difficult in such terrain. It remains important, however, to have a conceptual model of the system as a driving force behind the monitoring network design with a goal to quantify that model as knowledge of the system improves.

One should not infer that the simulation model in Figure 5 is always the same. Indeed, a stepwise approach may be used in which simpler analytic codes are used in the initial phases before constructing three-dimensional numerical models. Also, it is likely that further groundwater research will develop multiple models addressing different roles and objectives. Each model provides a means to reevaluate the monitoring network from a different perspective and to advance understanding of how the water balance of the aquifer system responds to human development.

Generalized long-term monitoring will provide critical information for many uses but will not offset the need for very specific monitoring to address more localized issues, such as the effects of pumping on the ecology of a stream reach. Ideally, the broader scale monitoring programs provide a hydrologic context for the design of such studies.

Conclusions

Groundwater monitoring data serve as a foundation that permits informed management decisions on many kinds of groundwater resource and sustainability issues. Unfortunately, data on groundwater conditions and trends are generally lacking worldwide: Groundwater is commonly undervalued, and there is a deceptive time lag between withdrawals and the resultant impacts of those withdrawals. Long-term groundwater data from individual wells are useful primarily as part of a broader analysis of aquifer systems; thus, the value of data from individual wells is often as invisible as the resource they represent.

Long-term water-level monitoring needs to be integrated with analysis of other monitoring data and an underlying model of the water budget of the aquifer system (typically a simulation model) as a means for interpreting monitoring results and guiding the design of monitoring networks. Regular reassessment of monitoring objectives is necessary to ensure that monitoring programs provide the information needed by groundwater users and those who manage water resources. To enhance the value of groundwater data, managers and policymakers must also ensure the continuity of data-collection programs over time.

Within the United States, one might summarize the current situation as one in which we have some ability to track groundwater levels and water use for many aquifers, generally have limited ability to place these data in the context of groundwater sustainability for most aquifers, and often lack an integrated approach with feedback among monitoring, simulation, scientific studies, and management approaches. Similar issues exist in managing groundwater resources throughout the world.[31] Fortunately, the ability to access groundwater data on the Internet and to portray them in a spatial context should continue to enhance their visibility and value in the coming years.

Notes

1. Single geologic units may define aquifers. Alternately, multiple aquifers and surrounding lower permeability units may be collectively referred to as "aquifer systems." The two terms are used somewhat interchangeably in this article, with "aquifer systems" used when an emphasis is placed on aquifers as hydrologic systems. The *Ground Water Atlas of the United States* describes many of the important aquifers of the nation and can be found at http://capp.water.usgs.gov/gwa/
2. W. M. Alley, R. W. Healy, J. W. LaBaugh, and T. E. Reilly, "Flow and Storage in Groundwater Systems," *Science* 296, 5575 (14 June 2002): 1985–90.
3. National Research Council, *Confronting the Nation's Water Problems: The Role of Research* (Washington, DC: National Academies Press, 2004), 187; and L. F. Konikow and E. Kendy, "Groundwater Depletion: A Global Problem," *Hydrogeology Journal* 13, no. 1 (2005): 317–20.
4. G. M. Hornberger, "A Water Cycle Initiative," *Ground Water* 43, no. 6 (2005): 771.
5. Saturated thickness is the vertical thickness of the aquifer in which the pore spaces are filled (saturated) with water. Specific yield is the ratio of the volume of water that a saturated rock will yield by gravity drainage to the volume of the rock. Specific yield typically ranges from 0.05 to 0.3.
6. B. B. Wilson, D. P. Young, and R. W. Buddemeier, *Exploring Relationships Between Water Table Elevations, Reported Water Use, and Aquifer Subunit Delineations,* Kansas Geological Survey Open File Report 2002-25D (Lawrence, KS, 2002).
7. For example, in a well-known study, researchers found that most of the water pumped from the confined Dakota sandstone aquifer in South Dakota has come from confining beds. J. D. Bredehoeft, C. E. Neuzil, and P. C. D. Milly, *Regional Flow in the Dakota Aquifer: A Study of the Role of Confining Layers,* U.S. Geological Survey Water-Supply Paper 2237 (Washington, DC, 1983).
8. G. L. Bertoldi, R. H. Johnston, and K. D. Evenson, *Ground Water in the Central Valley, California—A Summary Report,* U.S. Geological Survey Professional Paper 1401-A (Washington, DC, 1991). Land subsidence is a gradual settling or sudden sinking of the Earth's surface. Several different processes can cause it. Most water-related subsidence occurs as a result of compaction of aquifer materials (as in the Central Valley), drainage and oxidation of organic soils, and the dissolution and collapse of limestone and other susceptible rocks forming sinkholes and similar features.
9. Ibid., page 27.
10. J. D. Bredehoeft, "Safe Yield and the Water Budget Myth," *Ground Water* 35, no. 6 (1997): 929.
11. R. J. Glennon, *Water Follies: Groundwater Pumping and the Fate of America's Fresh Waters* (Washington, DC: Island Press, 2004).
12. U.S. Department of the Interior, *Water Management of the Regional Aquifer in the Sierra Vista Subwatershed, Arizona—2004 Report to Congress,* prepared in consultation with the Secretaries of Agriculture and Defense and in cooperation with the Upper San Pedro Partnership in response to Public Law 108-136, Section 321, 30 March 2005.
13. Arizona Department of Water Resources, *Upper San Pedro Basin Active Management Area Review Report* (Phoenix, AZ, 2005) 3–25, http://www.azwater.gov/dwr/Content/Publications/files/UpperSanPedro/UpperSanPedroBasinAMAReviewReport.pdf
14. M. A. Maupin and N. L. Barber, *Estimated Withdrawals from Principal Aquifers in the United States, 2000,* U.S. Geological Survey Circular 1279 (Reston, VA, 2005), http://pubs.water.usgs.gov/circ1279
15. M. R. Llamas and P. Martinez-Santos, "Intensive Groundwater Use: A Silent Revolution that Cannot be Ignored," *Water Science and Technology 51,* no. 8 (2005): 167–74.
16. B. L. Morris et al., *Groundwater and its Susceptibility to Degradation: A Global Assessment of the Problem and Options for Management,* United Nations Environment Programme (UNEP) Early Warning and Assessment Report Series, RS 03-3 (Nairobi, Kenya, 2001).
17. National Research Council, *Estimating Water Use in the United States* (Washington, DC: National Academy Press, 2002).
18. W. M. Alley, T. E. Reilly, and O. L. Franke, *Sustainability of Ground-Water Resources,* U.S. Geological Survey Circular 1186 (Denver, CO, 1999), http://pubs.water.usgs.gov/circ1186; and W. M. Alley and S. A. Leake, "The Journey from Safe Yield to Sustainability," *Ground Water* 42, no. 1 (2004): 12–16.
19. W. A. Abderrahman, "Should Intensive Use of Non-renewable Groundwater Resources Always Be Rejected?" in R. Llamas and E. Custodio, eds., *Intensive Use of Groundwater: Challenges and Opportunities* (Lisse, Netherlands: A. A. Balkema, 2002), 191–203.

20. D. L. Galloway, W. M. Alley, P. M. Barlow, T. E. Reilly, and P. Tucci, *Evolving Issues and Practices in Managing Ground-Water Resources: Case Studies on the Role of Science,* U.S. Geological Survey Circular 1247 (Reston, VA, 2003), http://pubs.water.usgs.gov/circ1247
21. C. J. Taylor and W. M. Alley, *Ground-Water-Level Monitoring and the Importance of Long-Term Water-Level Data,* U.S. Geological Survey Circular 1217 (Denver, CO, 2001), http://pubs.water.usgs.gov/circ1217
22. O. E. Meinzer, "Introduction" in R. M. Leggette, et al., *Report of the Committee on Observation Wells, United States Geological Survey,* (unpublished manuscript on file in Reston, VA, 1935), 3.
23. H. John Heinz III Center for Science, Economics and the Environment, *The State of the Nation's Ecosystems: Measuring the Lands, Waters, and Living Resources of the United States* (Cambridge, UK: Cambridge University Press, 2002), http://www.heinzctr.org/ecosystems/report.html
24. U.S. General Accountability Office, *Watershed Management: Better Coordination of Data Collection Efforts Needed to Support Key Decisions,* GAO-04-382 (Washington, DC, 2004).
25. M. Rodell and J. S. Famiglietti, "Detectability of Variations in Continental Water Storage from Satellite Observations of the Time Dependent Gravity Field," *Water Resources Research* 35 (1999): 2705–23.
26. For examples of the use of InSAR to understand groundwater systems, see G. W. Bawden, M. Sneed, S. V. Stork, and D. L. Galloway, *Measuring Human-Induced Land Subsidence from Space,* U.S. Geological Survey Fact Sheet 069-03 (Sacramento, CA, 2003), http://pubs.water.usgs.gov/fs-069-03/
27. National Ground Water Association, *Ground Water Level and Quality Monitoring* (2005) http://www.ngwa.org/pdf/monitoring7.pdf (accessed 8 February 2006).
28. P. M. Barlow et al., *Concepts for National Assessment of Water Availability and Use,* U.S. Geological Survey Circular 1223 (Reston, VA, 2002), http://pubs.water.usgs.gov/circ1223
29. G. E. Fogg, "Groundwater Quality Sustainability, Creeping Normalcy, and a Research Agenda," *Geological Society of America Abstracts with Programs* 37, no. 7 (2005): 247.
30. J. D. Bredehoeft, "The Conceptualization Model Problem—Surprise," *Hydrogeology Journal* 13, no. 1 (2005): 37–46.
31. Recently, the International Groundwater Resources Assessment Center (IGRAC) has been established to share data and information on groundwater resources worldwide, http://www.igrac.nl/

William M. Alley is chief of the Office of Ground Water at the U.S. Geological Survey. He is an active participant in groundwater conferences and has served on national and international committees for the American Geophysical Union, National Ground Water Association, UNESCO, and the National Research Council. He can be reached at walley@usgs.gov.

From *Environment,* Vol. 48, no. 3, April 2006, pp. 11–25. Public Domain.

Article 22

How Much Is Clean Water Worth?

A lot, say researchers who are putting dollar values on wildlife and ecosystems—and proving that conservation pays.

JIM MORRISON

The water that quenches thirsts in Queens and bubbles into bathtubs in Brooklyn begins about 125 miles north in a forest in the Catskill Mountains. It flows down distant hills through pastures and farmlands and eventually into giant aqueducts serving 9 million people with 1.3 billion gallons daily. Because it flows directly from the ground through reservoirs to the tap, this water—long regarded as the champagne of city drinking supplies—comes from what's often called the largest "unfiltered" system in the nation.

But that's not strictly true. Water percolating through the Catskills is filtered naturally—for free. Beneath the forest, fine roots and microorganisms break down contaminants. In streams, plants absorb nutrients from fertilizer and manure. And in meadows, wetlands filter nutrients while breaking down heavy metals.

New York City discovered how valuable these services were 15 years ago when a combination of unbridled development and failing septic systems in the Catskills began degrading the quality of the water that served Queens, Brooklyn and the other boroughs. By 1992, the U.S. Environmental Protection Agency (EPA) warned that unless water quality improved, it would require the city to build a filtration plant, estimated to cost between $6 and $8 billion and between $350 and $400 million a year to operate.

Instead, the city rolled the dice with nature in a historic experiment. Rather than building a filtration plant, officials decided to restore the health of the Catskills watershed, so it would do the job naturally.

What's this ecosystem worth to the city of New York? So far, $1.3 billion. That's what the city has committed to build sewage treatment plants upstate and to protect the watershed through a variety of incentive programs and land purchases. It's a lot of money. But it's a fraction of the cost of the filtration plant—a plant, city officials note, that wouldn't work as tirelessly or efficiently as nature.

"It was a stunning thing for the New York City council to think maybe we should invest in natural capital," says Stanford University researcher Gretchen Daily.

Daily is one of a growing number of academics—some from economics, some from ecology—who are putting dollar figures on the services that ecosystems provide. She and other "ecological economists" look not only at nature's products—food, shelter, raw materials—but at benefits such as clean water, clean air, flood control and storm mitigation, irreplaceable services that have been taken for granted throughout history. "Much of Mother Nature's labor has enormous and obvious value, which has failed to win respect in the marketplace until recently," Daily writes in the book *The New Economy of Nature: The Quest to Make Conservation Profitable.*

Ecological economist Geoffrey Heal, a professor of public policy and business responsibility at Columbia University, became interested in the field as an economist who was concerned about the environment. "The idea of ecosystem services is an interesting framework for thinking why the environment matters," says Heal, author of *Nature and the Marketplace: Capturing the Value of Ecosystem Services.* "The traditional argument for environmental conservation had been essentially aesthetic or ethical. It was beautiful or a moral responsibility. But there are powerful economic reasons for keeping things intact as well."

Daily notes that beyond providing clean water, the Catskills ecosystem has value for its beauty, as wildlife habitat and for recreation, particularly trout fishing. Such values are not inconsequential. While no one has assessed the total worth of the watershed, even a partial look reveals that habitat and wildlife are powerful economic engines.

Restored habitat for trout and other game fish, for example, attracts fishermen, and angling is big business in this state. According to a report by the U.S. Fish and Wildlife Service (FWS), more than 1.5 million people fished in New York during 2001, yielding an economic benefit to the state of more than $2 billion and generating the equivalent of 17,468 full-time jobs and more than $164 million in state, federal, sales and motor fuel taxes. Though not as easily measured, individual Catskills species also have value. Beavers, for instance, create wetlands that are vital to filtering water and to biodiversity.

Ecological economists maintain that ecosystems are capital assets that, if managed well, provide a stream of benefits just as any investment does. The FWS report, for example, notes that 66 million Americans spent more than $38 billion in 2001 observing, feeding or photographing wildlife. Those expenditures resulted in more than a million jobs with total wages and salaries of $27.8 billion. The analysis found that birders alone spent an estimated $32 billion on wildlife watching that year, generating $85 billion of economic benefits. In Yellowstone National Park, the reintroduction of gray wolves that began in 1995 has already increased revenues in surrounding communities by $10 million a year, with total benefits projected to reach $23 million annually as more visitors come to catch a glimpse of these charismatic predators.

When it comes to water quality, EPA projects that the United States will have to spend $140 billion over the next 20 years to maintain minimum required standards for drinking water quality. No wonder, then, that 140 U.S. cities have studied using an approach similar to New York's. Under that agreement, finalized in 1997, the city promised to pay farmers, landowners and businesses that abided by restrictions designed to protect the watershed. (The city owns less than 8 percent of the land in the 2,000-square-mile watershed; the vast majority is in private hands.) "In the case of the Catskills, it was a matter of coming up with a way to reward the stewards of the natural asset for something they had been providing for free," Daily says. "As soon as they got paid even a little bit, they were much happier and inclined to go about their stewardship." There's no guarantee this experiment will work, of course; it may be another decade before the city finds out.

Elsewhere, other governing bodies are also recognizing the value of ecosystem services. The U.S. Army Corps of Engineers, for example, bought 8,500 acres of wetlands along Massachusetts' Charles River for flood control. The land cost $10 million, a tenth of the $100 million the Corps estimated it would take to build the dam and levee originally proposed. To fight floods in Napa, California, county officials spent $250 million to reconnect the Napa River to its historical floodplains, allowing the river to meander as it once did. The cost was a fraction of the estimated $1.6 billion that would have been needed to repair flood damage over the next century without the project. Within a year, notes Daily, flood insurance rates in the county dropped 20 percent and real estate prices rose 20 percent, thanks to the flood protection now promised by nature.

Even insects supply vital ecosystem services. More than 218,000 of the world's 250,000 flowering plants, including 70 percent of all species of food plants, rely on pollinators for reproduction—and more than 100,000 of these pollinators are invertebrates, including bees, moths, butterflies, beetles and flies. Another 1,000 or more vertebrate species, including birds, mammals and reptiles, also pollinate plants. According to University of Arizona entomologist Stephen Buchmann, author of *The Forgotten Pollinators,* one of every three bites of food we eat comes courtesy of a pollinator.

A Cornell University study estimated the value of pollination by honeybees in the United States alone at $14.6 billion in 2000. Yet honeybee populations are dropping everywhere, as much as 25 percent since 1990, according to one study. Now many farms and orchards are paying to have the bees shipped in.

Today's interest in assigning dollar values to pollination and other ecosystem services was spawned by publication of a controversial 1997 report in *Nature* that estimated the total global contribution of ecosystems to be $33 trillion or more each year—roughly double the combined gross national product of all countries in the world. The study became a lightning rod. Detractors scoffed at the idea that one could put a dollar value on something people weren't willing to purchase. One report by researchers at the University of Maryland, Bowden College and Duke University called the estimate "absurd," noting that if taken literally, the figure suggests that a family earning $30,000 annually would pay $40,000 annually for ecosystem protection.

Other researchers, including Daily and Heal, charged that the $33 trillion figure greatly underestimates nature's value. "If you believe, as I do, that ecosystem services are necessary for human survival, they're invaluable really," Heal says. "We would pay anything we could pay."

Daily doesn't believe the absolute value of an ecosystem can ever be measured. Heal agrees, yet both scientists say that pricing ecosystem services is an important tool for making decisions about nature—and for making the case for conservation. "Valuation is just one step in the broader politics of decision making," she says. "We need to be creative and innovative in changing social institutions so we are aligning economic forces with conservation."

Indeed, as dollar values for nature's services become available, environmentalists increasingly use them to bolster arguments for conservation. One high-profile example is the contentious dispute over whether to tear down four dams on the lower Snake River in southeastern Washington to restore salmon habitat, and thus the region's lucrative salmon fishery. Ed Whitelaw, a professor of economics at the University of Oregon, notes that estimates of the economic impact of breaching

Nature's Services

How the Experts Categorize Them

- Ecosystem goods, the traditional measure of nature's products such as seafood, timber and agriculture
- Basic life support functions such as water purification, flood control, soil renewal and pollination
- Life-fulfilling functions, the beauty and inspiration we get from nature, including activities such as hiking and wildlife watching
- Basic insurance, the idea that nature's diversity contains something—like a new drug—the value of which isn't known today, but may be large in the future

Natural Capital

What's the Annual Dollar Value of . . .?

- Recreational saltwater fishing in the United States: $20 billion
- Wild bee pollinators to a single coffee farm in Costa Rica: $60,000
- Tourism to view bats in the city of Austin, Texas: $8 million
- Wildlife watching in the United States: $85 billion
- U.S. employment income generated by wildlife watching: $27.8 billion
- State and federal tax revenues from wildlife watching: $6.1 billion
- Natural pest control services by birds and other wildlife to U.S. farmers: $54 billion

the dams range from $300 million in net costs to $1.3 billion in net benefits, largely due to the wide range of projections about recreational spending.

A 2002 report by the respected, nonprofit think-tank RAND Corporation concluded the dams could be breached without hurting economic growth and employment. Energy lost as a result of the breaches could be replaced with more efficient sources, including natural gas, resulting in 15,000 new jobs. Further, the report noted that recreation, retail, restaurants and real estate would experience a marked growth. Recreational activities alone would increase by an estimated $230 million over 20 years.

There's no question that returning the salmon runs would have a major impact on the region. When favorable ocean conditions increased the runs in 2001, Idaho's Department of Fish and Game estimated the salmon season that year alone generated more than $90 million of revenue in the state, most of it in rural communities that badly needed the funds.

"Some people think it sounds crass to put a price tag on something that's invaluable, careening down the slippery slope of the market economy," says Daily. "In fact, the idea is to do something elegant but tricky: to finesse the economic system, the system that drives so much of our individual and collective behavior, so that without even thinking it makes natural sense to invest in and protect our natural assets, our ecosystem capital."

What Daily and other ecological economists want is to insinuate consideration of ecosystem services into daily decision making, whether it takes the form of financial incentive or penalty. "At a practical level, decisions are made at the margin, not at the 'should we sterilize the Earth' level," she says. "Earth's in all the little decisions—whether to farm here or leave a few trees, whether to build the shopping mall there or leave the wetland, whether to buy an SUV or a Prius—that ecosystem service values need to be incorporated."

Heal agrees. "Although ecosystem services have been with us for millennia," he says, "the scale of human activity is now sufficiently great that we can no longer take their continuation for granted."

Virginia journalist **Jim Morrison** wrote about polar bears and global warming in the February/March 2004 issue.

Searching for Sustainability

Forest Policies, Smallholders, and the Trans-Amazon Highway

EIRIVELTHON LIMA ET AL.

It is a powerful and disturbing image: loggers driving roads deep into the forest to remove a few mahogany trees, with slash-and-burn settlers following closely on their heels. However, it no longer captures the whole picture of logging in the Brazilian Amazon. So, then, what is the role of logging in the impoverishment or potential conservation of the Amazon rainforest? The answer to this question is deceptively complex: To achieve a sustainable future in Amazon forestry, policymakers and stakeholders must understand the physical, economic and political dimensions of competing land use options and economic interests. They must provide effective governance for multiple agendas that require individual oversight.

For simplicity's sake, suppose that forest governance can be approached from two angles: A preservation approach in which the land is tucked away, never to be used again; and a "use-it-or-lose-it" approach in which a well-managed forest estate becomes part of a sustainable economic development scenario and competes successfully with other land use options. In fact, 28 percent of the Brazilian Amazon is already listed as some form of park, or as a protected or indigenous area.[1] But what of the forest without protection, found mainly on private lands or on as-yet undesignated government lands? For many, selective logging of these forests is a form of forest impoverishment that is only slightly less devastating than forest clear-cutting.[2] For others, the selective harvest of timber is the best way to make the long-term protection of standing forests economically and politically viable.[3]

Opponents base their argument on two points: The long-term selective logging of primary tropical forest is financially impracticable, and selective logging is the first step in a vicious cycle of degradation that includes settlement and land clearing.[4] Advocates say that selective logging, when done well (called "reduced impact logging"), is renewable, economically viable, and may provide an important stream of revenue for government and private landowners that would encourage the maintenance of forest cover.[5] These proponents contend that if tropical forestry is to compete successfully with other land use options and essentially push back against the encroaching line of deforestation, some conditions must be met: The removal of subsidies to other land use options; the breakdown of barriers to entry, such as complex forest management plans; the dissemination of information on forestry to all potential market participants; and the elimination of perverse incentives for deforestation—in particular the establishment of land titles through clearing to demonstrate active use. Furthermore, if forestry is deemed the least cost-effective approach to maintaining forest cover outside parks and protected areas, then subsidies to forest management activities might also be appropriate.

There are vast forested areas in the Brazilian Amazon located outside parks and protected areas, and a multitude of landowners, including state and federal governments, are controlling that forest. As a result, policies to manage forest resources must necessarily be comprehensive, flexible, and appropriate to varying conditions and agents. The Brazilian government has recently identified, and is now beginning to implement, a strategy of timber concessions that should help to corral some part of the industry into a controlled region. This should make it easier to monitor and will hopefully reduce illegal logging. The policy, however, mostly ignores the sticky issue of forestry on private land, which, although complex, could provide the engine for sustainable economic development among the disenfranchised settlers of the Amazon frontiers. The settlers may straggle onto the frontier individually, but they eventually form communities, control large areas of land, and become an increasingly important component of the timber industry.

A major economic corridor in the Brazilian Amazon—the Trans-Amazon Highway—illustrates how logging can be transformed from a force driving forest impoverishment to one driving forest conservation, and how this transition, in turn, carries important potential benefits for the semi-subsistence farmers who live along this corridor. To fully describe this transformation, it is necessary to place it in light of the history and current context of the timber industry of the Amazon, and with the understanding that the complexity inherent in the largest and most diverse tropical forest in the world makes forest governance a mighty task.

A Brief History of the Amazon Timber Industry

Understanding logging along the Trans-Amazon Highway depends upon the historical context of the timber industry in the Amazon, which can be roughly divided into three periods.[6] The early production period lasted from the 1950s to the early 1970s and was followed by a transition or boom period, which lasted from the mid 1970s to the late 1980s. A third period, industry consolidation and migration to new frontiers, started in the early 1990s but is now coming to an end. The current timber industry is in such disarray from political mismanagement that in October 2005 the federal police temporarily suspended the transport of all logs from the Amazon.

Early Days (1950s to Mid-1970s)

In the 1950s, the island region of the Amazon delta in the state of Pará was the center of the wood industry in the Amazon. Through the 1960s, there were three large plywood mills and six large sawmills that controlled production. With no connection to the large domestic markets of southeastern Brazil and the dependence on fluvial transport to access raw materials and deliver products, these mills produced only for the export market. Limited shipping capacity and irregular delivery schedules hindered sales to ports in northeastern Brazil, which could be reached by ship along the Atlantic coast. The primary source of raw material was small-scale landowners who sold logs along the banks of rivers. The environmental impact of logging was minimal, as timber extraction was an integrated part of diverse small-scale family farming systems on the Amazon River floodplain. The two popular tree species harvested were Virola (*Virola surinamensis*) for plywood and Andiroba (*Carapa guianensis*) for sawnwood (the first stage of the log processing sequence in which logs are cut into boards, but not planed).

In the early and mid-1970s, a number of smaller sawmills began to appear in the island region and farther up along the upper Amazon River. Into the mid-1970s, the Amazon remained disconnected from domestic markets but the export market flourished. Estimated log consumption was in the region of 2.5 million cubic meters per year—all harvested by axe. Early reports on timber production in the Brazilian Amazon suggest this was a period of poor market access, poor quality of laborers, obsolete equipment, insufficient knowledge of local tree species, and scarce information on prices and markets for products.[7]

Transition Period (Late 1970s to Early 1990s)

A period of dramatic transition in the timber sector began in the late 1970s to early 1980s. Several highways were completed to link the Amazon to domestic southeastern and northeastern Brazilian markets. The states of Rondônia, Mato Grosso, and Pará became connected through the BR364, BR163, and BR010 highways. Large public investment programs for the construction of dams, hydropower plants, a railroad for the Carajás mining program, and the settlement of migrants from southern and northeastern Brazil changed the interfluvial forests of the Amazon, passively protected until that time by their inaccessibility.

Deforestation during this time was largely a response to government actions that either directly promoted or enabled land conversion from forests to other uses. The number and size of sawmills increased in response to the inexpensive primary resource and newly accessible markets, growing local demand, and the availability of cheap labor. Mechanization of harvesting, transport, and processing also contributed to the growth of sawnwood output.

By the early 1980s, Paragominas (a city in Pará) became the most important mill center in the Amazon, producing mostly for the domestic market. The state of Mato Grosso also produced lumber for the domestic market, with important logging centers appearing in the towns of Sinop and Alta Floresta, Meanwhile, the island region continued to produce for the export market. In all, the transition period during the 1970s and mid-to-late 1980s was a turning point in the timber industry of the Brazilian Amazon.

Consolidation and Migration (Mid-1990s to 2000s)

After the transition period, another (less dramatic) period of consolidation and expansion ensued along old and new logging frontiers.[8] Old frontiers can now be found in eastern Pará (Paragominas and Tailandia) and in northern Mato Grosso (Sinop). In these areas, virgin forests have become increasingly scarce, and the logging industry became more diverse and efficient. The more inefficient logging firms exited the market, and those that remained became vertically integrated in an effort to capture value added in downstream processing.

Access to the old frontiers is generally good given the high density of paved roads. In contrast, new frontiers are characterized by a rapid inflow of mills and producers from the old frontier, poor government regulation, and high transport costs. The notable new logging frontier is in western Pará along the northern section of the Santarém-Cuiaba Highway, the BR163.

The Industry Today

The current volume of wood produced in the Legal Amazon is between 20 and 30 million cubic meters, of which more than 50 percent is sold in the domestic Brazilian market.[9] Prior to 2003, legal timber harvest was possible through the preparation of a forest management plan submitted to the government agency and approved with a temporary land title. All that was required by IBAMA, the Brazilian government environmental agency, was proof that a firm or individual had initiated a land legalization process with Brazilian land titling institutions such as the Institute of Colonization and Agrarian Reform (INCRA).[10] Generally, land titling procedures took years, and they did not always result in legalization. By the time the land titling institution had made its decision, the harvest was already complete and the loggers had moved on to the next native forest stocks.

In 2003, the Brazilian government abruptly decided that management plans could no longer be approved on lands where property rights were not well established. That year, nearly all forest management plans were rejected.[11] The government, however, did not have an alternative readily available for the nearly 2,500 logging companies based in the Amazon, and an unintended side effect of the policy has been that more companies now simply operate illegally in such areas. Conflicts, protests, and widespread unregistered logging are now the norm.

To solve the problem of legalizing timber harvest and controlling the timber industry, the Brazilian government has proposed implementing forest concessions on public lands.[12] While this approach has some merit, and indeed has been debated extensively in the Brazilian public arena, large concessions controlled by a few companies may not be the best economic option in the regions where smallholders and other private landowners, including a large number of migrant settlers, are the predominant land users.[13]

The Case of the Trans-Amazon Highway

For two weeks in August 2003, the Trans-Amazon Highway was impassable: Angry loggers had blocked the road, stopping traffic to protest a government-imposed timber shortage. A similar display occurred outside the town of Santarém, Pará, in January 2005 and recently on the BR163 Highway in western Pará. Tragically, access to timber was also one of the underlying reasons for the murder of Sister Dorothy Stang in the municipality of Anapú.[14] Timber scarcity is a startling concept for the Amazon. How is it that a resource so apparently abundant can be the root cause of violent conflicts and protests?[15] The answer lies partially in the sudden requirement by the Brazilian government that loggers provide proper legal documentation for land rights in areas where logs are extracted. But who owns the forests and logs along this frontier highway?

Built by General Emílio Garrastazu Médici (president of Brazil from 1969–1974), the main part of the Trans-Amazon Highway stretches approximately 1,000 kilometers from the town of Marabá to Itaituba on the banks of the Tapajós River.[16] The highway is largely unpaved and virtually impassable for four months of the year during the rainy season. Homesteaders are usually allocated demarcated lots of 100 hectares apiece (approximately 250 acres) and then often battle the elements and wealthy land speculators to continue occupying the land.[17] Still, migration to the region is relentless, as a constant stream of formal and informal land control followed early colonization projects in the late 1970s.[18] INCRA, the federal land settlement agency, has formally settled approximately 30,000 families and an unknown number of informal squatters.

While it is commonly accepted that smallholders control vast areas of land along the Trans-Amazon Highway, the exact quantity of land is debatable. This question is taken up under the auspices of the Green Highways Project, an international multi-institutional project, led by the Brazilian nongovernmental organization Instituto de Pesquisa Ambiental da Amazônia (IPAM, Amazon Institute of Environmental Research) with the support of the Massachusetts-based Woods Hole Research Center (WHRC).[19] An area 100 kilometers (km) on either side of the Trans-Amazon Highway from the municipality of Itupiranga to Placas was mapped using satellite imagery and secondary data from Brazilian government sources. Land distribution was mapped and deforestation measured using 30-meter spatial resolution satellite images and secondary data from INCRA and the Brazilian Institute of Geography and Statistics (IBGE).[20] Images were classified into forest and non-forest classes by supervised classification and visual interpretation.[21] The objective was to identify where smallholders are located and where they will be located in the future.

Of the total 15.7 million hectares located within this buffer, 7.9 million are under the control of or are promised to smallholders. Of the total area within the 100-km study area, the land distribution is: 1.1 percent in demarcated settlements, 5.4 percent in current settlements, 11.4 percent as squatters (*posseiros*), 13.2 percent in old colonization projects, and 19.5 percent destined for future settlements by INCRA. Four percent of the land is in conservation areas, 7.6 percent in informal medium and large-scale land holdings, 15.4 percent in indigenous reserves, and a final 21.2 percent is unclaimed government land.[22] The number of smallholders currently residing in the 100-km zone was estimated by summing the area with active settlements, which includes current settlements, colonization, and squatters, and then dividing by an 82.6-hectare average lot size from survey results (see below), giving a total area of approximately 4.7 million hectares held by 57,000 smallholder families.

Given the observed distribution of smallholders from the spatial analysis, the next logical question for the Green Highways Project was whether these agents could potentially supply the timber industry with wood. Demand for timber in the area is strong; the demand for logs on the Trans-Amazon Highway more than doubled over 12 years, increasing from roughly 340,000 cubic meters in 1990 to approximately 840,000 cubic meters in 2002. To determine whether smallholders can provide this quantity it is important to first estimate the growing stock potential of the forest held by smallholders, assuming that smallholders will in fact sell wood (this assumption will be revisited below). Using conservative (high) deforestation assumptions (for example, a range of 60 percent deforested for old colonization areas to 15 percent deforested for INCRA land allocated to future settlement) and a conservative stand volume of ten cubic meters per hectare, forest stock in active settlement areas is estimated to be 25.8 million cubic meters.[23] Using a harvest cycle of 30 years, this would give a sustained harvest volume of approximately 860,000 cubic meters, which matches current demand. At an estimated stumpage price of 10 Reais (R$10) per cubic meter of standing trees (approximately US$3.33 per cubic meter), this volume would generate R$8.6 million per year.[24] To put this in perspective, if the smallholder forests within current settlements were used to their full potential right now, and the benefits distributed evenly to every family (recall there are an estimated 57,000), each smallholder household could receive R$150 per year—a large sum given the discussion below.

Assuming that smallholders will eventually settle in areas set aside by INCRA, there will be an estimated forest stock of 52.6 million cubic meters, which could render a sustainable harvest of approximately 1.7 million cubic meters per year, more than double the current regional demand. Thus, there appears to be sufficient potential forest stock to meet the demand, and a tremendous opportunity for a redistribution of wealth to the poor, should smallholders have an unhampered market to sell wood [see Table 1 next page].[25]

However, one needs to ask if these estimates based upon government census data are consistent with data on the ground. To answer this question we make use of data generated from a recent comprehensive socioeconomic survey of smallholders along the Trans-Amazon Highway. Between June and December 2003, a total of nearly 3,000 families were interviewed, of which 2,441 lived within the 100-km zone.[26]

In the survey, smallholders were asked about their forest production, and socioeconomic data were collected. The results add to the discussion above, showing that 26 percent had sold wood, and those sales had occurred largely within the last 5 years. There had been only one sale per lot. Ninety-six percent of the smallholders sold standing trees, and the average number of trees sold was 20 per smallholder, which corresponds to a harvest rate of approximately 1 tree per 5 hectares and, assuming an average volume of 5 cubic meters of log per tree, an average sale volume of 100 cubic meters. The average total sale value was R$173, which corresponds to R$8.65 per tree or R$1.73 per cubic meter.[27]

Comparing these observations with the results from the geospatial analysis above, based on timber produced through legal deforestation and harvest of legal forest reserves (the area of smallholder land prohibited from clearing for crops), smallholders are selling approximately 1 cubic meter per hectare, and only 26 percent of them actually sell wood. At this harvest volume, it would take the harvest from 10,000 families per year—about 18 percent of the estimated total smallholder families—to sustain current demand from the area industry at current prices. This amounts to a harvest volume that is only 4 percent of current estimates for Amazon timber production from other studies. At a harvest intensity of 10 cubic meters per hectare, this participation requirement would be drastically reduced to only 1,000 families per year (which represents 1.8 percent of all families). This level of participation could be easily achieved without undue change in the smallholder system by subcontracting the timber industry to do much of the technical work associated with logging. The production of logs is dramatically low on smallholder lots because smallholders have limited knowledge of the forest potential and limited access to the financial resources required to manage the forests. This barrier can be overcome with a partnership between smallholders and the timber industry. For the successful implementation of such a partnership, however, it will be important for smallholders to understand the logging process and have adequate access to production information so that they can maintain a check on their industrial partners.

Holding Back the Tide of Smallholder Forestry

From the perspective of community foresters, the current ideal is that individuals within the communities must work collectively and must control the entire chain of production through to sales of the final product. Formal interaction with the timber industry is anathema. Also, there is still the idea that forest management must happen in large, undisturbed, contiguous tracts of forests. This closely held and restrictive view has undermined the potential of community forestry in the Amazon. The reality is that there are more than 500,000 settlement families in the Brazilian Amazon who work individually or in community associations and who specialize (though perhaps not yet efficiently) in the supply of standing timber by working closely with logging companies.

However, most of the community-based forestry operations have two key problems. First, when dealing with smallholders on an individual basis, the loggers hold all the cards. They have more information about the species and value of timber, and they exploit the immediate financial needs of cash-poor smallholders. Second, logging on smallholder lots is legal only under two premises: Smallholders have deforestation licenses that allow the clearing of 3 hectares per year and the sale of 60 cubic meters per year (up to 20 percent of the land area owned), or they may have the option to develop a forest management plan that must be approved by IBAMA. Of the sales registered in the surveys, 26 percent came from deforestation permits, and a startling 79 percent came from the "legal reserve" on each plot.[28] Because no formal forest management plans have been developed for these smallholder systems, this would imply that nearly 80 percent of log sales from smallholders are currently illegal by government rules; in addition, few smallholders get legal deforestation permits. Why are there no formal plans? A forest management plan requires that the landowner hold legal title, and although 95 percent of smallholders surveyed claimed to be the landowner, we found only 26 percent held formal title; a statistic supported by previous research in the region.[29] This lack of coordination between agencies and resource users is a major barrier to overcoming illegal logging within smallholder systems and to the integration of smallholders into the formal timber market.

Small-Farm Family Forestry in the Amazon

Coordination between ministries is not an impossible task, however. For example, IBAMA, INCRA, and the Ministry of Public Works of the town of Santarém (in Pará) operating with limited resources but in partnership with loggers and smallholders, found a creative solution to this problem in the form of an equitable partnership between industry and smallholders. In this case, the community associations subcontract the loggers to plan and implement harvesting, while the government ministries have the responsibility of expediting title and management approval. The land is owned individually, and management plans are done for each private 100-hectare lot, but the negotiations are between the logger and the community association. The community can demand higher prices by selling as a group, and the logger is assured of a long-term supply of timber. As a result, legal forest operations are taking place

Table 1 Timber Potential from Smallholder Lots on the Trans-Amazon Highway

Smallholders	Total Area (hectares (ha))	Percent Land Area	Forest Cover (percent)	Total Forest Area (ha)	Timber Stock (m^3)	Potential Timber Flow (m^3/year)
Future settlement projects	3,055,000	19.5	85	2,596,000	25,965,000	865,000
Colonization projects	2,063,000	13.2	40	825,000	8,252,000	275,000
Informal settlement	1,792,000	11.4	60	1,075,000	10,750,000	358,000
INCRA settlements	852,000	5.4	80	682,000	6,815,000	227,000
Demarcated settlements	169,000	1.1	50	85,000	847,000	28,000
Total smallholders	7,931,000	50.6	–	5,263,000	52,629,000	1,753,000

Note. INCRA is Brazil's National Institute of Colonization and Agrarian Reform. The entire buffer area is 15,643,000 ha. The area not occupied by smallholders is comprised of unclaimed government land (21.2 percent), indigenous land (15.4 percent), medium and large informal settlement (7.6 percent), and conservation units (4.2 percent).

Source: Instituto de Pesquisa Ambiental da Amazônia (the Amazon Institute of Environmental Research).

André Da Silva Dias Reflects on the Forest Families Program

Forest management models that can contribute to the social, environmental, and economic development of smallholders and traditional populations have been the subject of many recent initiatives in the Amazon. The "Forest Families" program [in Santarém] works with a specific relationship that appears to be very common but little studied: smallholders and the timber industry. It is interesting to note some of the fundamental characteristics around which the program is built: the relationship between the smallholder and the industry already exists; its foundation is market-based; its actors are well-defined; and [it] is based on uncommonly strong legal and ethical rigor. The last characteristic alone makes one pay attention.

One can question whether this is community forest management or not, A pertinent doubt, but, in the end, there exists a forest and its resources and a people organized, or organizing, in communities. In fact, the smallholders are not directly managing their forests: they delegate this activity to a subcontractor and his team. And when they delegate they relinquish some personal control of the forest. However, they exercise their rights to the forest in a free manner, in a negotiation process that strengthens the local organization, generates collective responsibility, creates a commonly used infrastructure, provides income and, most importantly, gives value to the standing forest. All of which are the principles that underlie community forest management.

It is possible to imagine a scenario in which they should manage their own forests in accordance with their capacity, limitation, abilities, and interests. Perhaps this will happen one day. But for right now, the reality is different. No better and no worse, this is just different than many other community forest management initiatives where the local residents play the role of managers. The fact is that they, the owners, are who should say whether this is how it should be. And they seem to be making this [decision] in an informed way, understanding their limitations, and identifying opportunities. It is interesting to observe a community and its people (in this case Santo Antonio) started barely two years ago by families of different origin, who until this point never knew each other, but who already have solid development plans and a growing autonomy in the formulation of local projects, rather than just hope of better days.

I believe that one of the principal contributions that this program can lend to the discussion of local forest management is to define criteria and indicators of a healthy and egalitarian relationship between smallholders and the timber industry. To get there, some challenges that deserve more attention are:

- Improving local knowledge of good forest management practices.
- Identifying the impact of timber harvest on the supply of hunting and non-timber forest products.
- Analyzing the socioeconomic impact of the timber income on the smallholder systems.

Source: André da Silva Dias, Executive Manager, Fundação Floresta Tropical, December 2003. This box was translated from the Portuguese by Frank Merry and first published in a report by Instituto de Pesquisa Ambiental da Amazônia (IPAM) for the International Institute for Environment and Development (IIED) as part of the IIED Power Tools Initiative: Sharpening Policy Tools for Marginalized Managers of Natural Resources. F. Merry, E. Lima, G. Amacher, O. Almeida, A. Alves, and M. Guimares, *Overcoming Marginalization in the Brazilian Amazon Through Community Association: Case Studies of Forests and Fisheries,* (Edinburgh, UK, 2004). It is reprinted with permission.

and smallholders are capturing a fair share of the benefits from the timber harvest on their land [see box above].[30]

However, changes in government personnel and extreme inefficiency (the project industry coordinator has had management plans under review at IBAMA for more than a year) has made even this promising partnership tenuous. These types of projects are in danger of failing because government oversight is inefficient, inadequate, corrupt, and contradictory.[31] There may be a partial solution to be found in timber concessions, but even with successful concessions, the large-scale problems of illegal logging will not disappear. Indeed, the problems of illegal logging will never be solved if IBAMA cannot control the industry or support it effectively, but there is no indication so far that IBAMA can do it alone. It is reasonable to assume, however, that the economic benefits of timber production on their private landholdings will stimulate smallholders to manage their forests and help control illegal activities.

What Does the Future Hold?

What do the results of the Green Highways Project have to say about the issue of loggers and forest policies? As mentioned above, the main thrust of the new forest policy centers around timber concessions on public lands with some allowances given to communities. This is an effective program for a portion of the industry, but there are two problems with the idea. First, the evidence presented above indicates that this approach is inadequate for some major economic corridors where there are many smallholders, such as in the case of the region surrounding the Trans-Amazon Highway; of the 80 percent of land available for harvest (for example, excluding conservation units and indigenous areas), the Green Highways Project shows that 64 percent is under the control of, or is promised to, smallholders. Second, it also shows that forestry is highly underutilized in these smallholder systems. This and the fact that there are more than 500,000 families settled in the Amazon region mean these results imply a very large economic loss to Brazilian society from not capturing a potential timber supply that would almost do away with the need for timber concessions on public lands.

Further, by excluding smallholders from access to the timber industry through current management plan requirements, smallholders are denied what could amount to a substantial and vital source of economic development. In some settlements, research has shown that the value of a single harvest can equal more than 15 years of agricultural production.[32] And finally, even if only some portion of the demand for logs is met by concessions harvesting on public government lands, it may have a negative socioeconomic impact on the potential for small farm forestry by depressing overall prices.

To promote sustainable forestry, the evidence indicates that the government has to realistically deal with land titling, facilitate institutional coordination, and commit to stopping illegal logging through better enforcement. Invariably, the causes of policy failure and poor governance are related to corruption and political auction of important positions in government institutions. An intricate net of political obligation, to the detriment of technical decisions, is commonplace, and even those individuals fiercely committed to their tasks (and there are many) struggle to make quality strategies a reality. A lack of efficiency in government agencies, whether through poor coordination or delays, increases transaction costs and makes formal forest management difficult. Also, by neglecting secure property rights, or making these difficult and costly to obtain, the government inadvertently creates incentives for smallholders and loggers to engage in illegal logging.

Forest management projects on smallholder settlement lots in the Brazilian Amazon will, if widely adopted, help move the region toward equitable forest-based economic development and a peaceful resolution to the problems now facing migrant families. This is not the only solution for the Amazon, but it is a step forward and one well within the reach of the current administration. Without change, however, we can expect further illegal degradation of the forest and a continuing struggle for economic development and social justice on the Amazon frontier.

Notes

1. This approach has been plied very successfully by large conservation organizations, in particular Conservation International, which solicits funds to buy up biodiversity "hot-spots." For an analysis of the effects of parks and protected areas on fires in the Amazon, see D. Nepstad et al., "Inhibition of Amazon Deforestation and Fire by Parks and Indigenous Reserves," *Conservation Biology.* In press, expected publication February 2006.
2. I. Bowles, R. E. Rice, R. A. Mittermeier, and G. A. B. da Fonseca, "Logging On in the Rain Forests," Science, 4 September 1998, 1453–58; R. Rice, C. Sugal, and I. Bowles, Sustainable *Forest Management: A Review of the Current Conventional Wisdom.* (Washington, DC: Conservation International, 1998); R. Rice, R. Gullison, and J. Reid, "Can Sustainable Management Save Tropical Forests?" *Scientific American,* April 1997, 34–39.
3. D. Pearce, F. E. Putz, and J. Vanclay, "Sustainable Forestry in the Tropics: Panacea or Folly?" *Forest Ecology and Management* 172, no. 2 (2003): 229–247; M. Verissimo, A. Cochrane, and C. Sousa Jr., "National Forests in the Amazon," *Science,* 30 August 2002, 1478; F. E. Putz, K. H. Redford, J. G. Robinson, R. Fimbel, and G. Blate, *Biodiversity Conservation in the Context of Tropical Forest Management* (Washington, DC: Biodiversity Studies, The World Bank, 2000), http://world-bank.org/biodiversity
4. G. Asner, et al., "Selective Logging in the Amazon," Science, 21 October 2005: 480–481. The Asner study claimed that the selective logging of the Amazon is far more widespread than previously thought. The authors suggest that the source of logs in the Amazon is not slash-and-burn deforestation—those logs are simply burned—but conventional poor-quality selective logging and that this is the first step in the economic and ecological degradation of the forest. According to the data, this is more widely practiced and perhaps more damaging than previously thought.
5. Forest management and reduced impact logging (FM-RIL) guidelines are available from many sources: the Suriname Agricultural Training Center (CELOS); the International Tropical Timber Organization (ITTO); the Food and Agricultural Organization of the United Nations (FAO); the Institute of Humans and the Environment of the Amazon (IMAZON); and the Fundação Floresta Tropical (FFT, Tropical Forest Foundation). In addition, field models in Brazil demonstrate the improvements of FM-RIL practices over conventional selective logging. See the FFT website at http://www.fft.org.br and click "Research." There have been several studies on the economic benefits of reduced impact logging and comparisons with "conventional" selective logging. For a few examples see: S. Armstrong and C. J. Inglis, "RIL For Real: Introducing Reduced Impact Logging Techniques into a Commercial Forestry Operation in Guyana," *International Forestry Review* 2, (2000): 264–72; F. Boltz, D. R. Carter, T. P. Holmes, and R. Perreira Jr., "Financial Returns Under Uncertainty for Conventional and Reduced-Impact Logging in Permanent Production Forests of the Brazilian Amazon," *Ecological Economics* 39 (2001): 387–98; P. Barreto, P. Amaral, E. Vidal, and C. Uhl, "Costs and Benefits of Forest Management for Timber Production in Eastern Amazonia," *Forest Ecology and Management* 108, no. 1 (1998): 9–26; and T. P. Holmes et al., *Financial Costs and Benefits of Reduced Impact Logging Relative to Conventional Logging in the Eastern Amazon* (Washington, DC: Tropical Forest Foundation, 1999).
6. Thanks to Johan Zweede of the Instituto Florestal Tropical in Belém, Brazil, and Benno Pokorny of the University of Freiburg, Germany, for valuable comments on the history and context of the timber industry.
7. For an excellent review, see I. Sholtz, *Overexploitation or Sustainable Management: Action Patterns of the Tropical Timber Industry: The Case of Pará, Brazil,* 1960–1997 (London: Frank Cass Publishers, 2001).
8. By definition the term "frontier," when applied to forests, implies the point at which new logging occurs. It is, however, common in the literature of logging in the Amazon to differentiate frontiers by age. This is done partially out of custom, but also because logging on all "frontiers" is relatively new; even old frontiers are less than 30 years old.
9. The "Legal Amazon" is a geo-political definition of the Amazon region in Brazil and comprises the states of Amapá, Amazonas, Acre, Maranhão, Mato Grosso, Pará, Rondônia, and Tocantins. The volume of sawnwood destined for export is different across frontiers. More than 60 percent of logs from new frontiers are destined for the export market, whereas on the intermediate and old frontiers, that level dips to 50 and 15 percent, respectively, according to F. Merry et al., "Industrial Development on Logging Frontiers in the Brazilian Amazon," *International Journal of Sustainable Development,* in review. For a recent discussion of production volumes, see G. Asner et al., note 4 above.
10. The Instituto Brasileiro do Meio Ambiente e dos Recursos Naturais Renováveis (http://www.ibama.gov.br) is the Brazilian

government's environmental agency responsible for the forest sector and all issues of environmental control in the country. The federal land-titling agency is the Institute of Colonization and Agrarian Reform (INCRA). For more information, see http://www.incra.gov.br. Each state also has a local agency.

11. The forest management process includes a formal management plan that essentially states that the company intends to harvest in a given area (with accompanying maps and documentation) and subsequently an annual operating plan that delivers the details of each year's harvest operation. The term "forest management plan" includes both of these components of logging.
12. For more information on forest concessions, see A. Veríssimo, M. A. Cochrane, and C. Sousa Jr., National Forests in the Amazon," Science, 30 August 2002, 1478.
13. The forest concessions issue has long been debated in the scientific literature. Both sides of the argument for Brazil can be explored in F. D. Merry et al., "A Risky Forest Policy in the Amazon?" *Science,* 21 March 2003, 1843 and in F. D. Merry et al., "Some Doubts About Concessions in Brazil," *Tropical Forestry Update* 13, no. 3 (2003): 7–9 (see http://www.itto.or.jp/live/contents/download/tfu/TFU.2003.03.English.pdf). See also F. D. Merry and G. S. Amacher, "Forest Taxes, Timber Concessions, and Policy Choices in the Amazon," *Journal of Sustainable Forestry* 20, no. 2 (2005): 15–44; and Veríssimo, Cochrane, and Sousa, note 12 above. For earlier discussion on concessions see J.A. Gray, *Forestry Revenue Systems in Developing Countries,* FAO Forestry Paper 43 (Rome, 1983); R. Repetto and M. Gillis, eds., *Public Policies and the Misuse of Forest Resources,* (Cambridge, UK: Cambridge University Press, 1988); J. R. Vincent, "The Tropical Timber Trade and Sustainable Development," *Science,* 19 June, 1992, 1651–1655; and J. A. Gray, "Underpricing and Overexploitation of Tropical Forests: Forest Pricing in the Management, Conservation and Preservation of Tropical Forests," *Journal of Sustainable Forestry* 4, no. 1/2 (1997): 75–97. The Ministry of Environment has created a new law on public forest management (Law 4776/05), which was approved by Brazil's Chamber of Representatives in July, is still awaiting the vote of the Senate. This law would create the national forest service, the forest development fund, and would regulate timber harvest on public lands. Three kinds of harvest are sought for production forests: direct government management of conservation units (such as national forests); local community use (such as extractive reserves); and forest concessions.
14. Dorothy Stang, a 73-year-old nun from Dayton, Ohio, a practitioner of liberation theology, and an ardent supporter of local settlers, was assassinated in broad daylight in February 2005 in a remote farm community near her home of 25 years in Anapú on the Trans-Amazon Highway. Her battle for equal rights for the poor, including legal land and resource ownership, brought her in direct conflict with loggers and ranchers. Her death triggered an avalanche of government response. Two thousand soldiers were sent to the region to crack down on illegal loggers and land speculators, and five million hectares of forest (an area the size of Costa Rica) were designated as parks and reserves in what may be the world's single greatest act of tropical rainforest conservation.
15. The estimate of forest stock for the Amazon is approximately 60 billion cubic meters. There are varying estimates of the flow from the forest: The IBGE, which is the government institute of geography and statistics (http://ibge.gov.br), estimates log demand in the north of Brazil to be about 17 million cubic meters; IBAMA, the environmental regulation agency of Brazil, estimates it to be around 25 million; and IMAZON, a local nongovernmental research organization, estimates it at about 24 million—down from 28 million in 1999.
16. The entire Trans-Amazon Highway runs approximately 3,300 kilometers, connecting the state of Tocantins to the state of Acre near the Peruvian border. Continuing westward from Itaituba to the town of Humaitá (a stretch which lies to the west of the Tapajós River) is virtually uninhabited, but may be the future frontier on which this story is replayed some years hence.
17. For an excellent discussion on property rights, violence and settlement on the Trans-Amazon Highway see L. J. Alston, G. D. Libecap, and B. Mueller, *Titles, Conflict, and Land Use: The Development of Property Rights and Land Reform on the Brazilian Amazon Frontier* (Ann Arbor: The University of Michigan Press, 1999); and L. G. Alston, G. D. Libecap, and B. Mueller, "Land Reform Policies, the Sources of Violent Conflict in the Brazilian Amazon," *Journal of Environmental Economics and Management* 39, no. 2 (2000): 162–188.
18. For more discussion of smallholder settlement in new and old settlements and the roles of community associations in economic development in migrant settlements see F. Merry and D. J. Macqueen, Collective *Market Engagement* (Edinburgh, International Institute for Environment and Development, 2004), http://www.iied.org/docs/flu/PT7_collective_market_engagement.pdf
19. Other institutions working on the Trans-Amazon Highway within the Green Highways Project are the Fundação Viver, Produzir e Preservar (FVPP) and the Instituto Floresta Tropical (IFT).
20. The principal source of government statistics for Brazil is the Brazilian Institute of Geography and Statistics (Instituto Brasiliero de Geografia e Estatistica, IBGE). Their website can be accessed at http://www.ibge.gov.br
21. A supervised classification is a procedure for identifying spectrally similar areas on an image by pinpointing training sites of known targets and then extrapolating those spectral signatures to other areas of unknown targets. The signatures are quantitative measures of the spectral properties at one or several wavelength intervals. These measures include class maximum, minimum, mean and covariance matrix values. Training areas, usually small and discrete compared to the full image, are identified through visual interpretation and used to "train" the classification algorithm to recognize land cover classes based on their spectral signatures, as found in the image. The training areas for any one land cover class need to fully represent the variability of that class within the image.
22. The total area for squatters was 19 percent of the buffer zone, of which local extension agents estimated 60 percent to be smallholders. The remaining 40 percent were said to be medium- and large-size holdings.
23. The evidence also indicated that only one percent of the buffer area is currently deforested, so these estimates could be considered very conservative for deforestation.
24. The price of R$10 is based on a conservative estimate of a formal logging contract between smallholders and the industry near the town of Santarém and the example of a forest concession (3-year cutting contract) in the Tapajós

national forest—an ITTO project run by IBAMA—where the average stumpage fee for three price categories in 2003 was R$11.73. The exchange rate for the period of the survey was approximately R$3 per US$1, but is now at R$2.2 per US$1. For further commentary on the timber markets of Brazil, see A. Veríssimo and R. Smeraldi, *Acertando O Alvo: Consumo da Madeira no Mercado Interno Brasileiro a Promocao da Certificacao Florestal* (Finding the Target: Consumption of Wood in the Brazilian Domestic Market and the Promotion of Forest Certification) and M. Lentini, A. Verissimo, and L. Sobral, Fatos Florestais da Amazônia (Forest Facts of the Amazon) (Belém, Brazil: Imazon, 2003); E. Lima, and F. Merry, "Views of Brazilian Producers—Increasing and Sustaining Exports," in D. Macqueen, ed., *Growing Timber Exports: The Brazilian Tropical Timber Industry and International Markets* (London: IIED, 2003), 82–102.

25. For an economic model of smallholder decision-making, production, and labor allocation, see F. D. Merry and G. S. Amacher, "Emerging Smallholder Forest Management Contracts in the Brazilian Amazon: Labor Supply and Productivity Effects," Environment and Development Economics. Invited to revise and resubmit, expected publication 2006.
26. The preliminary results of the survey were presented in seminars to the smallholders in June 2004. Further details of this survey are available from the authors.
27. In comparison, the estimated price for logs at the mill gate in 2002 on the Trans-Amazon was R$58 per cubic meter, and an unadjusted five-year average price for logs from 1998 to 2002 was R$39 per cubic meter, but this is before accounting for harvest costs—which for intermediate frontiers such as the Trans-Amazon can run between 30 and 40 Reais per cubic meter and transportation costs; transport distances can run as far as 80 or 90 kilometers from log deck to mill.
28. The legal reserve (Reserva Legal) of a smallholder lot, or for that matter any private land holding in the Brazilian Amazon, is 80 percent of the total land area. This "reserve" area can only be used for forestry with approved forest management plans or the collection of non-timber forest products.
29. Alston, Libecap, and Mueller, note 17 above. Only 11 percent of land owners hold formal title. In our survey, individuals were asked whether they held "definitive title," not formal records.
30. This example is well documented. See D. Nepstad et al., "Managing the Amazon Timber Industry," *Conservation Biology* 18, no. 2 (2004): 575–577; D. Nepstad et al., "Governing the Amazon Timber Industry," in D. Zarin, J. R. R. Alavalapati, F. E. Putz, and M. Schmink, eds., *Working Forests in the American Tropics: Conservation through Sustainable Management?* (New York: Columbia University Press, 2004), 388–414.
31. Another example is the project Safra Legal (Legal Harvest) on the Trans-Amazon Highway. The objective of this project was to make use of the legal deforestation options available to smallholders. The idea of this project came from the forest management projects near Santarém and presented a wonderful alternative to smallholders who would have simply burned the trees where they planned to conduit agricultural activities. The project, however, has recently become embroiled in scandal as a condult of illegal logging, see L. Coutinho, "More Petista Mud in the Ibama," *VEJA,* 15 June 2005, 70. The problems behind the Safra Legal program were also described in L. Rohter, "Loggers, Scorning the Law, Ravage the Amazon Jungle," *The New York Times,* 16 October 2005. These articles illustrate the far-reaching negative effects of corrupt government on the sustainable management of natural resources.
32. F. Merry et al., "Collective Action Without Collective Ownership: The Role of Formal Logging Contracts in Community Associations on the Brazilian Amazon Frontier," International *Forestry Review*, in review. Drafts available from the authors.

Eirivelthon Lima is an associate researcher at the Instituto de Pesquisa Ambiental da Amazônia (IPAM, the Amazon Institute of Environmental Research), headquartered in Belém, Pará, Brazil, and doctoral student in Forest Economics at the Virginia Polytechnic Institute and State University (Virginia Tech). **Frank Merry** is an associate researcher at IPAM, research fellow in environmental studies at Dartmouth College, and visiting assistant scientist at the Woods Hole Research Center (WHRC). **Daniel Nepstad** is a senior researcher at IPAM and senior scientist at WHRC. **Gregory Amacher** is an associate professor of forest economics at Virginia Tech and associate researcher at IPAM. **Cláudia Azevedo-Ramos** is a senior researcher at IPAM. **Paul Lefebvre** is a senior research associate at WHRC. **Felipe Resque Jr.** is a GIS technician at IPAM. We gratefully acknowledge funding from (in alphabetical order) the European Union; the Gordon and Betty Moore Foundation; the William and Flora Hewlett Foundation; NASA Large Scale Biosphere and Atmosphere Project; the National Science Foundation; and the United States Agency for International Development–Brazil program.

Diet, Energy, and Global Warming

The energy consumption of animal- and plant-based diets and, more broadly, the range of energetic planetary footprints spanned by reasonable dietary choices are compared. It is demonstrated that the greenhouse gas emissions of various diets vary by as much as the difference between owning an average sedan versus a sport-utility vehicle under typical driving conditions. The authors conclude with a brief review of the safety of plant-based diets, and find no reasons for concern.

GIDON ESHEL AND PAMELA A. MARTIN

1. Introduction

As world population rises (2.5, 4.1, and 6.5 billion individuals in 1950, 1975, and 2005, respectively; United Nations 2005), human-induced environmental pressures mount. By some measures, one of the most pressing environmental issues is global climate change related to rising atmospheric concentrations of greenhouse gases (GHGs). The link between observed rising atmospheric concentrations of CO_2 and other GHGs, and observed rising global mean temperature and other climatic changes, is not unequivocally established. Nevertheless, the accumulating evidence makes the putative link harder to dismiss. As early as 2000, the United Nations–sponsored Intergovernmental Panel on Climate Change (Houghton et al. 2001) found the evidence sufficiently strong to state that "there is new and stronger evidence that most of the warming observed over the last 50 years is attributable to human activities" and that "[t]he balance of evidence suggests a discernible human influence on global climate."

If one views anthropogenic climate change as an undesirable eventuality, it follows that modifying the ways we conduct various aspects of our lives is required in order to reduce GHG emissions. Many changes can realistically only occur following policy changes (e.g., switching some transportation volume to less CO_2-intensive modes). However, in addition to policy-level issues, energy consumption is strongly affected by individual personal, daily-life choices. Perhaps the most frequently discussed such choice is the vehicle one drives, indeed a very important element of one's planetary footprint. As we show below, an important albeit often overlooked personal choice of substantial GHG emission consequences is one's diet. Evaluating the implications of dietary choices to one's planetary footprint (narrowly defined here as total personal GHG emissions) and comparing those implications to the ones associated with personal transportation choices are the purposes of the current paper.

2. Comparative Energy Consumption by Food Production

In 1999, Heller and Keoleian (2000) estimated the total energy used in food production (defined here as agricultural production combined with processing and distribution) to be 10.2×10^{15} BTU yr^{-1}. Given a total 1999 U.S. energy consumption of 96.8×10^{15} BTU yr^{-1} (Table 1.1 in U.S. Department of Energy 2004a), energy used for food production accounted for 10.5% of the total energy used. In 2002, the food production system accounted for 17% of all fossil fuel use in the United States (Horrigan et al. 2002). For example, Unruh (2002) states that delivered energy consumption by the food industry, 1.09×10^{18} J in 1998, rose to 1.16×10^{18} J in 2000 and is projected to rise by 0.9% yr^{-1}, reaching 1.39×10^{18} J in 2020. Unruh (2002) also reports that delivered energy consumption in the crops and other agricultural industries (the latter consisting of, e.g., animal and fishing) increases, on average, by 1% and 0.9% yr^{-1}, respectively. Thus, food production, a function of our dietary choices, represents a significant and growing energy user.

To place energy consumption for food production in a broader context, we compare it to the more often cited energy sink, personal transportation. The annual U.S. per capita vehicle miles of travel was 9848 in 2003 (Table PS-1 in U.S. Department of Transportation 2004). Using the same source, and focusing on cars (i.e., excluding buses and heavy commercial trucks), per capita vehicle miles traveled becomes 8332, of which an estimated 63% are traveled on highways (Table VM-1 U.S. Department of Transportation 2004). According to the U.S. Department of Energy's (2005) table of most and least efficient vehicles (http://www.fueleconomy.gov/feg/best/bestworstNF.shtml) and considering only highly popular models, the 2005 vehicle miles per gallon (mpg) range is bracketed by the Toyota Prius' 60:51 (highway:city) on the low end and by Chevrolet

Table 1 Energy Consumption for Personal Travel

	Miles per Gallon			Annual Consumption		
Model	**City**	**Highway**	**Weighted Average**[a]	**Gallons**	**10^7 BTU**[b]	**Ton CO_2**[c]
Prius	51	60	57	146	1.7	1.19
Camry	24	33	30	278	3.2	2.24
Suburban	11	15	14	595	6.8	4.76

[a]Based on 63% highway driving.
[b]The conversion of gallons consumed to BTU consumed is based on an average of 1 U.S. gallon of fuel = 115 000 BTU. Many sources report a conversion factor of 1 U.S. gallon of fuel = 125 000 BTU, but this assumes a so-called high heating value, which is not appropriate for motor vehicles' internal combustion engines [Oak Ridge National Laboratory bioenergy conversion factors; http://bioenergy.ornl.gov/papers/misc/energy_conv.html].
[c]The conversion of BTU consumed to CO_2 emissions is based on the total of the U.S. emissions, as described in the text.

Suburban's 11:15. At near average is the Toyota Camry Solara's 24:33 mpg. The salient transportation calculation (Table 1) demonstrates that, depending on the vehicle model, an American is likely to consume between 1.7×10^7 and 6.8×10^7 BTU yr^{-1} for personal transportation. This amounts to emissions of 1.19–4.76 ton CO_2 based on the estimated conversion factor of 7×10^{8} ton CO_2 BTU^{-1} derived from the 2003 U.S. total energy consumption, 98.6×10^{15} BTU (U.S. Department of Energy 2004a), and total CO_2 emissions of 6935.9×10^6 ton (U.S. Department of Energy 2004b).

Next, we perform a similar energetic calculation for food choices. Accounting for food exports, in 2002 the U.S. food production system produced 3774 kcal per person per day or 1.4×10^{15} BTU yr^{-1} nationwide (FAOSTAT 2005). (The difference between 3774 kcal per person per day and the needed average~2100 kcal per person per day is due to overeating and food discarded after being fully processed and distributed.) In producing those 1.4×10^{15} BTU yr^{-1}, the system used 10.2×10^{15} BTU yr^{-1}. That is, given both types of inefficiency, food production energy efficiency is 100(1.4/10.2) (2100/3774) $\approx$ 7.6%. Therefore, in order to ingest 2100 kcal day^{-1}, the average American uses $2100/0.076 \approx 72.6 \times 10^4$ kcal day^{-1} or

$$2100\frac{\text{kcal}}{\text{day}} \times \frac{1\ \text{BTU}}{0.252\ \text{kcal}} \times 365\frac{\text{day}}{\text{yr}} \times \frac{1}{0.076} \approx 4 \times 10^7\frac{\text{BTU}}{\text{yr}}. \quad (1)$$

In summary, while for personal transportation the average American uses 1.7×10^7 to 6.8×10^7 BTU yr^{-1}, for food the average American uses roughly 4×10^7 BTU yr^{-1}. Thus, there exists an order of magnitude parity in fossil energy consumption between dietary and personal transportation choices. This is relevant to climate because fossil fuel-based energy consumption is associated with CO_2 emissions. Note that both food production and transportation also release non-CO_2 GHGs produced during fossil fuel combustion (principally NO_x conversion to N_2O), but these are ignored below. This omission is irrelevant to the comparison between transportation and food production because these contributions are proportional to the mass of fossil fuel burned and thus scale with CO_2 emissions. They are noteworthy, however, as they render our bottom-line conclusion an underestimate of the range of GHG burden resulting from dietary choices.

The next logical step is quantifying the range of GHG emissions associated with various reasonable dietary choices. In exploring this question, we note that food production also releases non-CO_2 GHGs *un*related to fossil fuel combustion (e.g., methane emissions due to animal manure treatment). In comparing below the GHG burden exerted by various reasonable dietary choices we take note of both contributions.

3. Plant-Based versus Animal-Based Diets

To address the variability in energy consumption and GHG emissions for food, we focus on a principal source of such variability, plant- versus animal-based diets. To facilitate a quantitative analysis, we define and consider several semirealistic mixed diets: mean American, red meat, fish, poultry, and lacto-ovo vegetarian. To obtain the mean American diet, we use actual per capita food supply data summarized in the Food Balance Sheets for 2002 (FAOSTAT 2005). Those balance sheets report a total gross caloric consumption of 3774 kcal per person per day, of which 1047 kcal, or 27.7%, is animal based. Of those 1047 kcal day^{-1}, 41% are derived from dairy products, 5% from eggs, and the remaining 54% from various meats. For comparison, we let all diets, including the exclusively plant-based one ("vegan"), comprise the same total number of gross calories, 3774 kcal day^{-1}.

The red meat, fish, and poultry diets we consider share similar dairy and egg portions, 41% and 5% of the animal-based caloric fraction of the diet. The remaining 54% of the animal-based portion of the diet is attributed to the sole source given by the diet name. For example, the animal-based part of the red meat diet comprises 41%, 5%, and 54% of the animal-based calories from dairy, eggs, and red meat, respectively. For the purposes of this paper, we define red meat as comprising 35.6% beef, 62.6% pork, and 1.8% lamb, reflecting the proportions of these meats in the FAOSTAT data. In the lacto-ovo diet, we set the total animal-based energy derived from eggs and dairy to 15% and 85% based on values from Table 1 of Pimentel and Pimentel (2003).

Specific diets vary widely in the fraction of caloric input from animal sources (hereafter α). For example, Haddad and

Table 2 Energetic Efficiencies for a Few Representative Food Items Derived from Land Animals, Aquatic Animals, and Plants

Food item	$100 \times \frac{\text{kcal protein}^a}{\text{kcal input}}$	$\frac{\text{kcal total}^b}{\text{kcal protein}}$	$100 \times \frac{\text{kcal output}^c}{\text{kcal input}}$
Livestock			
Chicken	6.3	2.9	18.1
Milk	5.3	3.9	20.6
Eggs	3.6	3.1	11.2
Beef (grain fed)	2.9	2.3	6.4
Pork	1.5	2.5	3.7
Lamb	0.5	2.3	1.2
Fish			
Herring	50.0	2.2	110
Tuna	5.0	1.2	5.8
Salmon (farmed)	2.5	2.3	5.7
Shrimp	0.7	1.3	0.9
Plants			
Corn			250
Soy			415
Apple			110
Potatoes			123

[a]Pimentel and Pimentel (1996a,b); energy input refers to fossil fuels.
[b]Assuming 1 gram protein = 4 kcal and using U.S. Department of Agriculture (2005) values.
[c]For animal products, the product of the previous two columns.

Tanzman (2003) suggest that lacto-ovo vegetarian diets in the United States contain less than 15% of their calories from animal sources, well below the 27.7% derived from animal sources in the mean American diet. We therefore calculate the energy and GHG impact of each diet over a range of this fraction, $0\% \leq a \leq 50\%$, where $\alpha = 0$ corresponds to a vegan diet.

3.1. Greenhouse Effects of Direct Energy Consumption

This section addresses the greenhouse burden by agriculture that is directly exerted through (mostly fossil fuel) energy consumption and the subsequent CO_2 release. The fossil fuel inputs treated here are related to direct energy needs such as irrigation energy costs, fuel requirements of farm machinery, and labor. We are interested in the range of this burden affected by dietary choices, especially plant-versus animal-based diets.

We define energy efficiency as the percentage of fossil fuel input energy that is retrieved as edible energy [$e = 100 \times$ (output edible energy)/(fossil energy input); see Table 2]. We derive energy efficiency e of various animal-based food items by combining available estimates of (edible energy in protein output)/(fossil energy input) (Pimentel and Pimentel 1996a) and the total energy content relative to the energy from protein. The estimated energy efficiency of protein in animal products (Pimentel and Pimentel 1996a) varies from 0.5% for lamb to ~5% for chicken and milk to 3% for beef (second column of Table 2). This wide range reflects the different reproductive life histories of various animals, their feed, their genetic ability to convert nutrients and feed energy into body protein, fat, and offspring, the intensity of their rearing, and environmental factors (heat, humidity, severe cold) to which they are subjected, among other factors. Accounting for the total energy content relative to the energy from protein (Table 2; U.S. Department of Agriculture 2005), these numbers translate to roughly 1%, 20%, and 6% (e = 0.1, 0.2, and 0.06). The weighted mean efficiency of meat [red meat (consisting of beef, pork, and lamb, as previously defined), fish, and poultry] in the American diet is 9.32% (U.S. Department of Agriculture 2002; see Table 3). These efficiencies are readily comparable with the energy efficiency f of plant-based foods estimated by Pimentel and Pimentel (1996b,c): 60% for tomatoes, ~170% for oranges and potatoes, and 500% for oats. The wide range of f reflects differences in farming intensity, including labor, machinery operation, and synthetic chemical requirements.

Because of the wide range of efficiencies in both plant- and animal-based foods, we quantitatively compare plant-based diets with animal-based ones by considering

$$E = cd\left(\frac{\alpha}{e} + \frac{1-\alpha}{f}\right) \tag{2}$$

for the various hypothetical diets in (2),

$$c = 3774\frac{\text{kcal}}{\text{day}} \times 365\frac{\text{day}}{\text{yr}} \approx 1\,377\,510\frac{\text{kcal}}{\text{yr}}$$

Table 3 Weighted-Mean Energetic Efficiency of the Animal-Based Portion of the Hypothetical Mixed Diets Considered in This Paper

Diet	Component	Percent Efficiency	Caloric Fraction	Weighted Mean (%)
Lacto-ovo	Dairy	20.6	0.85α	19.19
	Eggs	11.2	0.15α	
	Dairy	20.6	0.41α	
Mean American	Eggs	11.2	0.05α	14.05
	Meat	9.3	0.54α	
	Dairy	20.6	0.41α	
Fish	Eggs	11.2	0.05α	11.52
	Fish	4.6	0.54α	
	Dairy	20.6	0.41α	
Red meat	Eggs	11.2	0.05α	11.52
	Meat	9.3	0.54α	
	Dairy	20.6	0.41α	
Poultry	Eggs	11.2	0.05α	18.76
	Poultry	9.3	0.54α	

is the U.S. per capita annual gross caloric consumption, and

$$d = \frac{1}{0.252}\frac{\text{kcal}}{\text{BTU}} \times (7 \times 10^{-8}) \frac{\text{ton CO}_2}{\text{BTU}} \approx 2.778 \times 10^{-7} \frac{\text{ton CO}_2}{\text{kcal}},$$

so that $cd \approx 0.383$ ton CO_2 yr^{-1} is the annual CO_2 emissions of a person consuming 3774 kcal day^{-1} using the BTU–CO_2 conversion factor introduced earlier and assuming perfect efficiency (the deviation from ideal efficiency is accounted for by the bracketed term). The parameter α is the fraction of the dietary caloric intake derived from animal sources. As defined above, e and f are the weighted mean caloric efficiencies of animal- and plant-based portions of a given diet. Those efficiencies for the five hypothetical diets considered here, shown in Table 3, are simply the weighted mean efficiencies derived from the characteristic caloric efficiency of each component of the diet and the caloric prevalence of those components.

The efficiencies are $e = 0.1152$ (fish), 0.1152 (red meat), 0.1405 (average American diet), 0.1876 (poultry), and 0.1919 (lacto-ovo). Recall that the red meat, mean American, fish, and poultry diets derive 41% and 5% of their animal-based calories from dairy and eggs; thus, the weighted-mean efficiency e of the diets reflects the higher efficiency of dairy and egg relative to fish or red meats. The specific (not weighted mean) efficiency of poultry production is between those of dairy and eggs (Table 2). The notable equality of fish and red meat efficiencies reflects: 1) the large energy demands of the long-distance voyages required for fishing large predatory fishes such as swordfish and tuna toward which western diets are skewed, and 2) the relatively low energetic efficiency of salmon farming. Note that similar e values for two or more diets (such as the poultry and lacto-ovo above) reflect similar overall energetic efficiency of the total diets only if those diets also share α, the animal-based caloric fraction of the diet. However, recall the aforementioned Haddad and Tanzman (2003) suggestion that American lacto-ovo vegetarians eat less than 15% of their calories from animal sources, indicating that the overall energetic efficiency of lacto-ovo diets is higher than that of the average poultry diet assumed here, with $\alpha = 0.277$, the same fraction as that of the mean American diet.

Equation (2) allows us to calculate the total CO_2 burden related to fossil fuel combustion for various diets characterized by specific α, e, and f values. However, our objective is to compute the difference between various mixed diets and an exclusively plant-based, vegan, diet. To facilitate such comparison, we get an expression for the difference in CO_2-based footprint between mixed diets and an exclusively plant-based one by subtracting from Equation (2) the expression for a purely plant-based diet. We get the latter by setting $\alpha = 0$ in Equation (2), yielding $E_{vegan} = cd/f$, with which

$$\delta E \equiv E - E_{vegan} = c\ d\ \alpha\left(\frac{1}{e} - \frac{1}{f}\right). \tag{3}$$

3.2. Greenhouse Effects in Addition to Energy Inputs

Of agriculture's various non-energy-related GHG emissions, we focus below on the two main non-CO_2 GHGs emitted by agriculture, methane, CH_4, and nitrous oxide, N_2O. In 2003, U.S. methane emissions from agriculture totaled 182.8×10^6 ton CO_2-eq, of which 172.2×10^6 ton CO_2-eq are directly due to livestock (U.S. Department of Energy 2004b). The same report also estimates the 2003 agriculture-related nitrous oxide emissions, 233.3×10^6 ton CO_2-eq, of which 60.7×10^6 ton CO_2-eq are due to animal waste. Thus, the production of livestock in the U.S. emitted methane and nitrous

Table 4 Non-CO_2 GHG Emissions Associated With the Production of Various Food Items. Units are 10^6 CO_2-eq yr^{-1}, Except Column 6

	CH_4^a				
Food	**Enteric Fermentation**	**Manure Management**	**N_2O* Manure Management**	**Sum**	**Approx Percentage of Total**
Eggs	—	2.08	0.62	2.70	1
Dairy	26.68	18.18	21.96	66.82	29
Beef	82.04	4.43	34.34	120.81	56
Pork	2.07	30.20	1.70	33.97	15
Poultry	—	2.31	0.68	2.99	1
Sheep	1.16	0.03	0.60	1.79	1
Goats	0.14	0.01	0.20	0.35	<1
				Total: 229.41	

*Sources: U.S. Department of Energy (2004b, Tables 21, 22, and 28).

oxide is equivalent to at least $172.2 \times 10^6 + 60.7 \times 10^6 = 232.9 \times 10^6$ ton CO_2 in 2003. With 291 million Americans in 2003, this amounts to 800 kg CO_2-eq per capita annually in excess of the emissions associated with a vegan diet.

One may reasonably argue that the ~0.8 ton CO_2-eq per person per year due to non-CO_2 GHGs does not accurately represent the difference between animal- and plant-based diets, which is our object of inquiry; if there were no animal-based food production at all, plant-based food production would have to increase. However, such a hypothetical transition will produce zero methane and nitrous oxide emissions in the categories considered above, animal waste management, and enteric fermentation by ruminants. Ignored categories, principally soil management, will indeed have to increase, but over an area far smaller than that vacated by eliminating feed production for animals. For example, Reijnders and Soret (2003) report that, per unit protein produced, meat production requires 6 to 17 times as much land as soy. Therefore, the net reduction in methane and nitrous oxide emissions will have to be larger than our estimate presented here.

Approximately 74% of the total nitrous oxide emissions from agriculture, ~173×10^6 ton CO_2-eq, are due to nitrogen fertilization of cropland, which supports production of both animal- and plant-based foods. The exact partitioning of nitrogen fertilization into animal feed and human food is a complex bookkeeping exercise beyond the scope of this paper. Consequently, we ignore this large contribution below. Nevertheless, simple analysis of the Food Balance Sheets (FAOSTAT 2005) and Agriculture Production Database (FAOSTAT 2005) data shows that the portion of those 173×10^6 ton CO_2-eq attributable to animal production is at least equal to, and probably larger than that attributable to plants, thereby rendering our estimate of the GHG burden exerted by animal-based food production a lower bound.

The value of 800 kg CO_2-eq yr^{-1} due to non-CO_2 emissions computed above represents the composition of the actual mean American diet. To calculate the added non-CO_2 burden of specific diets, we must first compute, from the mean American diet, the burden for individual food items.

This calculation requires intermediate steps, as available data are for specific farm animals, not individual food items. Using annual emissions reported by the U.S. Department of Energy (2004b), in Table 4 we sum the contributions of methane from enteric fermentation and manure management and the nitrous oxide from manure management for cattle, pigs, poultry, sheep, and goats. To partition cattle methane emissions from enteric fermentation [108.72 million ton CO_2-eq; Table 21 in U.S. Department of Energy (2004b)] among beef (75.46%) and dairy (24.54%) cattle, we use emission ratios derived from Table 5-3 of U.S. Environmental Protection Agency (2005) (we apply these ratios to the 2003 data, but we do not use the absolute values, because the table's latest entry is 2001). We similarly use Table 5-5 of U.S. Environmental Protection Agency (2005) to partition nitrous oxide emissions from cattle manure management, 56.3 million ton CO_2-eq [Table 28 in U.S. Department of Energy (2004b)], among dairy (39%) and beef (61%) cattle.

Table 28 in U.S. Department of Energy (2004b) reports emissions of 1.3 million ton CO_2-eq from N_2O due to poultry manure management. Because we do not have direct information on the partitioning of these emissions among eggs and poultry meat, we assume this partition in N_2O is proportional to total manure mass and thus is roughly similar to the partitioning of methane from poultry manure management, 47.38% and 52.62% for eggs and meat, respectively (Table 22 in U.S. Department of Energy 2004b). We thus partition the 1.3 million ton CO_2-eq from N_2O due to poultry manure management as 0.62 and 0.68 million ton CO_2-eq due to eggs and poultry meat, respectively.

Table 5 Non-CO_2 GHG Emissions Per Unit Food Consumed, Derived from the Actual Mean American Diet in the Food Balance Sheets (FAOSTAT 2005)

Food	Per Capita Emissions, gram CO_2-eq day^{-1}*	Per Capita Consumption in Mean American Diet, kcal day^{-1}	Emissions gram CO_2-eq kcal
Dairy + butter	629.1	429	1.47
Eggs	25.4	56	0.45
Beef	1137.4	120	9.48
Pork + fat	319.8	211	1.52
Poultry	28.2	196	0.14
Sheep	16.9	6	2.82
Fish	—	29	0.00
Total		1047	

*Results from dividing the "sum" column in Table 5 (FAOSTAT 2005) by 291 million Americans and 365 days.

To obtain the per capita daily emissions associated with food items, we divide the individual non-CO_2 GHG annual sums (Table 4, fourth numeric column) by the U.S. 2003 population, 291 million, and 365 days. The results, in grams of CO_2-eq per day, are shown in the first numeric column in Table 5. To calculate emissions per kcal associated with the consumption of individual food items, we divide the per capita daily emissions (Table 5, first numeric column) by the respective per capita consumptions (FAOSTAT 2005; Table 5, second numeric column). These divisions yield the non-CO_2 GHG emissions per kcal reported in the rightmost column in Table 5. Importantly, the non-CO_2 GHG emissions per kcal vary by as much as a factor of 70 for the animal-based food items considered, rendering some animal-based options (e.g., poultry meat) far more benign than other ones (most notably beef).

Using the emission associated with individual food items (Table 5, rightmost column), we calculate the weighted non-CO_2 GHG emissions for the ith hypothetical diet considered in this paper,

$$\beta_i = \sum_{j=1}^{M} \begin{bmatrix} \text{individual daily} \\ \text{emissions of} \\ \text{component } j \\ \text{in diet } i, \\ \text{grams } CO_2\text{-eq kcal}^{-1} \end{bmatrix} \times \begin{bmatrix} \text{daily} \\ \text{kcals of} \\ \text{component} \\ j \text{ in diet } i \end{bmatrix} \sim \frac{\text{gram } CO_2\text{-eq}}{\text{kcal}}, \quad (4)$$

where M is the number of food items diet i comprises. In calculating β for the red meat and mean American diets, we sum the emissions due to individual meat items. The composition of the mean American diet is detailed in Table 5. The composition of meat in the red meat diet is 35.6% beef, 62.6% pork, and 1.8% lamb, as defined in section 3. The βs of the various diets are computed, using Equation (4) in Table 6.

Using β we modify Equation (3) to take note of both CO_2 and non-CO_2 GHG emissions,

$$\delta E_i = c\,\alpha_i \left[d\left(\frac{1}{e_i} - \frac{1}{f_i} \right) + 10^{-6}\beta_i \right], \quad (5)$$

where the 10^{-16} factor is needed to convert from grams to tons. The GHG burden exerted by animal-based food production through both CO_2 emissions due to fossil fuel combustion and non-CO_2 (methane and nitrous oxide) emissions. Adding the non-CO_2 GHG emissions more than doubles the impact of the mean American diet at mean (27.7%) animal fraction, from 701 kg CO_2-eq per person per year based on fossil fuel input alone to nearly 1.5 ton CO_2-eq per person per year, taking note of fossil fuel inputs as well as non-CO_2 emissions. Recall that this is an underestimate of the actual radiative effect of an animal-based diet relative to a plant-based one because of the neglect of land management in the nitrous oxide budget and other conservative idealizations we have made.

In addition to amplifying the GHG burden of all mixed diets, the added inclusion of non-CO_2 GHGs reveals several consequences of dietary choices. First, red meat and fish diets, which previously coincided because the only consideration was caloric efficiency e, which is roughly 0.11 for both, are now clearly distinct. Second, with the effect of non-CO_2 GHGs included, the fish diet results in lower GHG emissions than both the red meat and mean American diets. This is partly attributable to our choice to ignore small non-CO_2 GHG emissions associated with fish consumption. Third, the lacto-ovo vegetarian diet appears to result in higher GHG emissions than the poultry diet. According to our calculations this is true for any α; however, if lacto-ovo vegetarians eat less than average animal products, as suggested by Haddad and Tanzman (2003), the relevant comparison is not

Table 6 Non-CO_2 GHG Emissions of the Hypothetical Diets Considered in This Paper

Diet	Component	Individual Emissions gram CO_2-eq / kcal	Caloric fraction	β, gram CO_2-eq / kcal
Lacto-ovo	Dairy	1.47	0.85α	
	Eggs	0.45	0.15α	1.317
Mean American	Dairy	1.47	0.41α	
	Eggs	0.45	0.05α	
	Meat*	2.67	0.54α	2.067
	Dairy	1.47	0.41α	
Fish	Eggs	0.45	0.05α	
	Fish	0.00	0.54α	0.625
Red meat	Dairy	1.47	0.41α	
	Eggs	0.45	0.05α	
	Meat**	4.43	0.54α	3.017
	Dairy	1.47	0.41α	
Poultry	Eggs	0.45	0.05α	
	Poultry	0.14	0.54α	0.701

*Beef 21.35%, pork 37.54%, lamb 1.07%, poultry 34.88%, fish 5.16%; the actual emissions are detailed in the third numeric column of Table 5.

**Beef 35.61%, pork 62.61%, and lamb 1.78%.

for a given α, but rather between one α for lacto-ovo diet, for example, $\alpha \approx 0.15$, and a higher one for poultry, for example, $\alpha \approx 0.27$.

To place the planetary consequences of dietary choices in a broader context, at mean U.S. caloric efficiency, it only requires a dietary intake from animal products of ~20%, well below the national average, 27.7%, to increase one's GHG footprint by an amount similar to the difference between an ultraefficient hybrid (Prius) and an average sedan (Camry). For a person consuming a red meat diet at ~35% of calories from animal sources, the added GHG burden above that of a plant eater equals the difference between driving a Camry and an SUV. These results clearly demonstrate the primary effect of one's dietary choices on one's planetary footprint, an effect comparable in magnitude to the car one chooses to drive.

4. Are Plant-Based Diets Safe?

The thrust of this paper has been that the United States bears a GHG burden for the animal-based portion of its collective diet. We can estimate this burden as roughly 1.485 ton CO_2-eq per person per year × 291 million Americans ≈ 432 million ton CO_2-eq yr^{-1} nationwide, or ~6.2% of the total [69 335.7 million ton CO_2-eq in 2003 (Table ES2 of U.S. Department of Energy 2004b)]. To the extent one subscribes to the notion that reducing GHG emissions is desirable, a corollary of this estimate is that it is advantageous to minimize the animal-based portion of the mean American diet. This raises the question of whether a plant-based diet is nutritionally adequate for public health. The following section addresses this question. The available evidence suggests that plant-based diets are safe, and are probably nutritionally superior to mixed diets deriving a large fraction of their calories from animals.

The adverse effects of dietary animal fat intake on cardiovascular diseases is by now well established (see Willett 2001 for a comprehensive review). Similar effects are also seen when meat, rather than fat, intake is considered (e.g., Key et al. 1999; Erlinger and Appel 2003). Less widely appreciated—despite being just as persuasively demonstrated, exhaustively researched, and robustly reproducible—are the links between animal protein consumption and cancer (for a thorough review, see Campbell and Campbell 2004).

The first studies linking dietary animal protein and cancer (e.g., Mgbodile and Campbell 1972; Preston et al. 1976) focused on cancer initiation, the brief process during which cancer-causing mutations first occur. Collectively, they documented numerous cellular mechanisms by which cancer initiation increases under high animal protein diets. Follow-up studies (e.g., Appleton and Campbell 1982; Dunaif and Campbell 1987) addressed cancer promotion after initiation, showing dramatically increased precancerous deformities in response to a given carcinogen dose under high animal protein diets. To unambiguously implicate animal protein in the observed enhanced cancer promotion, Schulsinger et al. (1989) compared induced carcinogenesis under high protein diets of animal and plant origins. Cancer promotion was significantly enhanced under animal-protein-rich diet. Youngman (1990) and Youngman and Campbell (1992) extended these results to clinical cancer (as opposed to cancer precursors), showing roughly an order of magnitude higher tumor incidence in rats on high animal-protein diets who lived their full natural life span. Similar results were also obtained with different species and carcinogens (e.g., Cheng et al. 1997; Hu et al. 1997). Note that the above laboratory results were all obtained at protein intakes per unit body mass routinely consumed by Westerners, suggesting the applicability of the results to humans (Campbell and Campbell 2004).

Human epidemiological evidence indeed corroborates the link between animal-based diet and cancer. For example, Larsson et al. (2004) show enhancement of ovarian cancer with dairy consumption in Swedish women; Sieri et al. (2002) show a strong association between animal protein intake and breast cancer in Italian women; Chao et al. (2005) show a tight positive relationship between meat consumption and colorectal cancer; and Fraser (1999) demonstrates an approximate halving of colon and prostate cancer risk among vegetarians. Barnard et al. (1995) documented the disease burden exerted by seven major diseases on the health care system directly related to meat consumption. Some of the above cited results may well be challenged in the future. Nevertheless, it is hard to avoid the conclusion, reached by, for example, Sabate (2003), that animal-based diets discernibly increase the likelihood of both cardiovascular diseases and certain types of cancer. To our knowledge, there is currently no credible evidence that plant-based diets actually undermine health; the balance of available evidence suggests that plant-based diets are at the very least just as safe as mixed ones, and are most likely safer.

5. Conclusions

We examine the greenhouse gas emissions associated with plant- and animal-based diets, considering both direct and indirect emissions (i.e., CO_2 emissions due to fossil fuel combustion, and methane and nitrous oxide CO_2-equivalent emissions due to animal-based food production). We conclude that a person consuming a mixed diet with the mean American caloric content and composition causes the emissions of 1485 kg CO_2-equivalent above the emissions associated with consuming the same number of calories, but from plant sources. Far from trivial, nationally this difference amounts to over 6% of the total U.S. greenhouse gas emissions. We conclude by briefly addressing the public health safety of plant-based diets, and find no evidence for adverse effects.

References

Appleton, B. S., and T. C. Campbell, 1982: Inhibition of aflatoxin-initiated preneoplastic liver lesions by low dietary protein. *Nutr. Cancer,* **3,** 200–206.

Barnard, N. D., A. Nicholson, and J. L. Howard, 1995: The medical costs attributable to meat consumption. *Prev. Med.,* **24,** 656–657.

Campbell, T. C., and T. M. Campbell II, 2004: *The China Study.* Benbella, 417 pp.

Chao, A., and Coauthors, 2005: Meat consumption and risk of colorectal cancer. *J. Amer. Med. Assoc.,* **293,** 172–182.

Cheng, Z., J. Hu, J. King, G. Jay, and T. C. Campbell, 1997: Inhibition of hepatocellular carcinoma development in hepatitis B virus transfected mice by low dietary casein. *Hepatology,* **26,** 1351–1354.

Dunaif, G., and T. C. Campbell, 1987: Dietary protein level and aflatoxin B_1-induced preneoplastic hepatic lesions in the rat. *J. Nutr.,* **117,** 1298–1302.

Erlinger, T. P., and L. J. Appel, 2003: The relationship between meat intake and cardiovascular disease. Review paper, Johns Hopkins Center for a Livable Future, 26 pp. [Available online at http://www.jhsph.edu/Environment/CLF]

FAOSTAT, cited 2005: Statistical database of the Food and Agricultural Organization of the United Nations. [Available online at http://faostat.fao.org/]

Fraser, G. E., 1999: Associations between diet and cancer, ischemic heart disease, and all-cause mortality in non-Hispanic white California Seventh-day Adventists. *Amer. J. Clin. Nutr.,* **70,** 532–538.

Haddad, E. H., and J. S. Tanzman, 2003: What do vegetarians in the United States eat? *Amer. J. Clin. Nutr.,* **78,** 626S–632S.

Heller, M. C., and G. A. Keoleian, 2000: Life cycle–based sustainability indicators for assessment of the U.S. food system. Rep. CSS0004, Center for Sustainable Systems, School of Natural Resources and Environment, University of Michigan, Ann Arbor, MI, 59 pp. [Available online at http://www.public.iastate.edu/~brummer/papers/FoodSystemSustainability.pdf]

Horrigan, L., R. S. Lawrence, and P. Walker, 2002: How sustainable agriculture can address the environmental and human health harms of industrial agriculture. *Environ. Health Persp.,* **110,** 445–456.

Houghton, J. T., Y. Ding, D. J. Griggs, M. Noguer, P. J. van der Linden, X. Dai, K. Maskell, and C. A. Johnson, Eds., 2001: *Climate Change 2001: The Scientific Basis: Contribution of Working Group I to the Third Assessment Report of the Intergovernmental Panel on Climate Change.* Cambridge University Press, 881 pp.

Hu, J.-F., Z. Cheng, F. V. Chisari, T. H. Vu, A. R. Hoffman, and T. C. Campbell, 1997: Repression of hepatitis B virus (HBV) transgene and HBV-induced liver injury by low protein diet. *Oncogene Rev.,* **15,** 2795–2801.

Key, T. J., and Coauthors, 1999: Mortality in vegetarians and nonvegetarians: Detailed findings from a collaborative analysis of 5 prospective studies. *Amer. J. Clin. Nutr.,* **70,** 516S–524S.

Larsson, S. C., L. Bergkvist, and A. Wolk, 2004: Milk and lactose intakes and ovarian cancer risk in the Swedish Mammography Cohort (1-3). *Amer. J. Clin. Nutr.,* **80,** 1353–1357.

Mgbodile, M. U. K., and T. C. Campbell, 1972: Effects of protein deprivation of male weanling rats on the kinetics of hepatic microsomal enzyme activity. *J. Nutr.,* **102,** 53–60.

Pimentel, D., and M. Pimentel, 1996a: Energy use in livestock production. *Food, Energy and Society,* D. Pimentel and M. Pimentel, Eds., University Press of Colorado, 77–84.

———, and ———, 1996b: Energy use in grain and legume production. *Food, Energy and Society,* D.

Pimentel and M. Pimentel, Eds., University Press of Colorado, 107–130.

———, and ———, 1996c: Energy use in fruit, vegetable and forage production. *Food, Energy and Society,* D. Pimentel and M. Pimentel, Eds., University Press of Colorado, 131–147.

———, and ———, 2003: Sustainability of meat-based and plant-based diets and the environment. *Amer. J. Clin. Nutr.*, **78,** 660S–663S.

Preston, R. S., J. R. Hayes, and T. C. Campbell, 1976: The effect of protein deficiency on the in vivo binding of aflatoxin B_1 to rat liver macromolecules. *Life Sci.,* **19,** 191–198.

Reijnders, L., and S. Soret, 2003: Quantification of the environmental impact of different dietary protein choices. *Amer. J. Clin. Nutr.,* **78** (Suppl.), 664S–668S.

Sabate, J., 2003: The contribution of vegetarian diets to health and disease: A paradigm shift? *Amer. J. Clin. Nutr.,* **78,** 502S–507S.

Schulsinger, D. A., M. M. Root, and T. C. Campbell, 1989: Effect of dietary protein quality on development of aflatoxin B1-induced hepatic preneoplastic lesions. *J. Nat. Cancer Inst.,* **81,** 1241–1245.

Sieri, S., V. Krogh, P. Muti, A. Micheli, V. Pala, P. Crosignani, and F. Berrino, 2002: Fat and protein intake and subsequent breast cancer risk in postmenopausal women. *Nutr. Cancer,* **42** (1), 10–17.

United Nations, 2005: World population prospects: The 2004 revision. United Nations Secretariat, Department of Economic and Social Affairs, Population Division, 91 pp.

Unruh, B., cited 2002: Delivered energy consumption projections by industry in the *Annual Energy Outlook 2002.* U.S. Department of Energy, Energy Information Administration. [Available online at http://www.eia.doe.gov/oiaf/analysispa]

U.S. Department of Agriculture, cited 2002: Agricultural Fact Book 2001–2002. [Available online at http://www.usda.gov/factbook/chapter2.htm]

———, cited 2005: USDA National Nutrient Database for Standard Reference. Release 18, Agricultural Research Service. [Available online at http://www.ars.usda.gov/main/site_main.htm?modecode=12354500]

U.S. Department of Energy, 2004a: Annual energy review 2003. Rep. DOE/EIA-0384(2003), Energy Information Administration, 390 pp.

———, 2004b: Emissions of greenhouse gases in the United States 2003. Energy Information Administration Rep. DOE/EIA-0573(2003), 108 pp. [Avaliable online at http://tonto.eia.doe.gov/FTPROOT/environment/057303.pdf]

U.S. Department of Transportation, 2004: Highway statistics 2003. Office of Highway Policy Information, Federal Highway Administration, 47 pp. [Available online at http://www.fhwa.dot.gov/policy/ohim/sh03/htm/ps1.htm]

U.S. Environmental Protection Agency, cited 2005: Inventory of U.S. greenhouse gas emissions and sinks: 1990–2001. EPA 430-R-03-004. [Available online at http://yosemite.epa.gov/oar/globalwarming.nsf/content/ResourceCenterPublications GHGEmissionsUSEmissionsInventory2003.html]

Willett, W. C., 2001: *Eat, Drink and Be Healthy.* Simon and Schuster, 299 pp.

Youngman, L. D., 1990: The growth and development of aflatoxin B1-induced preneoplastic lesions, tumors, metastasis, and spontaneous tumors as they are influenced by dietary protein level, type, and intervention. Ph.D. thesis, Cornell University, 203 pp.

———, and T. C. Campbell, 1992: The sustained development of preneoplastic lesions depends on high protein intake. *Nutr. Cancer,* **18,** 131–142.

Corresponding author address: **Gidon Eshel,** Dept. of the Geophysical Sciences, University of Chicago, 5734 S. Ellis Ave., Chicago, IL 60637. E-mail address: geshel@uchicago.edu

Acknowledgments—We thank two anonymous reviewers for their thoughtful, pertinent comments, and Editor Foley for handling the manuscript and suggesting numerous useful improvements.

Article 25

Landfill-on-Sea

Old plastic rubbish doesn't die; it just gets tossed away in far-off places that we rarely get to see. Daisy Dumas assesses its impact on the world's largest floating landfill, the Great Pacific Garbage Patch.

DAISY DUMAS

A challenge. Try, if you can, to spend at least five minutes without the company of plastic sometime today. I'm warning you, it won't be easy. We sit on it, wash in it, eat from it, drink from it, look through it, play with it and pay with it. It is more than likely that there is some residing inside you. Plastics are literally everywhere.

What was once seen as the durable, lightweight, cheap and easily manufactured answer to our needs and desires has now become an unwelcome ubiquity. We are only just beginning to understand the extent of damage caused by the uncontrolled, unparalleled and unexamined overproduction of plastics.

In the quest to produce a material that transports and stores effectively, we have unwittingly created a range of products made from a substance that is totally at odds with the environment. And having conquered the land, plastics are now taking over the planet's greatest oceans.

The Doldrums

The Central Pacific Gyre is the largest uniform ocean realm on the planet, stretching over a vast 10 million square miles. Subtropical highs cause the slow, clockwork rotation of the ocean, where a devastatingly calm core gently wanders with the currents. Once synonymous with a sailor's nemesis, the area has taken on a rather more sinister role as a site for the world's plastic trash. Trapped in these calm seas, a toxic dump of floating seaborne plastic waste swirls and grows, constantly accumulating substance.

At twice the size of France, this phenomenon was dubbed the Great Pacific Garbage Patch (GPGP) by leading flotsam expert Curtis Ebbesmeyer, and is perhaps the single largest body of pollution in the world; an aggregation of year upon year of discarded plastic entering the Pacific Ocean. In this place plastic waste can rotate and linger for over 16 years, its origin a multitude of shorelines, neighbouring waters and ocean vessels. The doldrums have always been an area where flotsam collected. Until the recent past, biodegradation has taken care of integrating much of this largely natural waste into the marine ecosystem. Nowadays, however, 90 percent of all marine debris is anything but natural. It is, instead, plastic. Defying even the most rapacious and stubborn bacteria, plastics slowly photo-degrade to a molecular level, at which point further degradation can only be achieved by burning.

Between 70 and 80 percent of the debris collecting in the Garbage Patch is post-consumer waste from the land, mostly swept into the marine ecosystem by storms and wind. Much of the remaining plastic is an unintended consequence of the mass-fishing industry, as vast trawling nets, broken buoys and mile upon mile of plastic cord and twine intermingle with plastic bottles, toys, trainers and cigarette lighters. A smaller but nonetheless significant fraction of the debris is pre-consumer, often in the form of nurdles—pre-manufacture pellets.

Given the nebulous nature of the GPGP, its rate of growth is hard to determine. I think it is growing faster than we can predict. At the moment it is enlarging at an exponential rate, increasing by a factor of 10 each year, says Captain Charles Moore, Founder of the Algalita Marine Research Foundation in California, who in 2006 found that in some areas of the GPGP, the ratio of plastic to plankton measured six to one. It is likely to be 100 times worse in six years' time and similar to rates found off the coast of Japan, where much of the waste originates.

Hideshige Takada, an environmental geochemist at Tokyo University, who is studying the problem off Japan's coastline, has measured a three-fold increase in plastic particulate

Plastic Facts

Almost every aspect of our lives is touched by plastics, so much so that:

- In 1979, the manufacture of plastic overtook that of steel.
- Today we use 20 times more plastic than we did 50 years ago.
- Each year, 100 million tonnes of plastic are used worldwide.
- We each dispose of 185lb of plastic every year.

pollution between 1989 and 1999, and tenfold increases in the past two to three years.

Today, particulate pollution in the GPGP is at least as high as 100,000 pieces per square mile.

Facts in the Water

The remote Midway Atoll lies at the north-eastern tip of the Hawaiian archipelago. Far from man, far from manufacturing plants and far from the prodigious demands of modern culture, Midway should, by definition, exemplify a storybook desert island.

It is anything but. Surrounded by the GPGP, Midway could be mistaken for a landfill site. Its beaches are littered with the harsh reality of extreme pollution, as carcasses jostle with coke bottles and clumps of fishing nets lie discarded like seaweed. An important albatross rookery, 40 percent of fledglings hatched on Midway never leave the island, instead dying from starvation.

Captain Moore's gruesome photo library bears macabre testimony to the first-hand effects of seaborne plastic. Decomposed albatross bodies, their bloated stomachs exposing horrific last meals of lids, nurdles and cigarette lighters, compete for space beside unrecognisable turtles, their shells disgustingly disfigured from a life with six-pack beer holders lodged tight around their middles.

Whether it be an algae-sifting whale or a fish-eating seal, small pieces of plastic are mistaken for food at all levels of the chain. Algalita researchers have seen styrofoam cups with bites taken out of them because they have the same texture as food. Indeed, recent media coverage of washed-up rubber ducks from a massive dump in the Pacific over a decade ago show telltale bite marks to their necks and abdomens. Nurdles of all colours and sizes fool jellyfish, birds and fish into ingesting them, blocking digestive and respiratory tracts and competing with scant nutrients for a place in their stomachs. Microplastics have even come to be known as plastic plankton—a befitting but twisted name to billions of indiscriminate filter feeders.

The figures speak for themselves—Greenpeace estimates that one million birds and 100,000 marine mammals die in the Garbage Patch each year. Individual species are quite literally on the brink of extinction, the onset of which can be attributed solely to plastic interference. We have counted more than 100,000 Laysan Albatross deaths in a single year and it won't be long until species become extinct—there is a whole list of endangered species and it is getting longer, says Moore. The species Captain Moore worries about most is the Hawaiian Monk Seal, which he says faces certain extinction if things don't change. It is not for lack of effort. But without removing plastic from oceans, or halting their entry into the marine environment in the first place, rescuers are fighting a losing battle. It is tragic. It is so sad to see hard working animal rescue centres treat animals and release them, only to find them washed up in nets a few months later, says Moore.

Toxic Sponges

Quite apart from physical implications, the biological impact is enormous. Not only can larger plastic objects entrap, entangle and entwine pelagic wildlife, they also act as floating islands and play a role in the colonisation of potentially poisonous new habitats. Man-made toxins freely migrate both in and out of plastics, and small plastic particles with high surface areas have the ability to absorb and transport a million times the concentration of hydrophobic toxic chemicals (such as DDT and PCBs) than that of ambient water.

Perhaps most disturbingly, plastics have the capacity to leach out the chemical compounds associated with their production. So much so that the US Food and Drug Administration used to term plastics indirect food additives. Plastics expert Paul Goettlich of mindfully.org is a harsh critic of the current regulatory structures (or lack thereof) for dealing with the production of plastics and their chemical components. Despite what we are led to believe, he explains, the [plastic production] process is never 100 percent perfect. Logically then, there are always toxicants available for migration into the many things they contact—whether these points of contact be seawater, fish, birds or mammals.

Ironically, where man has failed to clear these finegrained toxic sponges from the oceans, nature has erroneously stepped in. As Moore puts it an astronomical number of vectors for some of the most toxic pollutants known are being released into an ecosystem dominated by the most efficient natural vacuum cleaners nature ever invented—the jellies and salps living in the ocean. After those organisms ingest the toxins, they are eaten in turn by fish, and so the poisons pass into the food web that leads, in some cases, to human beings.

The most common group of such chemicals are proven endocrine disrupters. These substances interfere with the function of natural hormones, the most dangerous manifestations of which are reproductive disorders and cancer. On land, studies show that reproductive problems in sentinel species such as amphibians and birds—species that reflect the health of their ecosystem—are giving us all the warning signs we need, whilst the toxic effects of PCBs in humans is well-documented, going back to work-related exposure in the 1930s.

The plastic goods market is expanding at a far faster rate than the infrastructure to deal with waste plastic. Perhaps unsurprisingly, it may be that the lack of action can be traced to the relative economic dead-end posed by the problem. As Captain Moore puts it: There is no economic resource that would directly benefit from this process. We haven't yet learned how to factor the health of the environment into our economic paradigm. We need to get to work on this calculus quickly, because a stock market crash will pale in comparison to an ecological crash on an oceanic scale.

Short of filtering every drop of the planet's water, there is little we can do to turn the tide on the GPGP. Workable solutions must lie in reducing our need and desire for plastic and its subsequent entry into the environment, but plastic consumption in Western Europe alone is currently increasing by four percent each year.

Not in My Back Ocean

Though the Central Pacific may seem a million miles away, the GPGP is likely to exemplify the future of many marine areas. According to Richard Thompson of the University of Plymouth, while scales and densities differ, plastic pollution in Europe has increased sharply over the past 40 years. Locally, we find that patches of debris vary over time and depend on wind and tidal conditions. Concentrations of debris are found, but at smaller scales of resolution than the Garbage Patch. It is entirely possible that an accumulation similar to that could occur.

So, where can we expect to see the next Garbage Patch forming? It is not so much about specific debris sinkholes, Thompson warns. It is the fact that debris can collect in any number of hotspots around the world. Given that 40 percent of the world's oceans are subtropical gyres, not to mention the many smaller ebbs and flows of sea currents, potential hotspots are worryingly abundant.

Dive below the surface of the problem, and it becomes clear that there is yet another dimension to consider. A comprehensive study in Europe by Galgani et al, in 2000, recorded plastic debris during 27 oceanographic cruises and using submersibles down to 2,700m. The truth lurking in the depths was that some areas were contaminated with more than 100,000 items per square kilometre.

There is no prospect of plastic particulate pollution going away quickly. Rather, two trends are likely to increase. Firstly, fine-grained, smaller plastic particles will proliferate through photodegradation. Although the potential environmental impact of smaller debris and plastic plankton is relatively unknown, Algalita recently won a research grant, allowing the team to begin work on the effects of microplastics on zooplankton. Secondly, seabed, deep sea plastics will accumulate as larger objects are fouled and worn, altering their density and sinking. The UN estimates that 70 percent of all seaborne plastic will eventually sink, sequestered to the depths of oceans where a toxic graveyard will fester.

As Bill MacDonald of Algalita says, "People don't understand that what they do can affect the environment thousands of miles away. Perhaps they won't need to. The grim reality is that a plastic garbage patch may soon be coming to the waters near you."

So, is biodegradable plastic the answer? In short, no. While bio plastics have an application in modern life (especially in farming), they are limited in their effect. They require high temperatures, a very specific pH and high levels of light to decompose, but such conditions rarely occur in natural environments, let alone sea, where there are lower temperatures and levels of sunlight. In an ocean environment, as in a landfill, biodegradable plastic will remain intact, causing damage to wildlife and ecosystems for many years.

Daisy Dumas is a freelance journalist.

UNIT 6

The Politics of Climate Change

Unit Selections

26. **The Truth about Denial,** Sharon Begley
27. **Swift Boating, Stealth Budgeting, and Unitary Executives,** James Hansen
28. **The Myth of the 1970s Global Cooling Scientific Consensus,** Thomas C. Peterson, William M. Connolly, and John Fleck
29. **How to Stop Climate Change: The Easy Way,** Mark Lynas
30. **Environmental Justice for All,** Leyla Kokmen

Key Points to Consider

- What efforts are being made by fossil fuel-based energy companies to impede climate change regulation? How are politicians responding to the climate change debate?
- Rather than arguing, what can be done? Are global climate accords feasible? Can the United States be convinced to join in international legislation? What government legislation could be passed (rather than debated) that would reduce carbon emissions?
- Why is there so much confusion about climate change? Given that the scientific community is in overwhelming agreement on the long-term consequences of elevated greenhouse emissions, why is the public so unsure? How do politicians affect the debate?
- In what ways does environmental degradation affect communities of different socioeconomic levels differently? Why are lower-income regions so much more vulnerable?

Student Website

www.mhhe.com/cls

Internet References

Global Climate Change Facts: The Truth, The Consensus, and the Skeptics
http://www.climatechangefacts.info

The Climate Project
http://www.theclimateprojectus.org

NASA Global Climate Change site
http://climate.nasa.gov

Real Climate
http://www.realclimate.org

How we deal with our environmental issues ultimately comes down to politics. With a population just under seven billion people, the Earth is under great stress. A *laissez-faire* attitude toward the environment simply will not work. In spite of the good intentions of most people around the world, society must impose regulations in order to minimize environmental degradation. Once people understand the consequences of inaction, they are generally willing to make small or modest sacrifices to preserve their environment. And yet, much is to be gained by extracting wealth from the Earth and ignoring the consequences. There are many trillions of dollars to be had by extracting fossil fuels from the Earth. Can we really expect our society to simply leave all that "wealth" untouched forever? Why, when there is overwhelming consensus among climate scientists, do less than half of Americans believe that climate change is real? One major reason is that those who benefit from exploitation of our natural resources are working very hard to convince us that such activities are benign.

Sharon Begley's "The Truth about Denial" uncovers the efforts by the fossil fuel industries to lobby Congress and sow doubt as to the danger of global warming. In the next article, James Hansen, one of the early developers of the global warming hypothesis, comes out and states how the Bush administration actively attempted to keep the U.S. public in the dark on global warming. Hansen's reports would be altered so that the strength of his argument was lessened and factual statements were recast with a sense of uncertainty. Hansen also explains that the "climate contrarians" have an agenda other than scientific truth. They are interested in confusing the public and will say anything to move their agenda forward. That point is made clear in "The Myth of the 1970s Global Cooling Scientific Consensus." The authors debunk the common claim by "deniers" that climate scientists who now say that the Earth is heating were saying exactly the opposite some 30 years ago. The facts, clearly laid out in the article, point out that this statement is flat-out untrue.

So, are we getting anywhere, or will the back-and-forth go on as our planet degrades? In "How to Stop Climate Change: The Easy Way," Mark Lynas says it's time to stop debating and start acting. The final article in this compilation, "Environmental Justice for All," looks at another problem that requires government intervention. Where does it make most sense for a company to build a power plant that pollutes? In a upscale neighborhood, or the poorest in the city? Obviously, the latter, where the community does not have the financial resources to fight the large polluting companies. This concept is valid at the local level and at the global one, where rich countries have in the past and continue to pay poorer countries to take their toxic waste. Leyla Kokmen describes "toxic tours" into low-income, minority neighborhoods in order to show people firsthand the environmental hazards that so many poor urban people face. By opening our eyes to the living conditions of our fellow citizens, we become educated about the problems we face and are more likely to demand that something be done about it. This is a powerful and effective way of educating our citizenry. Ultimately, education is what will force our government to obey the wishes of its constituents, to act to curb environmental degradation. Let us all do our part and share our knowledge and passion with others so that we might bring about a cleaner world as soon as possible.

Oak Ridge National Laboratory, managed by U.S. Dept. of Energy

The Truth about Denial

SHARON BEGLEY

Sen. Barbara Boxer had been chair of the Senate's Environment Committee for less than a month when the verdict landed last February. "Warming of the climate system is unequivocal," concluded a report by 600 scientists from governments, academia, green groups and businesses in 40 countries. Worse, there was now at least a 90 percent likelihood that the release of greenhouse gases from the burning of fossil fuels is causing longer droughts, more flood-causing downpours and worse heat waves, way up from earlier studies. Those who doubt the reality of human-caused climate change have spent decades disputing that. But Boxer figured that with "the overwhelming science out there, the deniers' days were numbered." As she left a meeting with the head of the international climate panel, however, a staffer had some news for her. A conservative think tank long funded by ExxonMobil, she told Boxer, had offered scientists $10,000 to write articles undercutting the new report and the computer-based climate models it is based on. "I realized," says Boxer, "there was a movement behind this that just wasn't giving up."

If you think those who have long challenged the mainstream scientific findings about global warming recognize that the game is over, think again. Yes, 19 million people watched the "Live Earth" concerts last month, titans of corporate America are calling for laws mandating greenhouse cuts, "green" magazines fill newsstands, and the film based on Al Gore's best-selling book, "An Inconvenient Truth," won an Oscar. But outside Hollywood, Manhattan and other habitats of the chattering classes, the denial machine is running at full throttle—and continuing to shape both government policy and public opinion.

Since the late 1980s, this well-coordinated, well-funded campaign by contrarian scientists, free-market think tanks and industry has created a paralyzing fog of doubt around climate change. Through advertisements, op-eds, lobbying and media attention, greenhouse doubters (they hate being called deniers) argued first that the world is not warming; measurements indicating otherwise are flawed, they said. Then they claimed that any warming is natural, not caused by human activities. Now they contend that the looming warming will be minuscule and harmless. "They patterned what they did after the tobacco industry," says former senator Tim Wirth, who spearheaded environmental issues as an under secretary of State in the Clinton administration. "Both figured, sow enough doubt, call the science uncertain and in dispute. That's had a huge impact on both the public and Congress."

Just last year, polls found that 64 percent of Americans thought there was "a lot" of scientific disagreement on climate change; only one third thought planetary warming was "mainly caused by things people do." In contrast, majorities in Europe and Japan recognize a broad consensus among climate experts that greenhouse gases—mostly from the burning of coal, oil and natural gas to power the world's economies—are altering climate. A new NEWSWEEK Poll finds that the influence of the denial machine remains strong. Although the figure is less than in earlier polls, 39 percent of those asked say there is "a lot of disagreement among climate scientists" on the basic question of whether the planet is warming; 42 percent say there is a lot of disagreement that human activities are a major cause of global warming. Only 46 percent say the greenhouse effect is being felt today.

As a result of the undermining of the science, all the recent talk about addressing climate change has produced little in the way of actual action. Yes, last September Gov. Arnold Schwarzenegger signed a landmark law committing California to reduce statewide emissions of carbon dioxide to 1990 levels by 2020 and 80 percent more by 2050. And this year both Minnesota and New Jersey passed laws requiring their states to reduce greenhouse emissions 80 percent below recent levels by 2050. In January, nine leading corporations—including Alcoa, Caterpillar, Duke Energy, Du Pont and General Electric—called on Congress to "enact strong national legislation" to reduce greenhouse gases. But although at least eight bills to require reductions in greenhouse gases have been introduced in Congress, their fate is decidedly murky. The Democratic leadership in the House of Representatives decided last week not even to bring to a vote a requirement that automakers improve vehicle mileage, an obvious step toward reducing greenhouse emissions. Nor has there been much public pressure to do so. Instead, every time the scientific case got stronger, "the American public yawned and bought bigger cars," Rep. Rush Holt, a New Jersey congressman and physicist, recently wrote in the journal Science; politicians "shrugged, said there is too much doubt among scientists, and did nothing."

It was 98 degrees in Washington on Thursday, June 23, 1988, and climate change was bursting into public consciousness. The Amazon was burning, wildfires raged in the United States, crops in the Midwest were scorched and it was shaping up to be the hottest year on record worldwide. A Senate committee, including Gore, had invited NASA climatologist James

Hansen to testify about the greenhouse effect, and the members were not above a little stagecraft. The night before, staffers had opened windows in the hearing room. When Hansen began his testimony, the air conditioning was struggling, and sweat dotted his brow. It was the perfect image for the revelation to come. He was 99 percent sure, Hansen told the panel, that "the greenhouse effect has been detected, and it is changing our climate now."

The reaction from industries most responsible for greenhouse emissions was immediate. "As soon as the scientific community began to come together on the science of climate change, the pushback began," says historian Naomi Oreskes of the University of California, San Diego. Individual companies and industry associations—representing petroleum, steel, autos and utilities, for instance—formed lobbying groups with names like the Global Climate Coalition and the Information Council on the Environment. ICE's game plan called for enlisting greenhouse doubters to "reposition global warming as theory rather than fact," and to sow doubt about climate research just as cigarette makers had about smoking research. ICE ads asked, "If the earth is getting warmer, why is Minneapolis [or Kentucky, or some other site] getting colder?" This sounded what would become a recurring theme for naysayers: that global temperature data are flat-out wrong. For one thing, they argued, the data reflect urbanization (many temperature stations are in or near cities), not true global warming.

Shaping public opinion was only one goal of the industry groups, for soon after Hansen's sweat-drenched testimony they faced a more tangible threat: international proposals to address global warming. The United Nations had scheduled an "Earth Summit" for 1992 in Rio de Janeiro, and climate change was high on an agenda that included saving endangered species and rain forests. ICE and the Global Climate Coalition lobbied hard against a global treaty to curb greenhouse gases, and were joined by a central cog in the denial machine: the George C. Marshall Institute, a conservative think tank. Barely two months before Rio, it released a study concluding that models of the greenhouse effect had "substantially exaggerated its importance." The small amount of global warming that might be occurring, it argued, actually reflected a simple fact: the Sun is putting out more energy. The idea of a "variable Sun" has remained a constant in the naysayers' arsenal to this day, even though the tiny increase in solar output over recent decades falls far short of explaining the extent or details of the observed warming.

In what would become a key tactic of the denial machine—think tanks linking up with like-minded, contrarian researchers—the report was endorsed in a letter to President George H.W. Bush by MIT meteorologist Richard Lindzen. Lindzen, whose parents had fled Hitler's Germany, is described by old friends as the kind of man who, if you're in the minority, opts to be with you. "I thought it was important to make it clear that the science was at an early and primitive stage and that there was little basis for consensus and much reason for skepticism," he told Scientific American magazine. "I did feel a moral obligation."

Bush was torn. The head of his Environmental Protection Agency, William Reilly, supported binding cuts in greenhouse emissions. Political advisers insisted on nothing more than voluntary cuts. Bush's chief of staff, John Sununu, had a PhD in engineering from MIT and "knew computers," recalls Reilly. Sununu frequently logged on to a computer model of climate, Reilly says, and "vigorously critiqued" its assumptions and projections.

Sununu's side won. The Rio treaty called for countries to voluntarily stabilize their greenhouse emissions by returning them to 1990 levels by 2000. (As it turned out, U.S. emissions in 2000 were 14 percent higher than in 1990.) Avoiding mandatory cuts was a huge victory for industry. But Rio was also a setback for climate contrarians, says UCSD's Oreskes: "It was one thing when Al Gore said there's global warming, but quite another when George Bush signed a convention saying so." And the doubters faced a newly powerful nemesis. Just months after he signed the Rio pact, Bush lost to Bill Clinton—whose vice president, Gore, had made climate change his signature issue.

Groups that opposed greenhouse curbs ramped up. They "settled on the 'science isn't there' argument because they didn't believe they'd be able to convince the public to do nothing if climate change were real," says David Goldston, who served as Republican chief of staff for the House of Representatives science committee until 2006. Industry found a friend in Patrick Michaels, a climatologist at the University of Virginia who keeps a small farm where he raises prize-winning pumpkins and whose favorite weather, he once told a reporter, is "anything severe." Michaels had written several popular articles on climate change, including an op-ed in The Washington Post in 1989 warning of "apocalyptic environmentalism," which he called "the most popular new religion to come along since Marxism." The coal industry's Western Fuels Association paid Michaels to produce a newsletter called World Climate Report, which has regularly trashed mainstream climate science. (At a 1995 hearing in Minnesota on coal-fired power plants, Michaels admitted that he received more than $165,000 from industry; he now declines to comment on his industry funding, asking, "What is this, a hatchet job?")

The road from Rio led to an international meeting in Kyoto, Japan, where more than 100 nations would negotiate a treaty on making Rio's voluntary—and largely ignored—greenhouse curbs mandatory. The coal and oil industries, worried that Kyoto could lead to binding greenhouse cuts that would imperil their profits, ramped up their message that there was too much scientific uncertainty to justify any such cuts. There was just one little problem. The Intergovernmental Panel on Climate Change, or IPCC—the international body that periodically assesses climate research—had just issued its second report, and the conclusion of its 2,500 scientists looked devastating for greenhouse doubters. Although both natural swings and changes in the Sun's output might be contributing to climate change, it concluded, "the balance of evidence suggests a discernible human influence on climate."

Faced with this emerging consensus, the denial machine hardly blinked. There is too much "scientific uncertainty" to justify curbs on greenhouse emissions, William O'Keefe, then a vice president of the American Petroleum Institute and leader of the Global Climate Coalition, suggested in 1996. Virginia's Michaels echoed that idea in a 1997 op-ed in The

Washington Post, describing "a growing contingent of scientists who are increasingly unhappy with the glib forecasts of gloom and doom." To reinforce the appearance of uncertainty and disagreement, the denial machine churned out white papers and "studies" (not empirical research, but critiques of others' work). The Marshall Institute, for instance, issued reports by a Harvard University astrophysicist it supported pointing to satellite data showing "no significant warming" of the atmosphere, contrary to the surface warming. The predicted warming, she wrote, "simply isn't happening according to the satellite[s]." At the time, there was a legitimate case that satellites were more accurate than ground stations, which might be skewed by the unusual warmth of cities where many are sited.

"There was an extraordinary campaign by the denial machine to find and hire scientists to sow dissent and make it appear that the research community was deeply divided," says Dan Becker of the Sierra Club. Those recruits blitzed the media. Driven by notions of fairness and objectivity, the press "qualified every mention of human influence on climate change with 'some scientists believe,' where the reality is that the vast preponderance of scientific opinion accepts that human-caused [greenhouse] emissions are contributing to warming," says Reilly, the former EPA chief. "The pursuit of balance has not done justice" to the science. Talk radio goes further, with Rush Limbaugh telling listeners this year that "more carbon dioxide in the atmosphere is not likely to significantly contribute to the greenhouse effect. It's just all part of the hoax." In the new *NEWSWEEK* Poll, 42 percent said the press "exaggerates the threat of climate change."

Now naysayers tried a new tactic: lists and petitions meant to portray science as hopelessly divided. Just before Kyoto, S. Fred Singer released the "Leipzig Declaration on Global Climate Change." Singer, who fled Nazi-occupied Austria as a boy, had run the U.S. weather-satellite program in the early 1960s. In the Leipzig petition, just over 100 scientists and others, including TV weathermen, said they "cannot subscribe to the politically inspired world view that envisages climate catastrophes." Unfortunately, few of the Leipzig signers actually did climate research; they just kibitzed about other people's. Scientific truth is not decided by majority vote, of course (ask Galileo), but the number of researchers whose empirical studies find that the world is warming and that human activity is partly responsible numbered in the thousands even then. The IPCC report issued this year, for instance, was written by more than 800 climate researchers and vetted by 2,500 scientists from 130 nations.

Although Clinton did not even try to get the Senate to ratify the Kyoto treaty (he knew a hopeless cause when he saw one), industry was taking no chances. In April 1998 a dozen people from the denial machine—including the Marshall Institute, Fred Singer's group and Exxon—met at the American Petroleum Institute's Washington headquarters. They proposed a $5 million campaign, according to a leaked eight-page memo, to convince the public that the science of global warming is riddled with controversy and uncertainty. The plan was to train up to 20 "respected climate scientists" on media—and public—outreach with the aim of "raising questions about and undercutting the 'prevailing scientific wisdom' " and, in particular, "the Kyoto treaty's scientific underpinnings" so that elected officials "will seek to prevent progress toward implementation." The plan, once exposed in the press, "was never implemented as policy," says Marshall's William O'Keefe, who was then at API.

The GOP control of Congress for six of Clinton's eight years in office meant the denial machine had a receptive audience. Although Republicans such as Sens. John McCain, Jim Jeffords and Lincoln Chafee spurned the denial camp, and Democrats such as Congressman John Dingell adamantly oppose greenhouse curbs that might hurt the auto and other industries, for the most part climate change has been a bitterly partisan issue. Republicans have also received significantly more campaign cash from the energy and other industries that dispute climate science. Every proposed climate bill "ran into a buzz saw of denialism," says Manik Roy of the Pew Center on Climate Change, a research and advocacy group, who was a Senate staffer at the time. "There was no rational debate in Congress on climate change."

The reason for the inaction was clear. "The questioning of the science made it to the Hill through senators who parroted reports funded by the American Petroleum Institute and other advocacy groups whose entire purpose was to confuse people on the science of global warming," says Sen. John Kerry. "There would be ads challenging the science right around the time we were trying to pass legislation. It was pure, raw pressure combined with false facts." Nor were states stepping where Washington feared to tread. "I did a lot of testifying before state legislatures—in Pennsylvania, Rhode Island, Alaska—that thought about taking action," says Singer. "I said that the observed warming was and would be much, much less than climate models calculated, and therefore nothing to worry about."

But the science was shifting under the denial machine. In January 2000, the National Academy of Sciences skewered its strongest argument. Contrary to the claim that satellites finding no warming are right and ground stations showing warming are wrong, it turns out that the satellites are off. (Basically, engineers failed to properly correct for changes in their orbit.) The planet is indeed warming, and at a rate since 1980 much greater than in the past.

Just months after the Academy report, Singer told a Senate panel that "the Earth's atmosphere is not warming and fears about human-induced storms, sea-level rise and other disasters are misplaced." And as studies fingering humans as a cause of climate change piled up, he had a new argument: a cabal was silencing good scientists who disagreed with the "alarmist" reports. "Global warming has become an article of faith for many, with its own theology and orthodoxy," Singer wrote in The Washington Times. "Its believers are quite fearful of any scientific dissent."

With the Inauguration of George W. Bush in 2001, the denial machine expected to have friends in the White House. But despite Bush's oil-patch roots, naysayers weren't sure they could count on him: as a candidate, he had pledged to cap carbon dioxide emissions. Just weeks into his term, the Competitive Enterprise Institute heard rumors that the draft of a speech Bush was preparing included a passage reiterating that pledge. CEI's Myron Ebell called conservative pundit Robert Novak, who had booked Bush's EPA chief, Christie Todd Whitman, on CNN's "Crossfire." He asked her about the line, and within hours the possibility of a carbon cap was the talk of the Beltway. "We alerted anyone we thought could have influence and get the line, if it was in the speech, out," says CEI president Fred Smith, who counts this as another notch in CEI's belt. The White House declines to comment.

Bush not only disavowed his campaign pledge. In March, he withdrew from the Kyoto treaty. After the about-face, MIT's Lindzen told NEWSWEEK in 2001, he was summoned to the White House. He told Bush he'd done the right thing. Even if you accept the doomsday forecasts, Lindzen said, Kyoto would hardly touch the rise in temperatures. The treaty, he said, would "do nothing, at great expense."

Bush's reversal came just weeks after the IPCC released its third assessment of the burgeoning studies of climate change. Its conclusion: the 1990s were very likely the warmest decade on record, and recent climate change is partly "attributable to human activities." The weather itself seemed to be conspiring against the skeptics. The early years of the new millennium were setting heat records. The summer of 2003 was especially brutal, with a heat wave in Europe killing tens of thousands of people. Consultant Frank Luntz, who had been instrumental in the GOP takeover of Congress in 1994, suggested a solution to the PR mess. In a memo to his GOP clients, he advised them that to deal with global warming, "you need to continue to make the lack of scientific certainty a primary issue." They should "challenge the science," he wrote, by "recruiting experts who are sympathetic to your view." Although few of the experts did empirical research of their own (MIT's Lindzen was an exception), the public didn't notice. To most civilians, a scientist is a scientist.

Challenging the science wasn't a hard sell on Capitol Hill. "In the House, the leadership generally viewed it as impermissible to go along with anything that would even imply that climate change was genuine," says Goldston, the former Republican staffer. "There was a belief on the part of many members that the science was fraudulent, even a Democratic fantasy. A lot of the information they got was from conservative think tanks and industry." When in 2003 the Senate called for a national strategy to cut greenhouse gases, for instance, climate naysayers were "giving briefings and talking to staff," says Goldston. "There was a constant flow of information—largely misinformation." Since the House version of that bill included no climate provisions, the two had to be reconciled. "The House leadership staff basically said, 'You know we're not going to accept this,' and [Senate staffers] said, 'Yeah, we know,' and the whole thing disappeared relatively jovially without much notice," says Goldston. "It was such a foregone conclusion."

Especially when the denial machine had a new friend in a powerful place. In 2003 James Inhofe of Oklahoma took over as chairman of the environment committee. That summer he took to the Senate floor and, in a two-hour speech, disputed the claim of scientific consensus on climate change. Despite the discovery that satellite data showing no warming were wrong, he argued that "satellites, widely considered the most accurate measure of global temperatures, have confirmed" the absence of atmospheric warming. Might global warming, he asked, be "the greatest hoax ever perpetrated on the American people?" Inhofe made his mark holding hearing after hearing to suggest that the answer is yes. For one, on a study finding a dramatic increase in global temperatures unprecedented in the last 1,000 years, he invited a scientist who challenged that conclusion (in a study partly underwritten with $53,000 from the American Petroleum Institute), one other doubter and the scientist who concluded that recent global temperatures were spiking. Just as Luntz had suggested, the witness table presented a tableau of scientific disagreement.

Every effort to pass climate legislation during the George W. Bush years was stopped in its tracks. When Senators McCain and Joe Lieberman were fishing for votes for their bipartisan effort in 2003, a staff member for Sen. Ted Stevens of Alaska explained to her counterpart in Lieberman's office that Stevens "is aware there is warming in Alaska, but he's not sure how much it's caused by human activity or natural cycles," recalls Tim Profeta, now director of an environmental-policy institute at Duke University. "I was hearing the basic argument of the skeptics—a brilliant strategy to go after the science. And it was working." Stevens voted against the bill, which failed 43–55. When the bill came up again the next year, "we were contacted by a lot of lobbyists from API and Exxon-Mobil," says Mark Helmke, the climate aide to GOP Sen. Richard Lugar. "They'd bring up how the science wasn't certain, how there were a lot of skeptics out there." It went down to defeat again.

Killing bills in Congress was only one prong of the denial machine's campaign. It also had to keep public opinion from demanding action on greenhouse emissions, and that meant careful management of what federal scientists and officials wrote and said. "If they presented the science honestly, it would have brought public pressure for action," says Rick Piltz, who joined the federal Climate Science Program in 1995. By appointing former coal and oil lobbyists to key jobs overseeing climate policy, he found, the administration made sure that didn't happen. Following the playbook laid out at the 1998 meeting at the American Petroleum Institute, officials made sure that every report and speech cast climate science as dodgy, uncertain, controversial—and therefore no basis for making policy. Ex-oil lobbyist Philip Cooney, working for the White House Council on Environmental Quality, edited a 2002 report on climate science by sprinkling it with phrases such as "lack of understanding" and "considerable uncertainty." A short section on climate in another report was cut entirely. The White House "directed us to remove all mentions of it," says Piltz, who resigned in protest. An oil lobbyist faxed Cooney, "You are doing a great job."

The response to the international climate panel's latest report, in February, showed that greenhouse doubters have a lot of fight left in them. In addition to offering $10,000 to scientists willing to attack the report, which so angered Boxer, they are emphasizing a new theme. Even if the world is warming now, and even if that warming is due in part to the greenhouse gases emitted by burning fossil fuels, there's nothing to worry about. As Lindzen wrote in a guest editorial in *NEWSWEEK* International in April, "There is no compelling evidence that the warming trend we've seen will amount to anything close to catastrophe."

To some extent, greenhouse denial is now running on automatic pilot. "Some members of Congress have completely internalized this," says Pew's Roy, and therefore need no coaching from the think tanks and contrarian scientists who for 20 years kept them stoked with arguments. At a hearing last month on the Kyoto treaty, GOP Congressman Dana Rohrabacher asked whether "changes in the Earth's temperature in the past—all of these glaciers moving back and forth—and the changes that we see now" might be "a natural occurrence." (Hundreds of studies have ruled that out.) "I think it's a bit grandiose for us to believe . . . that [human activities are] going to change some major climate cycle that's going on." Inhofe has told allies he will filibuster any climate bill that mandates greenhouse cuts.

Still, like a great beast that has been wounded, the denial machine is not what it once was. In the *NEWSWEEK* Poll, 38 percent of those surveyed identified climate change as the nation's gravest environmental threat, three times the number in 2000. After ExxonMobil was chastised by senators for giving $19 million over the years to the Competitive Enterprise Institute and others who are "producing very questionable data" on climate change, as Sen. Jay Rockefeller said, the company has cut back its support for such groups. In June, a spokesman said ExxonMobil did not doubt the risks posed by climate change, telling reporters, "We're very much not a denier." In yet another shock, Bush announced at the weekend that he would convene a global-warming summit next month, with a 2008 goal of cutting greenhouse emissions. That astonished the remaining naysayers. "I just can't imagine the administration would look to mandatory [emissions caps] after what we had with Kyoto," said a GOP Senate staffer, who did not want to be named criticizing the president. "I mean, what a disaster!"

With its change of heart, ExxonMobil is more likely to win a place at the negotiating table as Congress debates climate legislation. That will be crucially important to industry especially in 2009, when naysayers may no longer be able to count on a friend in the White House nixing mandatory greenhouse curbs. All the Democratic presidential contenders have called global warming a real threat, and promise to push for cuts similar to those being passed by California and other states. In the GOP field, only McCain—long a leader on the issue—supports that policy. Fred Thompson belittles findings that human activities are changing the climate, and Rudy Giuliani backs the all-volunteer greenhouse curbs of (both) Presidents Bush.

Look for the next round of debate to center on what Americans are willing to pay and do to stave off the worst of global warming. So far the answer seems to be, not much. The *NEWSWEEK* Poll finds less than half in favor of requiring high-mileage cars or energy-efficient appliances and buildings. No amount of white papers, reports and studies is likely to change that. If anything can, it will be the climate itself. This summer, Texas was hit by exactly the kind of downpours and flooding expected in a greenhouse world, and Las Vegas and other cities broiled in record triple-digit temperatures. Just last week the most accurate study to date concluded that the length of heat waves in Europe has doubled, and their frequency nearly tripled, in the past century. The frequency of Atlantic hurricanes has already doubled in the last century. Snowpack whose water is crucial to both cities and farms is diminishing. It's enough to make you wish that climate change were a hoax, rather than the reality it is.

In *"The Truth About Denial"* (Aug. 13), we said that Congressman John Dingell "adamantly oppose[s] greenhouse curbs that might hurt the auto and other industries." While Dingell has long opposed greenhouse curbs, he is now a co-sponsor of a bill aimed at improving fuel efficiency.

With Eve Conant, Sam Stein and Eleanor Clift in Washington and Matthew Philips in New York.

Swift Boating, Stealth Budgeting, and Unitary Executives

James Hansen

The American Revolution launched the radical proposition that the commonest of men should have a vote equal in weight to that of the richest, most powerful citizen. Our forefathers devised a remarkable Constitution, with checks and balances, to guard against the return of despotic governance and subversion of the democratic principle for the sake of the powerful few with special interests. They were well aware of the difficulties that would be faced, however, placing their hopes in the presumption of an educated and honestly informed citizenry.

I have sometimes wondered how our forefathers would view our situation today. On the positive side, as a scientist, I like to imagine how Benjamin Franklin would view the capabilities we have built for scientific investigation. Franklin speculated that an atmospheric "dry fog" produced by a large volcano had reduced the Sun's heating of the Earth so as to cause unusually cold weather in the early 1780s; he noted that the enfeebled solar rays, when collected in the focus of a "burning glass," could "scarce kindle brown paper." As brilliant as Franklin's insights may have been, they were only speculation as he lacked the tools for quantitative investigation. No doubt Franklin would marvel at the capabilities provided by Earth-encircling satellites and super-computers that he could scarcely have imagined.

Yet Franklin, Jefferson, and the other revolutionaries would surely be distraught by recent tendencies in America, specifically the increasing power of special interests in our government, concerted efforts to deceive the public, and arbitrary actions of government executives that arise from increasing concentration of authority in a unitary executive, in defiance of the aims of our Constitution's framers. These tendencies are illustrated well by a couple of incidents that I have been involved in recently.

In the first incident, my own work was distorted for the purposes of misinforming the public and protecting special interests. In the second incident, the mission of the National Aeronautics and Space Administration (NASA) was altered surreptitiously by executive action, thus subverting constitutional division of power. These incidents help to paint a picture that reveals consequences for society far greater than simple enrichment of special interests. The effect is to keep the public in the dark about increasing risks to our society and our home planet.

The first incident prompted *New York Times* columnist Paul Krugman to argue not long ago that I must respond to "swift boaters"—those who distort the record to impugn someone's credibility. I have had reservations about doing so, stemming from the perceptive advice of Professor Henk van de Hulst, who said, when I was a post-doc at Leiden University, "Your success will depend upon choosing what not to work on." Unfortunately, given the shrinking fuse on the global warming time bomb, Krugman is probably right: we cannot afford the luxury of ignoring swift boaters and focusing only on science.

Pat Michaels, a swift boater to whom Krugman refers, is sometimes described as a "contrarian." Contrarians address global warming as if they were lawyers, not scientists. A lawyer's job often is to defend a client, not seek the truth. Instead of following Richard Feynman's dictum on scientific objectivity ("The only way to have real success in science . . . is to describe the evidence very carefully without regard to the way you feel it should be"), contrarians present only evidence that supports their desired conclusion.

Skepticism, an inherent aspect of scientific inquiry, should be carefully distinguished from contrarianism. Skepticism, and the objective weighing of evidence, are essential for scientific success. Skepticism about the existence of global warming and the principal role of human-made greenhouse gases has diminished as empirical evidence and our understanding have advanced. However, many aspects of global warming need to be understood better, including the best ways to minimize climate change and its consequences. Legitimate skepticism will always have an important role to play.

However, hard-core global warming contrarians have an agenda other than scientific truth. Their target is the public. Their goal is to create an impression that global warming or its causes are uncertain. Debating a contrarian leaves an impression with today's public of an argument among theorists. Sophistical contrarians do not need to win the scientific debate to advance their cause.

Science Fiction

Consider, for example, Pat Michaels' deceit (in a 2000 article in *Social Epistemology*) in portraying climate "predictions" that I made in 1988 as being in error by "450 percent." This distortion is old news, but by sheer repetition has become received wisdom among climate-change deniers. In fact, science fiction writer Michael Crichton was duped by Michaels, although Crichton reduced my "error" to "wrong by 300 percent" in his 2004 novel *State of Fear.*

People acquainted with this topic are aware that Michaels, in comparing global warming predictions made with the GISS (Goddard Institute for Space Studies) climate model with observations, played a dirty trick by showing model calculations for only one of the three scenarios (not predictions!) that I presented in 1988. Here's why this trick has a big impact.

The three scenarios (see figure, next page) were intended to bracket the range of likely future climate forcings (changes imposed on the Earth's energy balance that tend to alter global temperature either way). Scenario C had the smallest greenhouse gas forcing: it extended recent greenhouse gas growth rates to the year 2000 and thereafter kept greenhouse gas amounts constant, i.e., it assumed that after 2000 human sources of these gases would be just large enough to balance removal of these gases by the "sinks." Scenario B continued approximately linear growth of greenhouse gases beyond 2000. Scenario A showed exponential growth of greenhouse gases and included a substantial allowance for trace gases that were suspected of increasing but were unmeasured.

Scenarios A, B, and C also differed in their assumptions about future volcanic eruptions. Scenarios B and C included occasional eruptions of large volcanoes, at a frequency similar to that of the real world in the previous few decades. Scenario A, intended to yield the largest plausible warming, included no volcanic eruptions, as it is not uncommon to have no large eruptions for extended periods, such as the half century between the Katmai eruption in 1912 and the Agung eruption in 1963.

Multiple scenarios are used to provide a range of plausible climate outcomes, but also so that we can learn something by comparing real-world outcomes with model predictions. How well the model succeeds in simulating the real world depends upon the realism of both the assumed forcing and the climate sensitivity (the global temperature response to a standard climate forcing) of the model.

As it turned out, in the real world the largest climate forcing in the decade after 1988, by far, was caused by the Mount Pinatubo volcanic eruption, the greatest volcanic eruption of the past century. Forcings are measured in watt-years per square meter ($W\text{-}yr/m^2$) averaged over the surface of the Earth (1 $W\text{-}yr/m^2$ is a heating of 1 W/m^2 over the entire planet maintained for one year). The small particles injected into the Earth's stratosphere by Pinatubo reflected sunlight back to space, causing a negative (cooling) climate forcing of about -5 $W\text{-}yr/m^2$. In contrast, the added greenhouse gas climate forcings ranged from about $+1.6 W\text{-}yr/m^2$ in scenario C to about $+2.3 W\text{-}yr/m^2$ in scenario A.

So of the four scenarios (A, B, C, and the real world) only scenario A had no large volcanic eruption. The volcanic activity modeled in scenarios B and C was somewhat weaker than in the real world and was misplaced by a few years, but by good fortune it was such as to have a cooling effect pretty similar to that of Pinatubo. Despite the fact that scenario A omitted the largest climate forcing, Michaels chose to compare scenario A—and *only* scenario A—with the real world. Is this a case of scientific idiocy or is there something else at work? Perhaps Michaels is just not very interested in learning about the real world.

Although less important for the temperature change between 1997 and 1988 that Michaels examined, measured real-world greenhouse gas changes in carbon dioxide (CO_2), methane (CH_4), nitrous oxide (N_2O), and chlorofluorocarbons (CFCs) yielded a forcing similar to those in scenarios B and C. The reason for the slow real-world growth rate was that both CH_4 and CO_2 growth rates decreased in the early 1990s (the slowdowns may have been associated with Pinatubo; in any case the CO_2 growth rate has subsequently accelerated rapidly).

An astute reader may wonder why the world showed any warming during the period 1988–97, given that the negative (cooling) forcing by Pinatubo exceeded the positive (warming) forcing by greenhouse gases added in that period. The reason is that the climate system was also being pushed by the planetary "energy imbalance" that existed in 1988. The climate system had not yet fully responded to greenhouse gases added to the atmosphere before then. The observed continued decadal warming, despite the very large negative volcanic forcing, provides some confirmation of that planetary energy imbalance.

Noise and Distortion

Michaels' trick of comparing the real world only with the inappropriate scenario A accounts for his specious, incorrect conclusions. However, a second unscientific aspect of his method is also worth pointing out.

Scientists seek to learn something by comparing the real world with climate model calculations. Climate sensitivity is of special interest, as future climate change depends strongly upon it. In principal, we can extract climate sensitivity if we have accurate knowledge of the net forcing that drove climate change, and the global temperature change that occurred in response to that change. However, even if these demanding conditions are met, it is necessary to compare the magnitude of the calculated changes with the magnitude of "noise," including errors in the measurements and chaotic (unforced) variability in the model and real-world climate changes.

If Michaels had examined the noise question he would have realized that a nine-year change is insufficient to determine the real-world temperature trend or distinguish among the model runs. Even the period 1988–2005 is too brief for most purposes. Within several years the differences among scenarios A, B, and C, and comparisons with the real world, will become more meaningful.

Michaels' latest tomfoolery, repeated on several occasions, is the charge that I approve of exaggeration of potential consequences of future global warming. This is more unadulterated hogwash. Michaels quotes me as saying, "Emphasis on extreme scenarios may have been appropriate at one time, when the public and decision-makers were relatively unaware of the global warming issue."

What trick did Michaels use to create the impression that I advocate exaggeration? He took the above sentence out of context from a paragraph in which I was being gently critical of a tendency of Intergovernmental Panel on Climate Change climate simulations to emphasize only cases with very large increases of climate forcings. My entire paragraph (from a June 2003 presentation to the Council on Environmental Quality) reads as follows:

> *Summary opinion on scenarios. Emphasis on extreme scenarios may have been appropriate at one time, when the public and decision-makers were relatively unaware of the global warming issue, and energy sources such as "synfuels," shale oil, and tar sands were receiving strong consideration. Now, however, the need is for demonstrably objective climate forcing scenarios consistent with what is realistic under current conditions. Scenarios that accurately fit recent and near future observations have the best chance of bringing all of the important players into the discussion, and they also are what is needed for the purpose of providing policy-makers the most effective and efficient options to stop global warming.*

Would an intelligent reader who read the entire paragraph (or even the entire sentence; by chopping off half of the sentence Michaels brings quoting-out-of-context to a new low) infer that I was advocating exaggeration? On the contrary. Perhaps I should take it as a compliment that anyone would search my writing so hard to find something that can be quoted out of context.

Having taken this trouble to refute Michaels' claims, I still wonder about the wisdom of arguing with contrarians as a strategy. Many of them, including Michaels, receive support from special interests such as fossil fuel and automotive companies. It is understandable that special interests gravitated, early on, to scientists who had a message they preferred to hear. But now that global warming and its impacts are clearer, it is time for business people to reconsider their position—and scientists, rather than debating contrarians, may do better to communicate with business leaders. The latter did not attain their positions without being astute and capable of changing. We need to make clear to them the legal and moral liabilities that accrue with continued denial of global warming. It is time for business leaders to chuck contrarians and focus on the business challenges and opportunities.

Stealth Budgets & Unitary Executives

The second incident involved NASA's budget. Many people are aware that something bad happened to the NASA Earth Science budget this year, yet the severity of the cuts and their long-term implications are not universally recognized. In part this is because of a stealth budgeting maneuver.

When annual budgets for the coming fiscal year are announced, the differences in growth from the previous year, for agencies and their divisions, are typically a few percent. An agency with +3 percent growth may crow happily, in comparison to agencies receiving +1 percent. Small differences are important because every agency has fixed costs (civil service salaries, buildings, other infrastructure), so new programs or initiatives are strongly dependent upon any budget growth and how that growth compares with inflation.

When the administration announced its planned fiscal 2007 budget, NASA science was listed as having typical changes of 1 percent or so. However, Earth Science research actually had a staggering reduction of about 20 percent from the 2006 budget. How could that be accomplished? Simple enough: reduce the 2006 research budget retroactively by 20 percent! One-third of the way into fiscal year 2006, NASA Earth Science was told to go figure out how to live with a 20-percent loss of the current year's funds.

The Earth Science budget is almost a going-out-of-business budget. From the taxpayers' point of view it makes no sense. An 80-percent budget must be used mainly to support infrastructure (practically speaking, you cannot fire civil servants; buildings at large facilities such as Goddard Space Flight Center will not be bulldozed to the ground; and the grass at the centers

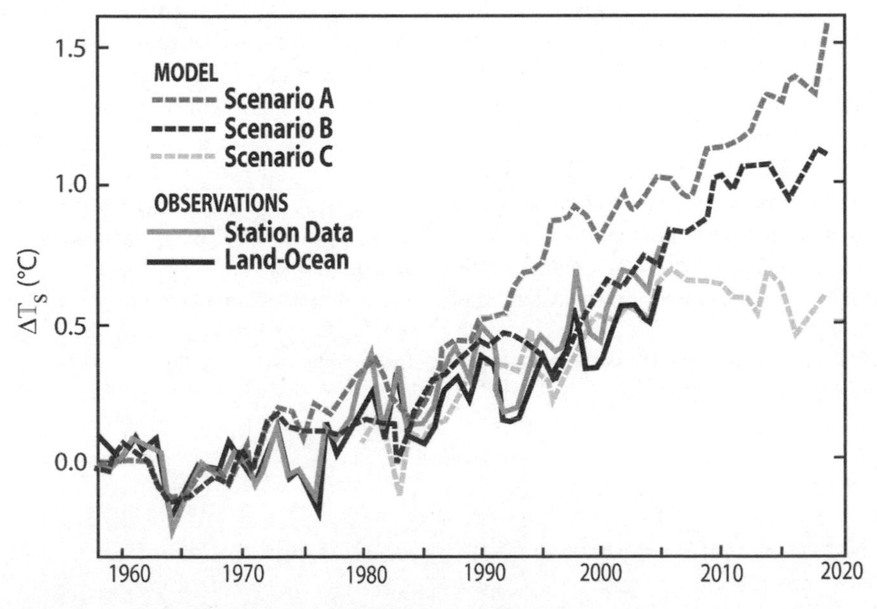

Figure 1 Annual mean global temperature change.

must continue to be cut). But the budget cuts wipe off the books most planned new satellite missions (some may be kept on the books, but only with a date so far in the future that no money needs to be spent now), and support for contractors, young scientists, and students disappears, with dire implications for future capabilities.

Bizarrely, this is happening just when NASA data are yielding spectacular and startling results. Two small satellites that measure the Earth's gravitational field with remarkable precision found that the mass of Greenland decreased by the equivalent of 200 cubic kilometers of ice in 2005. The area on Greenland with summer melting has increased 50 percent, the major ice streams on Greenland (portions of the ice sheet moving most rapidly toward the ocean and discharging icebergs) have doubled in flow speed, and the area in the Arctic Ocean with summer sea ice has decreased 20 percent in the last 25 years.

One way to avoid bad news: stop the measurements! Only hitch: the first line of the NASA mission is "to understand and protect our home planet." Maybe that can be changed to ". . . protect special interests' backside."

I should say that the mission statement *used* to read "to understand and protect our home planet." That part has been deleted—a shocking loss to me, as I had been using the phrase since December 2005 to justify speaking out about the dangers of global warming. The quoted mission statement had been constructed in 2001 and 2002 via an inclusive procedure involving representatives from the NASA Centers and e-mail interactions with NASA employees. In contrast, elimination of the "home planet" phrase occurred in a spending report delivered to Congress in February 2006, the same report that retroactively slashed the Earth Science research budget. In July 2006 I asked dozens of NASA employees and management people (including my boss) if they were aware of the change. Not one of them was. Several expressed concern that such management changes by fiat would have a bad effect on organization morale.

The budgetary goings-on in Washington have been noted, e.g., in editorials of *The Boston Globe:* "Earth to NASA: Help!" (June 15, 2006) and "Don't ask; don't ask" (June 22), both decrying the near-termination of Earth measurements. Of course, the *Globe* might be considered "liberal media," so their editorials may not raise many eyebrows.

But it is conservatives and moderates who should be most upset, and I consider myself a moderate conservative. When I was in school we learned that Congress controlled the purse strings; it is in the Constitution. But it does not really seem to work that way, not if the Bush administration can jerk the science budget the way they have, in the middle of a fiscal year no less. It seems more like David Baltimore's "Theory of the Unitary Executive" (the legal theory that the president can do pretty much whatever he wants) is being practiced successfully. My impression is that conservatives and moderates would prefer that the government work as described in the Constitution, and that they prefer to obtain their information on how the Earth is doing from real observations, not from convenient science fiction.

Congress is putting up some resistance to the budget manipulation. The House restored a fraction of the fiscal year 2007 cuts to science and is attempting to restore planning for some planetary missions. But the corrective changes are moderate. You may want to check your children's textbooks for the way the U.S. government works. If their books still say that Congress controls the purse strings, some updating is needed.

The NASA Mission
To understand and protect our home planet,
To explore the universe and search for life,
To inspire the next generation of explorers
. . . as only NASA can.

But may it be that this is all a bad dream? I will stand accused of being as wistful as the boy who cried out, "Joe, say it ain't so!" to the fallen Shoeless Joe Jackson of the 1919 Chicago Black Sox, yet I maintain the hope that NASA's dismissal of "home planet" is not a case of either shooting the messenger or a too-small growth of the total NASA budget, but simply an error of transcription. Those who have labored in the humid, murky environs of Washington are aware of the unappetizing forms of life that abound there. Perhaps the NASA playbook was left open late one day, and by chance the line "to understand and protect our home planet" was erased by the slimy belly of a slug crawling in the night. For the sake of our children and grandchildren, let us pray that this is the true explanation for the devious loss, and that our home planet's rightful place in NASA's mission will be restored.

James Hansen is an adjunct professor at the Columbia University Earth Institute and director of NASA's Goddard Institute for Space Studies in New York. He expresses his opinions here as a private citizen under the protection of the First Amendment.

The Myth of the 1970s Global Cooling Scientific Consensus

There was no scientific consensus in the 1970s that the Earth was headed into an imminent ice age. Indeed, the possibility of anthropogenic warming dominated the peer-reviewed literature even then.

THOMAS C. PETERSON, WILLIAM M. CONNOLLEY, AND JOHN FLECK

The Myth. When climate researcher Reid Bryson stood before the members of the American Association for the Advancement of Science in December 1972, his description of the state of scientists' understanding of climate change sounded very much like the old story about the group of blind men trying to describe an elephant. The integrated enterprise of climate science as we know it today was in its infancy, with different groups of scientists feeling blindly around their piece of the lumbering climate beast. Rigorous measurements of increasing atmospheric carbon dioxide were available for the first time, along with modeling results suggesting that global warming would be a clear consequence. Meanwhile, newly created global temperature series showed cooling since the 1940s, and other scientists were looking to aerosols to explain the change. The mystery of waxing and waning ice ages had long entranced geologists, and a cohesive explanation in terms of orbital solar forcing was beginning to emerge. Underlying this discussion was a realization that climate could change on time scales with the potential for significant effects on human societies, and that human activities could trigger such changes (Bryson 1974).

Bryson laid out the following four questions that still stand today as being central to the climate science enterprise:

1. How large must a climate change be to be important?
2. How fast can the climate change?
3. What are the causal parameters, and why do they change?
4. How sensitive is the climate to small changes in the causal parameters?

Despite active efforts to answer these questions, the following pervasive myth arose: there was a consensus among climate scientists of the 1970s that either global cooling or a full-fledged ice age was imminent (see the "Perpetuating the myth" sidebar). A review of the climate science literature from 1965 to 1979 shows this myth to be false. The myth's basis lies in a selective misreading of the texts both by some members of the media at the time and by some observers today. In fact, emphasis on greenhouse warming dominated the scientific literature even then. The research enterprise that grew in response to the questions articulated by Bryson and others, while considering the forces responsible for cooling, quickly converged on the view that greenhouse warming was likely to dominate on time scales that would be significant to human societies (Charney et al. 1979). However, perhaps more important than demonstrating that the global cooling myth is wrong, this review shows the remarkable way in which the individual threads of climate science of the time—each group of researchers pursuing their own set of questions—was quickly woven into the integrated tapestry that created the basis for climate science as we know it today.

Recognition of a Problem: The Potential for Warming

In 1965, when U.S. President Lyndon Johnson asked the members of his President's Science Advisory Committee (PSAC) to report on the potential problems of environmental pollution, climate change was not on the national agenda. The polluting effects of detergents and municipal sewage, the chronic problems associated with urban air pollution, and the risks associated with pesticides dominated public discourse about humanity's impact on the environment. However, in a 23-page appendix, which today appears prescient, the committee's Environmental Pollution Panel laid out the following stark scenario: emissions of carbon dioxide from the burning of fossil fuels could rapidly reshape Earth's climate (Revelle et al. 1965).

Perpetuating the Myth

The following are examples of modern writers perpetuating the myth of the 1970s global cooling scientific consensus.

Citing Singer (1998) as their source of information, Singer and Avery (2007) indicate that the National Academy of Science (1975) experts exhibited "hysterical fears" about a "finite possibility" that a serious worldwide cooling could befall the Earth, and that Ponte (1976) captured the "then-prevailing mood" by contending that the Earth may be on the brink of an ice age.

Balling (1992) Posits

> Could the [cold] winters of the late 1970s be the signal that we were returning to yet another ice age? According to many outspoken climate scientists in the late 1970s, the answer was absolutely yes—and we needed action now to cope with the coming changes . . . However, some scientists were skeptical, and they pointed to a future of global *warming,* not cooling, resulting from a continued build up of greenhouse gases. These scientists were in the minority at the time.

According to Horner (2007), the massive funding of climate change research was prompted by "'consensus' panic over 'global cooling'." This was "three decades ago—when the media were fanning frenzy about global cooling" (Will 2008) or, as Will (2004) succinctly put it, "the fashionable panic was about global cooling." "So, before we take global warming as a scientific truth, we should note that the opposite theory was once scientific verity" (Bray 1991).

In a Narrative, Crichton (2004) Put It This Way:

> "Just think how far we have come!" Henley said. "Back in the 1970s, all the climate scientists believed an ice age was coming. They thought the world was getting colder. But once the notion of global *warming* was raised, they immediately recognized the advantages. Global warming creates a crisis, a call to action. A crisis needs to be studied, it needs to be funded . . ."

According to Michaels (2004)

> Thirty years ago there was much scientific discussion among those who believed that humans influenced the . . . reflectivity [which would] cool the earth, more than . . . increasing carbon dioxide, causing warming. Back then, the "coolers" had the upper hand because, indeed, the planet was cooling . . . But nature quickly shifted gears . . . Needless to say, the abrupt shift in the climate caused almost as abrupt a shift in the balance of scientists who predictably followed the temperature.

Giddens (1999) States

> Yet only about 25 or so years ago, orthodox scientific opinion was that the world was in a phase of global cooling. Much the same evidence that was deployed to support the hypothesis of global cooling is now brought into play to bolster that of global warming—heat waves, cold spells, unusual types of weather.

The panel's members had two new tools at their disposal that had not been available just a few years before. The first up-to-date global temperature reconstructions had recently become available, allowing them to consider the twentieth century's somewhat confusing temperature trends (Somerville et al. 2007). More importantly, they had access to carbon dioxide data that Charles David Keeling and his colleagues had been collecting since 1957 on Mauna Loa, Hawaii, and in Antarctica (Pales and Keeling 1965; Brown and Keeling 1965). The data showed "clearly and conclusively," in the panel's words, that atmospheric carbon dioxide was rising as a result of fossil fuel burning. Human activities, the panel concluded, were sufficient in scale to impact not just the immediate vicinity where those activities were taking place. Industrial activities had become a global, geophysical force to be recognized and with which to be reckoned. With estimated recoverable fossil fuel reserves sufficient to triple atmospheric carbon dioxide, the panel wrote, "Man is unwittingly conducting a vast geophysical experiment." With the emission of just a fraction thereof, emissions by the year 2000 could be sufficient to cause "measurable and perhaps marked" climate change, the panel concluded (Revelle et al. 1965).

The Global Temperature Records: A Cooling Trend

Efforts to accumulate and organize global temperature records began in the 1870s (Somerville et al. 2007). The first analysis to show long-term warming trends was published in 1938. However, such analyses were not updated very often. Indeed, the Earth appeared to have been cooling for more than 2 decades when scientists first took note of the change in trend in the 1960s. The seminal work was done by J. Murray Mitchell, who, in 1963, presented the first up-to-date temperature reconstruction showing that a global cooling trend had begun in the 1940s. Mitchell used data from nearly 200 weather stations, collected by the World Weather Records project under the auspices of the World Meteorological Organization, to calculate latitudinal average temperature. His analysis showed that global temperatures had increased fairly steadily from the 1880s, the start of his record, until about 1940, before the start of a steady multidecade cooling (Mitchell 1963).

By the early 1970s, when Mitchell updated his work (Mitchell 1972), the notion of a global cooling trend was widely accepted, albeit poorly understood. The first satellite records showed increasing snow and ice cover across the Northern Hemisphere from the late 1960s to the early 1970s. This trend was capped by unusually severe winters in Asia and parts of North America in 1972 and 1973 (Kukla and Kukla 1974), which pushed the issue into the public consciousness (Gribbin 1975). The new data about global temperatures came amid growing concerns about world food supplies, triggering fears that a planetary cooling trend might threaten humanity's ability to feed itself (Thompson 1975). It was not long, however, before scientists teasing apart the details of Mitchell's trend found that it was not necessarily a global phenomenon. Yes, globally averaged temperatures were cooling, but this was largely due to changes in the Northern Hemisphere. A closer examination of Southern Hemisphere data revealed thermometers heading in the opposite direction (Damon and Kunen 1976).

New Revelations about the Ice Ages

While meteorologists were collecting, analyzing, and trying to explain the temperature records, a largely separate group of scientists was attacking the problem from a paleoclimate perspective, assembling the first detailed understanding of the Earth's ice age history. The fact that parts of the Northern Hemisphere had once been covered in ice was one of the great realizations of nineteenth-century geology. Even more remarkable was the realization that the scars on the landscape had been left by not one but several ice ages. Climate clearly was capable of remarkable variability, beyond anything humanity had experienced in recorded history.

It was not until the mid-twentieth century that scientists finally assembled the details of the coming and going of the last ice ages. The geologists' classic story had suggested four short ice ages over the Quaternary, with long warm periods between them. However, analysis of coral, cores from ice caps and the ocean floor, along with the application of newly developed radiometric techniques, forced a radical re-evaluation. Climate was far more variable, with long ice ages punctuated by short interglacial periods (Broecker et al. 1968; Emiliani 1972). The new work went beyond filling in gaps in scientists' knowledge of the past. It laid the foundation of an explanation for why ice age cycles occurred. Building on earlier work (e.g., Adhémar 1842; Croll 1875), Serbian engineer and geophysicist Milutin Milankovitch calculated that highly regular changes in the tilt of Earth's axis and the eccentricity of its orbit around the sun would change the distribution of sunlight hitting the Earth's surface, leading to the waxing and waning of ice ages (Milankovitch 1930). Milankovitch's work won few converts, in part because it did not match geologists' understanding of the history of the ice ages. However, the new dating of the ice's ebbs and flows led to new interest in Milankovitch's ideas (e.g., Ericson et al. 1964; Damon 1965). "The often-discredited hypothesis of Milankovitch must be recognized as the number-one contender in the climatic sweepstakes," Wallace Broecker wrote (Broecker et al. 1968). It took the rest of the science world a while to catch up with Broecker, but by the late 1970s they had (Hays et al. 1976; Kerr 1978; Weart 2003).

Because Milankovitch's astronomical metronome was predictable over thousands of years, climate scientists could now begin talking about predicting the onset of the next ice age. And they did. Members of the Climate: Long-range Investigation, Mapping and Prediction (CLIMAP) project lived up to their project's name with a "prediction" of sorts; in the absence of possible anthropogenic warming, "the long-term trend over the next several thousand years is toward extensive Northern Hemisphere glaciation" (Hays et al. 1976).

Carbon Dioxide

Mid-nineteenth-century British naturalist John Tyndall was fascinated by the new emerging evidence of past ice ages, and believed he had found a possible explanation for such dramatic changes in Earth's climate: changes in the composition of the atmosphere. Some molecules, he realized, could absorb thermal radiation, and as such could be the cause for "all the mutations of climate which the researches of geologists reveal" (Weart 2003; Tyndall 1861; Somerville et al. 2007). In 1896 Swedish scientist Svante Arrhenius calculated that a doubling of atmospheric carbon dioxide would raise global temperatures 5°–6° C. However, he figured it would take 3,000 yr of fossil fuel burning to do it (Weart 2003). Thus continued what would be a century of scientific debate and uncertainty, both about the effect of such so-called "greenhouse gases" and the possibility that the burning of fossil fuels could contribute substantially to their concentration (Landsberg 1970). It was not until the second half of the twentieth century that scientists finally had the tools to begin measuring the concentrations of those greenhouse gases in sufficient detail to begin evaluating their effects.

Using funding available through the International Geophysical Year, Charles David Keeling was able to overcome problems of local interference in carbon dioxide measurements in 1957 by establishing stations in Antarctica and atop Mauna Loa. By 1965, his data were sufficient to show an unambiguous trend. Keeling's observation also showed that atmospheric carbon dioxide was increasing far faster than Arrhenius's 70-yr-old estimate. That was enough for members of the U.S. President's Scientific Advisory Committee to pronounce the possibility that increasing carbon dioxide could "modify the heat balance of the atmosphere to such an extent that marked changes in climate, not controllable through local or even national efforts, could occur" (Revelle et al. 1965).

The PSAC scientists had a new tool for understanding the implications—the first preliminary results of newly developing climate models. The same year the PSAC report came out, Syukuro Manabe and Richard Wetherald developed the first true three-dimensional climate model. The results were raw at the time the PSAC report was written, but within 2 yr, the first seminal modeling results from the Geophysical Fluid Dynamics Laboratory team were published. Given their simplifying constraints, they found that a doubling of atmospheric carbon dioxide would raise global temperature 2°C (Manabe and Wetherald 1967). Within a decade, the models' sophistication had grown dramatically, enough for Manabe and Wetherald to conclude that high latitudes were likely to see greater warming in a doubled-CO_2 world, and that the intensity of the hydrologic cycle could be expected to increase significantly (Manabe and Wetherald 1975). The accumulating evidence of the new carbon dioxide record and the modeling results was enough for Wallace Broecker to ask in 1975, "Are we on the brink of a pronounced global warming?" Broecker's answer was a resounding "yes" (Broecker 1975).

Aerosols

In December 1968, a group of scientists convened in Dallas, Texas, for a "Symposium on Global Effects of Environmental Pollution" (Singer 1970). Reid Bryson showed the panel a remarkable graph illustrating the correlation between rising levels of dust in the Caucasus and the rising output of the Russian economy over the previous three decades. It was the foundation for an argument leading from human activities to dust to changing climate. Atmospheric pollution caused by humans was sufficient, Bryson argued, to explain the decline in global temperatures identified earlier in the decade by J. Murray Mitchell (Bryson and Wendland 1970).

Also on the symposium panel was Mitchell himself, and he disagreed. Mitchell's calculations suggested that particulates added to the atmosphere were insufficient to explain the cooling seen in his temperature records. However, he raised the possibility that, over time, cooling caused by particulates could overtake warming caused by what he called the "the CO_2 effect" (Mitchell 1970).

In 1971, S. Ichtiaque Rasool and Stephen Schneider wrote what may be the most misinterpreted and misused paper in the story of global cooling (Rasool and Schneider 1971). It was the first foray into climate science for Schneider, who would become famous for his work on climate change. Rasool and Schneider were trying to extend the newly developed tool of climate modeling to include the effects of aerosols, in an attempt to sort out two potentially conflicting trends—the warming brought about by increasing carbon dioxide and the cooling potential of aerosols emitted into the Earth's atmosphere by industrial activity.

The answer proposed by Rasool and Schneider to the questions posed by Bryson and Mitchell's disagreement was stark. An increase by a factor of 4 in global aerosol concentrations, "which cannot be ruled out as a possibility," could be enough to trigger an ice age (Rasool and Schneider 1971). Critics quickly pointed out flaws in Rasool and Schneider's work, including some they acknowledged themselves (Charlson et al. 1972; Rasool and Schneider 1972). Refinements, using data on aerosols from volcanic eruptions, showed that while cooling could result, the original Rasool and Schneider paper had overestimated cooling while underestimating the greenhouse warming contributed by carbon dioxide (Schneider and Mass 1975; Weart 2003). Adding to the confusion at the time, other researchers concluded that aerosols would lead to warming rather than cooling (Reck 1975; Idso and Brazel 1977).

It was James Hansen and his colleagues who found what seemed to be the right balance between the two competing forces by modeling the aerosols from Mount Agung, a volcano that erupted in Bali in 1963. Hansen and his colleagues fed data from the Agung eruption into their model, which got the size and timing of the resulting pulse of global cooling correct. By 1978, the question of the relative role of aerosol cooling and greenhouse warming had been sorted out. Greenhouse warming, the researchers concluded, had become the dominant forcing (Hansen et al. 1978; Weart 2003).

Media Coverage

When the myth of the 1970s global cooling scare arises in contemporary discussion over climate change, it is most often in the form of citations not to the scientific literature, but to news media coverage. That is where U.S. Senator James Inhofe turned for much of the evidence to support his argument in a U.S. Senate floor speech in 2003 (Inhofe 2003). Chief among his evidence was a frequently cited *Newsweek* story: "The cooling world" (Gwynne 1975). The story drew from the latest global temperature records, and suggested that cooling "may portend a drastic decline for food production." Citing the Kuklas' work on increasing Northern Hemisphere snow and ice, and Reid Bryson's concerns about a long-term cooling trend, the *Newsweek* story juxtaposes the possibility of cooling temperatures and decreasing food production with rising global populations. Other articles of the time featured similar themes (see "Popular literature of the era" sidebar).

Even cursory review of the news media coverage of the issue reveals that, just as there was no consensus at the time among scientists, so was there also no consensus among journalists. For example, these are titles from two *New York Times* articles: "Scientists ask why world climate is changing; major cooling may be ahead" (Sullivan 1975a) and "Warming trend seen in climate; two articles counter view that cold period is due" (Sullivan 1975b). Equally juxtaposed were *The Cooling* (Ponte 1976), which was published the year after *Hothouse Earth* (Wilcox 1975).

However, the news coverage of the time does reflect what *New York Times* science writer Andrew Revkin calls "the tyranny of the news peg," based on the idea that reporters need a "peg" on which to hang a story. Developments that are dramatic or new tend to draw the news media's attention, Revkin argues, rather than the complexity of a nuanced discussion within the scientific community (Revkin 2005). A handy peg for climate stories during the 1970s was the cold weather.

Survey of the Peer-Reviewed Literature

One way to determine what scientists think is to ask them. This was actually done in 1977 following the severe 1976/77 winter in the eastern United States. "Collectively," the 24 eminent climatologists responding to the survey "tended to anticipate a slight global warming rather than a cooling" (National Defense University Research Directorate 1978). However, given that an opinion survey does not capture the full state of the science of the time, we conducted a rigorous literature review of the

Popular Literature of the ERA

There are too many potential newspaper articles to adequately assess and, because they report on current events, even articles in the same paper by the same author separated by only a few months can be quite different. For example, the following are titles from two *New York Times* articles: "Scientists ask why world climate is changing; major cooling may be ahead" (Sullivan 1975a) and "Warming trend seen in climate; two articles counter view that cold period is due" (Sullivan 1975b). The most *frequently* cited magazine articles are described below. While these articles described the past climate and a distant future of another ice age, the following is a review only of their decadal-to-century-scale global temperature projections.

Science Digest's 1973 article "Brace yourself for another Ice Age" (Colligan 1973) primarily focused on ice ages and global cooling, with the warning that "the end of the present interglacial period is due 'soon." However, it clarified that "'soon' in the context of the world's geological time scale could mean anything from two centuries to 2,000 years, but not within the lifetime of anyone now alive." The article also mentioned that "scientists seem to think that a little more carbon dioxide in the atmosphere could warm things up a good deal."

Time Magazine (1974) ominously worried that "climatological Cassandras are becoming increasingly apprehensive, for the weather aberrations they are studying may be the harbinger of another ice age." However, only one scientist was indicated by name issuing any sort of projection: "Some scientists like Donald Oilman, chief of the National Weather Service's long-range-prediction group, think that the cooling trend may be only temporary."

Science News' 1975 article "Climate change: Chilling possibilities" (Douglas 1975) mainly discussed the new findings that raised the possibility of "the approach of a full-blown 10,000-year ice age." However, it also put these results into perspective with statements such as "the cooling trend observed since 1940 is real enough . . . but not enough is known about the underlying causes to justify any sort of extrapolation," and "by the turn of the century, enough carbon dioxide will have been put into the atmosphere to raise the temperature of earth half a degree."

The 1975 *Newsweek* article (Gwynne 1975) quotes four scientists by name and none of them offered a projection of the future; three discussed observations of the recent cooling and one the relationship between climate and agriculture. The article did, however, state that "seemingly disparate [weather] incidents represent the advance signs of fundamental changes in the world's weather," though "meteorologists disagree about the cause and extent of the cooling trend." The article states that there was an "almost unanimous" view that the cooling trend would "reduce agricultural productivity for the rest of the century," and it even discussed possible solutions such as spreading black soot on the Arctic ice cap.

In 1976, *National Geographic Magazine* published an article entitled "What's happening to our climate?" In this article, Mathews (1976) discusses projections on the relevant time frame from four different scientists. Reid Bryson of the University of Wisconsin believed that the critical factor was cooling caused by aerosols generated by an exploding population. If Willi Dansgaard of the University of Copenhagen is correct—that western Europe's climate lags 250 yr behind Greenland's—"Europe could be in for a cooler future," although he cautions that man-made atmospheric pollution "may completely change the picture." The "cooling trend of world climate" was documented in the 1960s by J. Murray Mitchell Jr., of the National Oceanic and Atmospheric Administration (NOAA). Now, he notes, "carbon dioxide pollution may be contributing to an opposite, or warming, tendency." And last, "it is possible that we are on the brink of a several-decade-long period of rapid warming," observes Dr. Wallace S. Broecker of Columbia University's Lamont-Doherty Geological Observatory. "If the natural cooling trend bottoms out . . . global temperature would begin a dramatic rise . . . this warming would, by the year 2000, bring average global temperatures beyond the range experienced during the past 1,000 years."

There were also lay books on climate change, some of which received rather scathing reviews in the scientific literature. For example, discussing *The Climatic Threat: What's Wrong with our Weather?* (Gribbin 1978a). Wigley (1978) wrote that the average reader "cannot possibly know how incompletely the author reviews the field he discusses, how uncritical and selective are his references to the scientific literature, how much he has mixed sound well accepted work with controversial opinion and speculation, and how often the cautious, tentative words of others are represented as established fact." Note also, "A casual reader" of *Climates of Hunger: Mankind and the World's Changing Weather* (Bryson and Murray 1977) "will not get a balanced picture of the current climatic debate" (Gribbin 1978b). Kellogg's (1979) review of Halacy (1978) that also comments on Calder (1974), stated that

> Halacy, in *Ice and Fire?* Like Calder, has chosen to write a book whose central theme is the prediction of a global cooling as the beginning of a new ice age—perhaps occurring very quickly. . . . Furthermore, even a non-expert will notice that he has blurred his timescales cleverly (as did Nigel Calder, whom he quotes extensively), giving the impression that the advent of an ice age could occur in a matter of a decade or so—perhaps it will take a century if we are lucky.

Landsberg (1976) also took Calder's book. *The Weather Machine,* to task, stating that "he quotes his favorite scientists at length, and then covers himself by a sentence at the end that there are others with diverging opinions . . . The amount of half-digested meteorology, such as the potential dust effect in the atmosphere, is formidable."

A common feature of the popular articles and books is the probable negative impacts of climate variability on agriculture, which was felt to be stressed already by population pressures. The book, *The Genesis Strategy* (Schneider 1976) takes this further and argues for a policy resilient to

(continued)

Popular Literature of the ERA *(continued)*

any future changes in climate, though without predicting either warming or cooling. A more extreme book, *The Cooling* (Ponte 1976), predicts that cooling could lead to billions of deaths by 2050, but struggles to find any good source for predictions of such a cooling; it is also somewhat undermined by its own preface by Reid Bryson, which states that "there are very few pages that, as a scientist, I could accept without questions of accuracy, of precision, or of balance." On the other side, the book *Hothouse Earth* (Wilcox 1975) has both polar ice caps melting due to anthropogenic global warming (Landsberg 1976) and the 1973 Charlton Heston film *Soylent Green* "imagines the Earth of 2022 as a dried-up wasteland where the greenhouse effect, brought about by an exponentially growing population and unchecked industry, has led to the destruction of the environment" (Bertram 2006).

American Meteorological Society's electronic archives as well as those of *Nature* and the scholarly journal archive *Journal Storage* (*JSTOR*). To capture the relevant topics, we used global temperature, global warming, and global cooling, as well as a variety of other less directly relevant search terms. Additionally, in order to make the survey more complete, even at the expense of no longer being fully reproducible by electronic search techniques, many references mentioned in the papers located by these searches were evaluated, as were references mentioned in various history-of-science documents. Because the time period attributed to the global cooling consensus is typically described as the 1970s, the literature search was limited to the period from 1965 through 1979. While no search can be 100% complete, this methodology offers a reasonable test of the hypothesis that there was a scientific consensus in the 1970s regarding the prospect of imminent global cooling. Such a consensus would be easily shown by both the presence of many articles describing global cooling projections and the absence of articles projecting global warming.

One measure of the relevance of a paper to a developing scientific consensus is the number of citations it receives. For that reason, a citation analysis of the papers found in our survey was undertaken. Not all of the citations may be supportive of the paper in question, but they do help indicate which papers dominated the thinking of the day. Because the period assessed ended in 1979 and it takes time for citations to start appearing, the citation count was extended through 1983. The gray literature of conference proceedings were not authoritative enough to be included in the literature search. However, a few prestigious reports that may not have been peer reviewed have been included in this literature survey because they clearly represent the science of their day.

Our literature survey was limited to those papers projecting climate change on, or even just discussing an aspect of climate forcing relevant to, time scales from decades to a century. While some of these articles make clear predictions of global surface temperature change by the year 2000, most of these articles do not. Many of the articles simply examined some aspect of climate forcing. However, it was generally accepted that both CO_2 and anthropogenic aerosols were increasing. Therefore, for example, articles that estimated temperature increases resulting from doubling CO_2 or temperature decreases resulting from anthropogenic aerosols would be listed in Table 1 as warming or cooling articles, respectively. The neutral category in Table 1 includes papers that project no change, that discuss both warming and cooling influences without specifically indicating which are likely to be dominant, or that state not enough is known to make a sound prediction. Articles were not included in the survey if they examined the climate impacts of factors that did not have a clear expectation of imminent change, such as increases in volcanic eruptions or the creation of large fleets of supersonic aircraft.

The survey identified only 7 articles indicating cooling compared to 44 indicating warming. Those seven cooling articles garnered just 12% of the citations. Graphical representations of this survey are shown in Fig. 1 for the number of articles and Fig. 2 for the number of citations. Interestingly, only two of the articles would, according to the current state of climate science, be considered "wrong" in the sense of getting the wrong sign of the response to the forcing they considered—one cooling (Bryson and Dittberner 1976) and one warming (Idso and Brazel 1977) paper—and both were immediately challenged (Woronko 1977; Herman et al. 1978). As climate science and the models progressed over time, the findings of the rest of the articles were refined and improved, sometimes significantly, but they were not reversed.

Given that even a cursory examination of Fig. 1 reveals that global cooling was never more than a minor aspect of the scientific climate change literature of the era, let alone the scientific consensus, it is worth examining the ways in which the global cooling myth persists. One involves the simple misquoting of the literature. In a 2003 *Washington Post* op-ed piece, former Energy Secretary James Schlesinger quoted a 1972 National Science Board report as saying, "Judging from the record of the past interglacial ages, the present time of high temperatures should be drawing to an end . . . leading into the next glacial age" (Schlesinger 2003). The quote repeatedly appeared other places in the political debate over climate change, including the floor of the U.S. Senate where Inhofe (2003) followed up that quote by stating, "That was the same timeframe that the global-warming alarmists are concerned about global warming." The actual report, however, shows that the original context, rather than supporting the global cooling myth, discusses the full state of the science at the time, as described earlier. The words not extracted by Schlesinger and Inhofe are highlighted with italics:

Table 1 Cooling, Neutral, and Warming Papers as Defined in the Text Followed by the Number of Times They Have Been Cited up through 1983.

Year	Cooling Papers	Neutral Papers	Warming Papers
1965			Revelle et al. (1965)
1966			
1967	McCormick and Ludwig (1967): 67		Manabe and Weatherald (1967): 306
1968			
1969			Sellers (1969): 191
1970		Landsberg (1970): 83	Benton (1970): 0; Report of the Study of Critical Environmental Problems (1970): 130
1971	Barrett (1971): 14; Rasool and Schneider (1971): 144		Mitchell (1971): 81
1972	Hamilton and Seliga (1972): 12	Charlson et al. (1972): 0; Lowry (1972): 0; National Science Board (1972): 0; Rasool and Schneider (1972): 0	Budyko (1972): 36; Machta (1972): 31; Mitchell (1972): 36; Sawyer (1972): 8
1973		Sellers (1973): 104	
1974	Chylek and Coakley (1974): 38	Bryson (1974): 113; Hobbs et al. (1974): 22; Weare et al. (1974): 12; Willett (1974): 0	Federal Council for Science and Technology Interdepartmental Committee for Atmospheric Sciences (1974): I; Kellogg and Schneider (1974): 30; Sellers (1974): 33
1975		National Academy of Sciences (1975): 0	Broecker (1975): 54; Manabe and Wetherald (1975): 211; Ramanathan (1975): 63; Reck (1975): 13; Schneider and Mass (1975): 82; Schneider (1975): 94; Thompson (1975): 49
1976	Bryson and Dittberner (1976): 31	Shaw (1976): 6	Budyko and Vinnikov (1976): 0; Damon and Kunen (1976): 29; Mitchell (1976): 50; Wang et al. (1976): 89
1977	Twomey (1977): 19	Bryson and Dittberner (1977): 0	Flohn (1977): 7; Idso and Brazel (1977): 1; Lee and Snell (1977): 8; National Academy of Sciences (1977): 1; Nordhaus (1977): 13; Panel on Energy and Climate (1977): 78; Woronko (1977): 1
1978		Herman et al. (1978): 0; Mason (1978a): 0; Miles (1978): 8; Ramanathan and Coakley (1978): 44; Shutts and Green (1978): 3	Budyko et al. (1978): 0; Cooper (1978): 0; Gilchrist (1978): 5; Idso and Brazel (1978): 2; Mason (1978b): 0; Mercer: (1978): 48; Ohring and Adler (1978): 25; Stuiver (1978): 101
1979		Choudhury and Kukla (1979): 4; Sagan et al. (1979): 25	Berger (1979): 6; Charney et al. (1979): 50; Houghton (1979): 0; Hoyt (1979): 13; Rotty (1979): 1

Judging from the record of the past interglacial ages, the present time of high temperatures should be drawing to an end, *to be followed by a long period of considerably colder temperatures* leading to the next glacial age *some 20,000 years from now. However, it is possible, or even likely, that human interference has already altered the environment so much that the climatic pattern of the near future will follow a different path. For instance, widespread deforestation in recent centuries, especially in Europe and North America, together with increased atmospheric opacity due to man-made dust storms and industrial wastes, should have increased the Earth's reflectivity. At the same time increasing concentration of industrial carbon dioxide in the atmosphere should lead to a temperature increase by absorption of infrared radiation from the Earth's surface. When these human factors are added to such other natural factors as volcanic eruptions, changes in solar activity, and resonances within the hydro-atmosphere, their effect can only be estimated in terms of direction, not of amount* (National Science Board 1972).

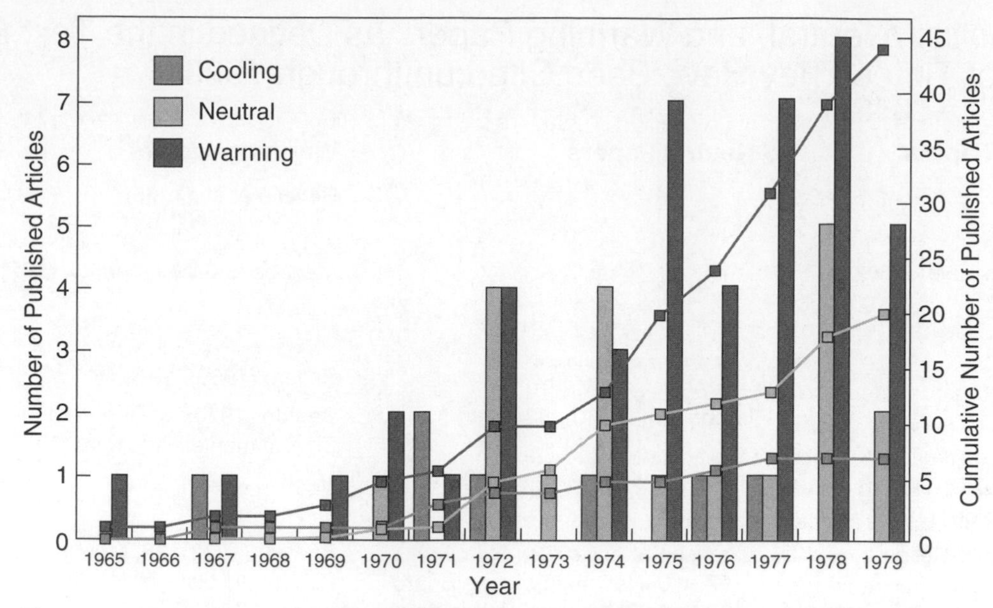

Figure 1 The number of papers classified as predicting, implying, or providing supporting evidence for future global cooling, warming, and neutral categories as defined in the text and listed in Table 1. During the period from 1965 through 1979, our literature survey found 7 cooling, 20 neutral, and 44 warming papers.

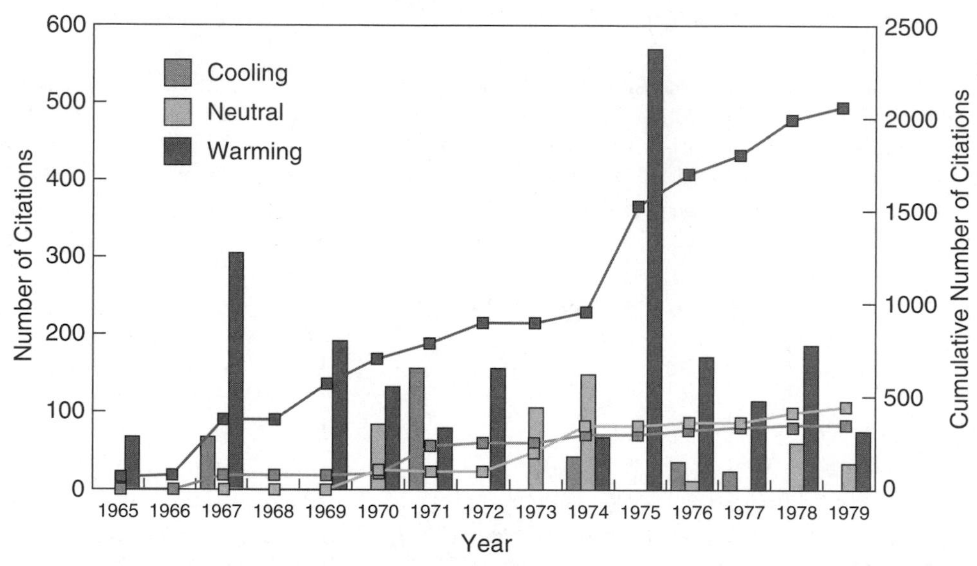

Figure 2 The number of citations for the articles shown in Fig. 1 and listed in Table 1. The citation counts were from the publication date through 1983 and are graphed on the year the article was published. The cooling papers received a total of 325 citations, neutral 424, and warming 2,043.

Underlying the selective quotation of the past literature is an example of what political scientist Daniel Sarewitz calls "scientization" of political debate: the selective emphasis on particular scientific "facts" to advance a particular set of political values (Sarewitz 2004). In this case, the primary use of the myth is in the context of attempting to undermine public belief in and support for the contemporary scientific consensus about anthropogenic climate change by appeal to a past "consensus" on a closely related topic that is alleged to have been wrong (see "Perpetuating the myth" sidebar).

Integrating Climate Science in the Late 1970s

When James D. Hays and colleagues published their landmark 1976 paper linking variations in the Earth's orbit to the ice ages, they offered the following two caveats:

> Such forecasts must be qualified in two ways: First, they apply only to the natural component of future climatic trends—and not to anthropogenic effects such as those

due to the burning of fossil fuels. Second, they describe only the long-term trends, because they are linked to orbital variations with periods of 20,000 years and longer. Climatic oscillations at higher frequencies are not predicted (Hays et al. 1976).

As the various threads of climate research came together in the late 1970s into a unified field of study—ice ages, aerosols, greenhouse forcing, and the global temperature trend—greenhouse forcing was coming to be recognized as the dominant term in the climate change equations for time scales from decades to centuries. That was the message from B. John Mason of the British Meteorological Office when he stood before members of the Royal Society in London on 27 April 1978 to deliver a review lecture on the state of the science. Taking his audience through the details of how the new computer climate models worked and what they showed, Mason ticked off the following now-familiar list of climate variables: variations in the Earth's orbit, aerosols, and the rapid increase in greenhouse gases. The effect of the latter, he said, was by far the largest, and more detailed study of the issue "now deserves high priority" (Mason 1978b).

In July 1979 in Woods Hole, Massachusetts, Jule Charney, one of the pioneers of climate modeling, brought together a panel of experts under the U.S. National Research Council to sort out the state of the science. The panel's work has become iconic as a foundation for the enterprise of climate change study that followed (Somerville et al. 2007). Such reports are a traditional approach within' the United States for eliciting expert views on scientific questions of political and public policy importance (Weart 2003). In this case, the panel concluded that the potential damage from greenhouse gases was real and should not be ignored. The potential for cooling, the threat of aerosols, or the possibility of an ice age shows up nowhere in the report. Warming from doubled CO_2 of 1.5°–4.5°C was possible, the panel reported. While there were huge uncertainties, Verner Suomi, chairman of the National Research Council's Climate Research Board, wrote in the report's foreword that he believed there was enough evidence to support action: "A wait-and-see policy may mean waiting until it is too late" (Charney et al. 1979). Clearly, if a national report in the 1970s advocates urgent action to address global warming, then the scientific consensus of the 1970s was not global cooling.

References

Adhémar, J. A., 1842: *Révolutions de la Mer: Déluges Périodiques.* Carilian-Goeury et V. Dalmont, 184 pp.

Balling, R. C., Jr., 1992: *The Heated Debate: Greenhouse Prediction versus Climate Reality.* Pacific Research Institute for Public Policy, 195 pp.

Barrett, E. W., 1971: Depletion of short-wave irradiance at the ground by particles suspended in the atmosphere. *Sol. Energy,* **13,** 323–337.

Benton, G. S., 1970: Carbon dioxide and its role in climate change. *Proc. Natl. Acad. Sci.,* **67,** 898.

Berger, A., 1979: Spectrum of climatic variations and their causal mechanisms. *Geophys. Surv.,* **3,** 351–402.

Bertram, D., 2006: More than just one inconvenient truth. *The Australian,* 14 September. [Available on-line at www.news.com.au/story/0,23599,20407484-5007146,00.html.]

Bray, A. J., 1991: The Ice Age cometh. *Policy Review,* No. 58, 82–84.

Broecker, W. S., 1975: Climate change: Are we on the brink of a pronounced global warming? *Science,* **189,** 460–463.

———, D. L. Thurber, J. Goddard, T.-L. Ku, R. K. Matthews, and K. J. Mesolella, 1968: Milankovitch hypothesis supported by precise dating of coral reefs and deep-sea sediments. *Science,* **159,** 297–300.

Brown, C. W., and C. D. Keeling, 1965: The concentration of atmospheric carbon dioxide in Antarctica. *J. Geophys. Res.,* **70,** 6077–6085.

Bryson, R. A., 1974: A perspective on climatic change. *Science,* **184,** 753–760.

———, and W. M. Wendland, 1970: Climatic effects of atmospheric pollution. *Global Effects of Environmental Pollution,* S. F. Singer, Ed., Springer-Verlag/D. Reidel, 130–138.

———, and G. J. Dittberner, 1976: A non-equilibrium model of hemispheric mean surface temperature. *J. Atmos. Sci.,* **33,** 2094–2106.

———, and ———, 1977: Reply. *J. Atmos. Sci.,* **34,** 1821–1824.

———, and T. J. Murray, 1977: *Climates of Hunger: Mankind and the World's Changing Weather.* American University Publishers Group, 171 pp.

Budyko, M. I., 1972: The future climate. *Eos, Trans. Amer. Geophys. Union,* **53,** 868–874.

———, and K. Y. Vinnikov, 1976: Global warming. *Sov. Meteor. Hydrol.,* **7,** 12–20.

———, ———, O. A. Drozdov, and N. A. Yefimova, 1978: Impending climatic change (in Russian). *Izv. Acad. Sci., USSR, Ser. Geogr.,* **6,** 5–20; English translation: 1979, *Sov. Geogr.,* **7,** 395–411.

Calder, N., 1974: *The Weather Machine.* BBC, 143 pp.

Charlson, R. J., H. Harrison, and G. Witt, 1972: Aerosol concentrations: Effects on planetary temperatures. *Science,* **175,** 95–96.

Charney, J. G., and Coauthors, 1979: *Carbon Dioxide and Climate: A Scientific Assessment.* National Academy of Science, 22 pp.

Choudhury, B., and G. Kukla, 1979: Impact of CO_2 on cooling of snow and water surfaces. *Nature,* **280,** 668–671.

Chylek, P., and J. A. Coakley Jr., 1974: Aerosols and climate. *Science,* **183,** 75–77.

Colligan, D., 1973: Brace yourself for another ice age. *Sci. Digest,* **73** (2), 57–61.

Cooper, C. F., 1978: What might man-induced climate change mean? *Foreign Affairs,* **56,** 500–520.

Crichton, M., 2004: *State of Fear.* Avon Books, 672 pp.

Croll, J., 1875: *Climate and Time in Their Geological Relations.* Appleton, 186 pp.

Damon, P. E., 1965: Pleistocene time scales. *Science,* **148,** 1037–1039.

———, and S. M. Kunen, 1976: Global cooling? *Science,* **193,** 447–453.

Douglas, J.H., 1975: Climate changes: Chilling possibilities. *Science News,* **107,** 138–140.

Emiliani, C., 1972: Quaternary paleotemperatures and the duration of the high-temperature intervals. *Science,* **178,** 398–401.

Ericson, D. B., M. Ewing, and G. Wollin, 1964: The Pleistocene Epoch in deep-sea sediments. *Science,* **146,** 723–732.

Federal Council for Science and Technology, Interdepartmental Committee for Atmospheric Sciences, 1974: *Report of the Ad Hoc Panel on the Present Interglacial.* 22 pp.

Flohn, H., 1977: Climate and energy: A scenario to a 21st century problem. *Climatic Change,* **1,** 5–20.

Giddins, A., 1999: *Runaway World.* BBC Reith Lectures. [Available online at http://news.bbc.co.uk/hi/english/static/events/reith_99/week2/week2.htm.]

Gilchrist, A., 1978: Numerical simulation of climate and climatic change. *Nature,* **276,** 342–245.

Gribbin, J., 1975: Cause and effects of global cooling. *Nature,* **254,** 14.

———, 1978a: *The Climatic Threat: What's Wrong with Our Weather?* Fontana, 206 pp.

———, 1978b: Climatic shifts. *Nature,* **271,** 785.

Gwynne, P., 1975: The cooling world. *Newsweek,* 28 April.

Halacy, D. S., 1978: *Ice or Fire? Surviving Climatic Change.* Harper and Row, 212 pp.

Hamilton, W. L., and T. A. Seliga, 1972: Atmospheric turbidity and surface temperature on the polar ice sheets. *Nature,* **235,** 320–322.

Hansen, J. E., W.-C. Wang, and A. A. Lacis, 1978: Mount Agung eruption provides test of a global climatic perturbation. *Science,* **199,** 1065–1068, doi:10.1126/science.199.4333.1065.

Hays, J. D., J. Imbrie, and N. J. Shackleton, 1976: Variations in the Earth's orbit: Pacemaker of the ice ages. *Science,* **194,** 1121–1132.

Herman, B. M., S. A. Twomey, and D. O. Staley, 1978: Atmospheric dust: Climatological consequences. *Science,* **201,** 378.

Hobbs, P. V., H. Harrison, and E. Robinson, 1974: Atmospheric effects of pollutants. *Science,* **183,** 909–915.

Horner, C. C., 2007: *The Politically Incorrect Guide to Global Warming and Environmentalism.* Regnery Publishing, 350 pp.

Houghton, J. T., 1979: Greenhouse effects of some atmospheric constituents. *Philos. Trans. Roy. Soc. London, Math. Phys. Sci.,* **290A,** 515–521.

Hoyt, D. V., 1979: An empirical determination of the heating of the Earth by the carbon dioxide greenhouse effect. *Nature,* **282,** 388–390.

Idso, S. B., and A. J. Brazel, 1977: Planetary radiation balance as a function of atmospheric dust: Climatological consequences. *Science,* **198,** 731–733.

———, and ———, 1978: Atmospheric dust: Climatological consequences. *Science,* **201,** 378–379.

Inhofe, J., cited 2003: Science of climate change. *Congressional Record,* No. 149, S10012-S10023. [Available online at wais.access.gpo.gov.]

Kellogg, W. W., 1979: Prediction of a global cooling. *Nature,* **280,** 615.

———, and S. H. Schneider, 1974: Climate stabilization: For better or for worse? *Science,* **186,** 1163–1172.

Kerr, R. A., 1978: Climate control: How large a role for orbital variations? *Science,* **201,** 144–146.

Kukla, G. J., and H. J. Kukla, 1974: Increased surface albedo in the Northern Hemisphere. *Science,* **183,** 709–714.

Landsberg, H. E., 1970: Man-made climatic changes. *Science,* **170,** 1265–1274.

———, 1976: Whence global climate: Hot or cold? An essay review. *Bull. Amer. Meteor. Soc.,* **57,** 441–443.

Lee, P. S., and F. M. Snell, 1977: An annual zonally averaged global climate model with diffuse cloudiness feedback. *J. Atmos. Sci.,* **34,** 847–853.

Lowry, W. P., 1972: Atmospheric pollution and global climatic change. *Ecology,* **53,** 908–914.

Machta, L., 1972: Mauna Loa and Global trends in air quality. *Bull. Amer. Meteor. Soc.,* **53,** 402–420.

Manabe, S., and R. T. Weatherald, 1967: Thermal equilibrium of the atmosphere with a given distribution of relative humidity. *J. Atmos. Sci.,* **24,** 241–259.

———, and ———, 1975: The effects of doubling the CO_2 concentration on the climate of a general circulation model. *J. Atmos. Sci.,* **32,** 3–15.

Mason, B. J., 1978a: The World Climate Programme, *Nature,* **276,** 339–342.

———, 1978b: Review lecture: Recent advances in the numerical prediction of weather and climate. *Proc. Roy. Soc. London, Math. Phys. Sci.,* **363A,** 297–333.

Mathews, S. W., 1976: What's happening to our climate? *Natl. Geogr. Mag.,* **150,** 576–615.

McCormick, R. A., and J. H. Ludwig, 1967: Climate modification by atmospheric aerosols. *Science,* **156,** 1358–1359.

Mercer, J. H., 1978: West Antarctic ice sheet and CO_2 greenhouse effect: A threat of disaster. *Nature,* **271,** 321–325.

Michaels, P. J., 2004: *Meltdown: The Predictable Distortion of Global Warming by Scientists, Politicians, and the Media.* Cato Institute, 271 pp.

Milankovitch, M., 1930: *Mathematische Klimalehre und Astronomische Theorie der Klimaschwankungen.* Vol. I, Part A, *Handbuch der Klimatologie,* W. Koppen and R. Geiger, Eds., Gebruder Borntrager, 176 pp.

Miles, M. K., 1978: Predicting temperature trend in the Northern Hemisphere to the year 2000. *Nature,* **276,** 356.

Mitchell, J. M., Jr., 1963: On the world-wide pattern of secular temperature change. *Changes of Climate: Proceedings of the Rome Symposium Organized by UNESCO and the World Meteorological Organization,* UNESCO, 161–181.

———, 1970: A preliminary evaluation of atmospheric pollution as a cause of the global temperature fluctuation of the past century. *Global Effects of Environmental Pollution,* S. F. Singer, Ed., Springer-Verlag/D. Reidel, 139–155.

———, 1971: The effects of atmospheric aerosols on climate with special reference to temperature near the Earth's surface. *J. Appl. Meteor.,* **10,** 703–714.

———, 1972: The natural breakdown of the present interglacial and its possible intervention by human activities. *Quart. Res.,* **2,** 436–445.

———, 1976: An overview of climatic variability and its causal mechanisms. *Quat. Res.,* **6,** 481–494.

National Academy of Sciences, 1975: *Understanding Climatic Change.* U.S. Committee for the Global Atmospheric Research Program, National Academy Press, 239 pp.

———, 1977: *Climate, Climatic Change, and Water Supply.* Panel on Water and Climate of the National Research Council, 132 pp.

National Defence University, Research Directorate, 1978: *Climate Change to the Year 2000: A Survey of Expert Opinion.* U.S. Government Printing Office, 109 pp.

National Science Board, 1972: *Patterns and Perspectives in Environmental Science.* National Science Foundation, 426 pp.

Nordhaus, W. D., 1977: Economic growth and climate: The carbon dioxide problem. *Amer. Econ. Rev.,* **67,** 341–346.

Ohring, G., and S. Adler, 1978: Some experiments with a zonally averaged climate model. *J. Atmos. Sci.,* **35,** 186–205.

Pales, J. C., and C. D. Keeling, 1965: The concentration of atmospheric carbon dioxide in Hawaii. *J. Geophys. Res.,* **70,** 6053–6076.

Panel on Energy and Climate, 1977: *Energy and Climate.* National Research Council, National Academy of Sciences, 158 pp.

Ponte, L., 1976: *The Cooling.* Prentice-Hall, 306 pp.

Ramanathan, V., 1975: Greenhouse effect due to chloroflurocarbons: Climatic implications. *Science,* **190,** 50–52.

———, and J. A. Coakley Jr., 1978: Climate modeling through radiative convective models. *Rev. Geophys. Space Phys.,* **16,** 465–489.

Rasool, S. I., and S. H. Schneider, 1971: Atmospheric carbon dioxide and aerosols: Effects of large increases on global climate. *Science,* **173,** 138–141.

———, and ———, 1972: Aerosol concentrations: Effect on planetary temperatures. *Science,* **175,** 96.

Reck, R. A., 1975: Aerosols and polar temperature change. *Science,* **188,** 728–730.

Report of the Study of Critical Environmental Problems, 1970: *Man's Impact on the Global Environment.* Massachusetts Institute of Technology Press, 319 pp.

Revelle, R., W. Broecker, H. Craig, C. D. Kneeling, and J. Smagorinsky, 1965: *Restoring the Quality of Our Environment: Report of the Environmental Pollution Panel.* President's Science Advisory Committee, The White House, 317 pp.

Revkin, A. C., 2005: The daily planet: Why the media stumble over the environment. *A Field Guide for Science Writers,* D. Blum, M. Knudson, and R. M. Henig, Eds., Oxford University Press, 222–228.

Rotty, R. M., 1979: Atmospheric CO_2 consequences of heavy dependence on coal. *Environ. Health Perspect.,* **33,** 273–283.

Sagan, C., O. B. Toon, and J. B. Pollack, 1979: Anthropogenic albedo changes and the Earth's climate. *Science,* **206,** 1363–1368.

Sarewitz, D., 2004: How science makes environmental controversies worse. *Environ. Sci. Policy,* **7,** 385–403.

Sawyer, J. S., 1972: Man-made carbon dioxide and the "greenhouse" effect. *Nature,* **239,** 23–26.

Schlesinger, J., 2003: Climate change: The science isn't settled. *Washington Post,* 7 July, A17.

Schneider, S. H., 1975: On the carbon dioxide-climate confusion. *J. Atmos. Sci.,* **32,** 2060–2066.

———, 1976: *The Genesis Strategy: Climate and Global Survival.* Plenum Press, 419 pp.

———, and C. Mass, 1975: Volcanic dust, sunspots, and temperature trends. *Science,* **190,** 741–746.

Sellers, W. D., 1969: A global climatic model based on the energy balance of the earth-atmosphere system. *J. Appl. Meteor.,* **8,** 392–400.

———, 1973: A new global climatic model. *J. Appl. Meteor.,* **12,** 241–254.

———, 1974: A reassessment of the effect of CO_2 variation on a simple global climatic model. *J. Appl. Meteor.,* **13,** 831–833.

Shaw, G. E., 1976: Properties of the background global aerosol and their effects on climate. *Science,* **192,** 1334–1336.

Shutts, G. J., and J. S. A. Green, 1978: Mechanisms and models of climatic change. *Nature,* **276,** 339–342.

Singer, S. F., Ed., 1970: *Global Effects of Environmental Pollution.* Springer-Verlag/D. Reidel, 218 pp.

———, 1998: Scientists add to heat over global warming. *Washington Times,* 5 May.

———, and D. T. Avery, 2007: *Unstoppable Global Warming Every 1,500 Years.* Rowman and Littlefield, 260 pp.

Somerville, R., H. Le Treut, U. Cubasch, Y. Ding, C. Mauritzen, A. Mokssit, T. Peterson, and M. Prather, 2007: Historical overview of climate change. *Climate Change 2007: The Physical Science Basis,* S. Solomon et al., Eds., Cambridge University Press, 93–127.

Stuiver, M., 1978: Atmospheric carbon dioxide and carbon reservoir changes. *Science,* **199,** 253–258.

Sullivan, W., 1975a: Scientists ask why world climate is changing; major cooling may be ahead. *New York Times,* 21 May.

———, 1975b: Warming trend seen in climate; two articles counter view that cold period is due. *New York Times,* 14 August.

Thompson, L. M., 1975: Weather variability, climate change, and grain production. *Science,* **188,** 535–541.

Time Magazine, 1974: Another ice age? 24 June. [Available online at www.time.com/time/magazine/article/0,9171,944914,00.html.]

Twomey, S., 1977: The influence of pollution on the shortwave albedo of clouds. *J. Atmos. Sci.,* **34,** 1149–1152.

Tyndall, J., 1861: On the absorption and radiation of heat by gases and vapours, and on the physical connection. *Philos. Mag.,* **22,** 277–302.

Wang, W. C., Y. L. Yung, A. A. Lacis, T. Mo, and J. E. Hansen, 1976: Greenhouse effects due to man-made perturbations of trace gases. *Science,* **194,** 685–690.

Weare, B. C., R. L. Temkin, and F. M. Snell, 1974: Aerosols and climate: Some further considerations. *Science,* **186,** 827–828.

Weart, S., cited 2003: The Discovery of Global Warming. [Available online at www.aip.org/history/climate.]

Wigley, T. M. L., 1978: Climatic change. *Nature,* **272,** 788.

Wilcox, H. A., 1975: *Hothouse Earth.* Praeger Publishers, 181 pp.

Will, G. F., 2004: Global warming? Hot air. *Washington Post,* 23 December.

———, 2008: March of the polar bears. *Washington Post,* 22 May.

Willett, H. C., 1974: Do recent climatic fluctuations portend an imminent ice age? *Geofis. Int.,* **14,** 265–302.

Woronko, S. F., 1977: Comments on "A non-equilibrium model of hemispheric mean surface temperature." *J. Atmos. Sci.,* **34,** 1820–1821.

Thomas C. Peterson—NOAA/National Climatic Data Center. Asheville, North Carolina; **William M. Connolley**—British Antarctic Survey, National Environment Research Council, Cambridge, United Kingdom; **John Fleck**—*Albuquerque Journal,* Albuquerque, New Mexico

Corresponding Author—Thomas C. Peterson, NOAA/National Climatic Data Center, 151 Patton Avenue, Asheville, NC 28803

Acknowledgments—We appreciate Byron Gleason for his detailed review of the article; librarian Mara Sprain for her invaluable assistance; André Berger, Spencer Weart, and Neville Nicholls for providing valuable insights and additional references; Mike Wallace for his cogent review, which precipitated considerable additional and much needed work; and two anonymous reviewers whose comments improved the paper. This project prompted us to reread many articles by the great climatologists of past decades such as Mikhail Budyko, Charles David Keeling, Helmut Landsberg, Syukuro Manabe, B. John Mason, and J. Murray Mitchell, which made us realize the debt of gratitude we owe to these pioneers.

How to Stop Climate Change: The Easy Way

Changing your light bulbs may not be enough to save a single polar bear, but there *are* things we can do collectively—and easily—that will really make a measurable difference in the battle against global warming.

MARK LYNAS

We have about 100 months left. If global greenhouse gas emissions have not begun to decline by the end of 2015, then our chances of restraining climate change to within the two degrees "safety line"—the level of warming below which the impacts are severe but tolerable—diminish day by day thereafter. This is what the latest science now demands: the peaking of emissions within eight years, worldwide cuts of 60 percent by 2030, and 80 percent or more by 2050. Above two degrees, our chances of crossing "tipping points" in the earth's system—such as the collapse of the Amazon rainforest, or the release of methane from thawing Siberian permafrost—is much higher.

Despite this urgent timetable, our roads continue to heave with traffic. Power companies draft blueprints for new coal-fired plants. The skies over England are criss-crossed with vapour trails from aircraft travelling some of the busiest routes in the world. Global emissions, far from decreasing, remain on a steep upward curve of almost exponential growth.

Sure, there are some encouraging signs. Media coverage of climate change remains high, and a worldwide popular movement—now perhaps upwards of a million people—is mobilising. But with so little time left, we must recognise that most people won't do anything to save the planet unless we make it much, much easier for them. This essay outlines my three-part strategy for stopping climate change—the easy way.

Step One: Stop Debating, Start Doing

Although there is now a very broad consensus on climate in the media and politics, opinion polls show that many people still harbour doubts about climate change. One of the peculiarities of the climate debate is that although more than 99 percent of international climate change scientists agree on the causes of global warming, the denial lobby still only has to produce one contrarian to undermine the consensus in the public mind. Similarly, changes in our understanding can be magnified and distorted to suggest that, because we don't know everything, therefore we must know nothing. Thus, data from one glacier that apparently bucks the global trend can be wielded as a trump card against all the accumulated knowledge of climate science.

This partly reflects a perhaps healthy scepticism in the public mind about believing "experts". But there is also a darker force at work: doubt undermines responsibility for action. If you don't know for sure that global warming isn't caused by sunspots or cosmic rays, then it's OK to go on driving and flying without feeling as if you're doing something bad. When it comes to global warming, many people—subconsciously at least—actually *want* to be lied to.

This is where the psychology gets interesting. Most green campaigners assume that information leads to action, and that deeper knowledge will undermine denial. Actually, the reverse may well be true: the more disempowered that people feel about a huge, scary issue like climate change, the more unwilling they may be to believe it is a problem. This sounds illogical, but it makes sense. If people don't feel they can do very much about climate change, they will prefer to cling to any tempting doubts that are dangled their way. Presenting people with more gloom-and-doom scenarios, however true they might be, may thus serve to reinforce denial.

Most campaigners try to mitigate this by also offering people easy things they can do: the "just change your light bulbs" approach. However, most people intuitively understand that an enormous problem cannot be solved by a tiny solution; that changing your light bulbs will not save a single polar bear. They are right, of course. So how can we mobilise collective action on a sufficiently grand scale to make a measurable contribution to solving the problem?

The American political strategists Ted Nordhaus and Michael Shellenberger make a specific proposal in a recent paper, and this forms the first plank of my three-part strategy to tackle global warming. Stop debating, they say, and start doing. Instead of confronting deeply established patterns of behaviour head on, let's start focusing on preparing for the impacts of global warming that are already inevitable. That means working on flood defences for vulnerable towns, helping to drought-proof agriculture and population centres, and adapting to sea-level rise in low-lying areas.

By sidestepping the tedious causality argument (is it us or natural cycles?), focusing on global warming preparedness can also help reopen the mitigation agenda. Shifting sandbags is empowering because you feel as if you're doing something tangible and useful. But accepting the need for adaptation and preparation implicitly involves accepting the reality of global warming, and therefore the eventual need to cut emissions. Many more people may be prepared to accept the change—the introduction of personal carbon allowances, for example—that this will inevitably mean.

In any case, adaptation is now essential because of the one degree or so of additional global warming that is already locked into the system thanks to past emissions. With proper planning, we can not only save thousands of human lives, but also try to protect natural ecosystems by establishing new "refuge" coral reefs in cooler waters or helping species to migrate as temperature zones shift.

Step Two: Focus on the Big Wins

But this is a long-term agenda, and we don't have much time. Hence my second proposal, which is for a much clearer focus on win-win strategies for immediate emissions reductions. These are things we would want to be doing anyway, even if global warming had never been thought of. Reducing deforestation in the tropics is a big win-win. Inherently desirable, this by itself would reduce global carbon emissions by 10 percent or more. All it takes is money: we have to pay countries such as Brazil and Indonesia to leave their forests alone rather than chop them down to sell to us as plywood and furniture.

There are obvious win-win strategies in the domestic sector. Better insulation makes living conditions more comfortable and reduces fuel bills. Even without climate change we'd still want to be getting cars out of town centres to reduce air pollution and improve the urban experience. Getting more children to walk and cycle to school improves their physical health and helps to tackle obesity. Enforcing speed limits (and reducing them further) would save hundreds of lives a year, and give some respite from the incessant noise pollution of speeding traffic.

Quality-of-life issues are by their nature subjective, so we need to focus on things that most people will agree on. Partly, this depends on how an issue is framed: most people don't want motorists to be unjustifiably hounded, but nor are they likely to oppose a measure that is about saving children's lives. The ban on smoking in public, for instance, was accepted precisely because the issue was correctly framed, and quickly became imbued with a sense of inevitability.

There is also a high degree of consensus about the desirability of localisation: protecting and encouraging small shops and local businesses, privileging farmers' markets over supermarkets, helping build stronger and more cohesive communities by reducing the need for travel, and so on. The fact that all of these measures will also reduce carbon emissions simply underlines the need for a more determined approach to their implementation. A much longer-term agenda here might be the reconnecting of people with their place and surroundings, helping them feel more rooted in their communities and proud of what is distinctive about their own areas. We are bringing up children who often have no direct experience of nature any more. Tree houses are replaced with Nintendos, the unsupervised exercise of playing outdoors replaced with structured exercise of sporting events. The author Richard Louv terms this "nature deficit disorder" and asks whether this disconnection might have something to do with the alienation and boredom that many youngsters feel today.

Step Three: Use Technology

But there are some areas of high-carbon behaviour that people will always be reluctant to give up, and this brings me to the third and final part of my strategy to deal with global warming—technology.

Today we face a situation where a global population of potentially nine billion or so by 2050 continues to demand a steadily increasing consumer lifestyle. There is nothing we can do to stop this, and nor should we try. But it does put humanity on a very real collision course with the planet, so we are going to have to throw every technological tool we have at the problem to try to meet people's aspirations without worsening our climatic predicament. Some of this will involve technology leapfrogging: helping developing countries skip over our dirty phase of industrialisation, by installing solar power in remote, off-grid areas of Africa and Asia, for example. We also need to help developing countries make choices that put fossil fuels at the bottom of the energy shopping list, by helping them use carbon capture and storage technology as well as nuclear power. Both have obvious drawbacks, but I would rather see China building two nuclear reactors a week than two coal-fired plants.

The localisation agenda can only go so far: in an age of carbon-fuelled globalisation, we need to figure out ways to transport people and goods long distances without increasing emissions. Aviation in particular is crying out for a techno-fix. Humanity went from the first manned flight in 1903 to putting a man on the moon in 1969. I think we should give the aviation industry 15 years to find a low-carbon way to shuttle people between continents—or get taxed out of existence. I believe with this kind of incentive, designers would come up with ideas none of us today could even conceive of.

The technological challenge is not just to come up with new inventions, but—in the words of Robert Socolow and Stephen Pacala from Princeton University—"to scale up what we already know how to do". In their concept of "stabilisation wedges," each wedge represents a billion tonnes of carbon shaved off the upward trend of emissions over the next

50 years. Building two million one-megawatt wind turbines, for example, is a wedge, as are two million hectares of solar panels, a 700-fold increase from today's deployment. There are many more wedges in the fields of transport, power generation and energy efficiency. As the two researchers say, this reduces a "heroic challenge" merely to a set of "monumental tasks". No one said it would be easy.

Perhaps the most controversial technological option of all is one that we need to keep strictly in reserve for real emergencies—geo-engineering. Here, some proposals have more merit than others, whether they be seeding the oceans with iron filings or putting up solar mirrors in space. None of them is an alternative to reducing emissions, but one just might be a valuable piece of insurance against the worst-case climate change scenarios. Believe me, pretty much anything is better than five or six degrees of global warming.

This may seem like a depressing conclusion, but it's really an optimistic one. If we fail to reduce emissions quickly enough and find ourselves frying, we must throw everything we possibly can at the problem to counteract the warming process, however temporarily. At no point—I repeat, at no point—do we give up and admit that all is lost. If we go over two degrees, then we have to try and stop ourselves going over three. If we fail to stabilise emissions by 2015, then we have to try and stabilise them by 2016 or 2020. If people continue to demand economic growth, then we have to try to deliver than growth in a low-carbon way. It will never be too late. As long as people and nature remain alive on this planet, we will still have everything to fight for.

Mark Lynas is the New Statesman's environment correspondent, and author of "*Six Degrees: Our Future on a Hotter Planet,*" published by Fourth Estate.

Environmental Justice for All

LEYLA KOKMEN

Manuel Pastor ran bus tours of Los Angeles a few years back. These weren't the typical sojourns to Disneyland or the MGM studios, though; they were expeditions to some of the city's most environmentally blighted neighborhoods—where railways, truck traffic, and refineries converge, and where people live 200 feet from the freeway.

The goal of the "toxic tours," explains Pastor, a professor of geography and of American studies and ethnicity at the University of Southern California (USC), was to let public officials, policy makers, and donors talk to residents in low-income neighborhoods about the environmental hazards they lived with every day and to literally see, smell, and feel the effects.

"It's a pretty effective forum," says Pastor, who directs USC's Program for Environmental and Regional Equity, noting that a lot of the "tourists" were eager to get back on the bus in a hurry. "When you're in these neighborhoods, your lungs hurt."

Like the tours, Pastor's research into the economic and social issues facing low-income urban communities highlights the environmental disparities that endure in California and across the United States. As stories about global warming, sustainable energy, and climate change make headlines, the fact that some neighborhoods, particularly low-income and minority communities, are disproportionately toxic and poorly regulated has, until recently, been all but ignored.

A new breed of activists and social scientists are starting to capitalize on the moment. In principle they have much in common with the environmental justice movement, which came of age in the late 1970s and early 1980s, when grassroots groups across the country began protesting the presence of landfills and other environmentally hazardous facilities in predominantly poor and minority neighborhoods.

In practice, though, the new leadership is taking a broader-based, more inclusive approach. Instead of fighting a proposed refinery here or an expanded freeway there, all along trying to establish that systematic racism is at work in corporate America, today's environmental justice movement is focusing on proactive responses to the social ills and economic roadblocks that if removed would clear the way to a greener planet.

The new movement assumes that society as a whole benefits by guaranteeing safe jobs, both blue-collar and white-collar, that pay a living wage. That universal health care would both decrease disease and increase awareness about the quality of everyone's air and water. That better public education and easier access to job training, especially in industries that are emerging to address the global energy crisis, could reduce crime, boost self-esteem, and lead to a homegrown economic boon.

That green rights, green justice, and green equality should be the environmental movement's new watchwords.

"This is the new civil rights of the 21st century," proclaims environmental justice activist Majora Carter.

A lifelong resident of Hunts Point in the South Bronx, Carter is executive director of Sustainable South Bronx, an eight-year-old nonprofit created to advance the environmental and economic future of the community. Under the stewardship of Carter, who received a prestigious MacArthur Fellowship in 2005, the organization has managed a number of projects, including a successful grassroots campaign to stop a planned solid waste facility in Hunts Point that would have processed 40 percent of New York City's garbage.

Her neighborhood endures exhaust from some 60,000 truck trips every week and has four power plants and more than a dozen waste facilities. "It's like a cloud," Carter says. "You deal with that, you're making a dent."

The first hurdle Carter and a dozen staff members had to face was making the environment relevant to poor people and people of color who have long felt disenfranchised from mainstream environmentalism, which tends to focus on important but distinctly nonurban issues, such as preserving Arctic wildlife or Brazilian rainforest. For those who are struggling to make ends meet, who have to cobble together adequate health care, education, and job prospects, who feel unsafe on their own streets, these grand ideas seem removed from reality.

That's why the green rights argument is so powerful: It spans public health, community development, and economic growth to make sure that the green revolution isn't just for those who can afford a Prius. It means cleaning up blighted communities like the South Bronx to prevent potential health problems and to provide amenities like parks to play in, clean trails to walk on, and fresh air to breathe. It also means building green industries into the local mix, to provide healthy jobs for residents in desperate need of a livable wage.

Historically, mainstream environmental organizations have been made up mostly of white staffers and have focused more on the ephemeral concept of the environment rather than on the people who are affected. Today, though, as climate change and gas prices dominate public discourse, the concepts driving the new environmental justice movement are starting to catch on. Just recently, for instance, *New York Times* columnist Thomas Friedman dubbed the promise of public investment in the green economy the "Green New Deal."

Van Jones, whom Friedman celebrated in print last October, is president of the Ella Baker Center for Human Rights in Oakland, California. To help put things in context, Jones briefly sketches the history of environmentalism:

The first wave was conservation, led first by Native Americans who respected and protected the land, then later by Teddy Roosevelt, John Muir, and other Caucasians who sought to preserve green space.

The second wave was regulation, which came in the 1970s and 1980s with the establishment of the Environmental Protection Agency (EPA) and Earth Day. Increased regulation brought a backlash against poor people and people of color, Jones says. White, affluent communities sought to prevent environmental hazards from entering their neighborhoods. This "not-in-my-backyard" attitude spurred a new crop of largely grassroots environmental justice advocates who charged businesses with unfairly targeting low-income and minority communities. "The big challenge was NIMBY-ism," Jones says, noting that more toxins from power plants and landfills were dumped on people of color.

The third wave of environmentalism, Jones says, is happening today. It's a focus on investing in solutions that lead to "eco-equity." And, he notes, it invokes a central question: "How do we get the work, wealth, and health benefits of the green economy to the people who most need those benefits?"

There are a number of reasons why so many environmental hazards end up in the poorest communities.

Property values in neighborhoods with environmental hazards tend to be lower, and that's where poor people—and often poor people of color—can afford to buy or rent a home. Additionally, businesses and municipalities often choose to build power plants in or expand freeways through low-income neighborhoods because the land is cheaper and poor residents have less power and are unlikely to have the time or organizational infrastructure to evaluate or fight development.

"Wealthy neighborhoods are able to resist, and low-income communities of color will find their neighborhoods plowed down and [find themselves] living next to a freeway that spews pollutants next to their schools," USC's Manuel Pastor says.

Moreover, regulatory systems, including the EPA and various local and state zoning and environmental regulatory bodies, allow piecemeal development of toxic facilities. Each new chemical facility goes through an individual permit process, which doesn't always take into account the overall picture in the community. The regulatory system isn't equipped to address potentially dangerous cumulative effects.

In a single neighborhood, Pastor says, you might have toxins that come from five different plants that are regulated by five different authorities. Each plant might not be considered dangerous on its own, but if you throw together all the emissions from those static sources and then add in emissions from moving sources, like diesel-powered trucks, "you've created a toxic soup," he says.

In one study of air quality in the nine-county San Francisco Bay Area, Pastor found that race, even more than income, determined who lived in more toxic communities. That 2007 report, "Still Toxic After All These Years: Air Quality and Environmental Justice in the San Francisco Bay Area," published by the Center for Justice, Tolerance & Community at the University of California at Santa Cruz, explored data from the EPA's Toxic Release Inventory, which reports toxic air emissions from large industrial facilities. The researchers examined race, income, and the likelihood of living near such a facility.

More than 40 percent of African American households earning less than $10,000 a year lived within a mile of a toxic facility, compared to 30 percent of Latino households and fewer than 20 percent of white households.

As income rose, the percentages dropped across the board but were still higher among minorities. Just over 20 percent of African American and Latino households making more than $100,000 a year lived within a mile of a toxic facility, compared to just 10 percent of white households.

The same report finds a connection between race and the risk of cancer or respiratory hazards, which are both associated with environmental air toxics, including emissions both from large industrial facilities and from mobile sources. The researchers looked at data from the National

Air Toxics Assessment, which includes estimates of such ambient air toxics as diesel particulate matter, benzene, and lead and mercury compounds. The areas with the highest risk for cancer had the highest proportion of African American and Asian residents, the lowest rate of home ownership, and the highest proportion of people in poverty. The same trends existed for areas with the highest risk for respiratory hazards.

According to the report, "There is a general pattern of environmental inequity in the Bay Area: Densely populated communities of color characterized by relatively low wealth and income and a larger share of immigrants disproportionately bear the hazard and risk burden for the region."

Twenty years ago, environmental and social justice activists probably would have presented the disparities outlined in the 2007 report as evidence of corporations deliberately targeting minority communities with hazardous waste. That's what happened in 1987, when the United Church of Christ released findings from a study that showed toxic waste facilities were more likely to be located near minority communities. At the 1991 People of Color Environmental Leadership Summit, leaders called the disproportionate burden both racist and genocidal.

In their 2007 book *Break Through: From the Death of Environmentalism to the Politics of Possibility,* authors Ted Nordhaus and Michael Shellenberger take issue with this strategy. They argue that some of the research conducted in the name of environmental justice was too narrowly focused and that activists have spent too much time looking for conspiracies of environmental racism and not enough time looking at the multifaceted problems facing poor people and people of color.

"Poor Americans of all races, and poor Americans of color in particular, disproportionately suffer from social ills of every kind," they write. "But toxic waste and air pollution are far from being the most serious threats to their health and well-being. Moreover, the old narratives of intentional discrimination fail to explain or address these disparities. Disproportionate environmental health outcomes can no more be reduced to intentional discrimination than can disproportionate economic and educational outcomes. They are due to a larger and more complex set of historic, economic, and social causes."

Today's environmental justice advocates would no doubt take issue with the finer points of Nordhaus and Shellenberger's criticism—in particular, that institutional racism is a red herring. Activists and researchers are acutely aware that they are facing a multifaceted spectrum of issues, from air pollution to a dire lack of access to regular health care. It's because of that complexity, however, that they are now more geared toward proactively addressing an array of social and political concerns.

"The environmental justice movement grew out of putting out fires in the community and stopping bad things from happening, like a landfill," says Martha Dina Argüello, executive director of Physicians for Social Responsibility—Los Angeles, an organization that connects environmental groups with doctors to promote public health. "The more this work gets done, the more you realize you have to go upstream. We need to stop bad things from happening."

"We can fight pollution and poverty at the same time and with the same solutions and methods," says the Ella Baker Center's Van Jones.

Poor people and people of color have borne all the burden of the polluting industries of today, he says, while getting almost none of the benefit from the shift to the green economy. Jones stresses that he is not an environmental justice activist, but a "social-uplift environmentalist." Instead of concentrating on the presence of pollution and toxins in low-income communities, Jones prefers to focus on building investment in clean, green, healthy industries that can help those communities. Instead of focusing on the burdens, he focuses on empowerment.

With that end in mind, the Ella Baker Center's Green-Collar Jobs Campaign plans to launch the Oakland Green Jobs Corps this spring. The initiative, according to program manager Aaron Lehmer, received $250,000 from the city of Oakland and will give people ages 18 to 35 with barriers to employment (contact with the criminal justice system, long-term unemployment) opportunities and paid internships for training in new energy skills like installing solar panels and making buildings more energy efficient.

The concept has gained national attention. It's the cornerstone of the Green Jobs Act of 2007, which authorizes $125 million annually for "green-collar" job training that could prepare 30,000 people a year for jobs in key trades, such as installing solar panels, weatherizing buildings, and maintaining wind farms. The act was signed into law in December as part of the Energy Independence and Security Act.

While Jones takes the conversation to a national level, Majora Carter is focusing on empowerment in one community at a time. Her successes at Sustainable South Bronx include the creation of a 10-week program that offers South Bronx and other New York City residents hands-on training in brownfield remediation and ecological restoration. The organization has also raised $30 million for a bicycle and pedestrian greenway along the South Bronx waterfront that will provide both open space and economic development opportunities.

As a result of those achievements, Carter gets calls from organizations across the country. In December she traveled to Kansas City, Missouri, to speak to residents, environmentalists, businesses, and students. She mentions exciting work being done by Chicago's Blacks in Green collective, which aims to mobilize the African American community around environmental issues. Naomi Davis, the collective's founder, told Chicago Public Radio in November that the group plans to develop environmental and economic opportunities—a "green village" with greenways, light re-manufacturing, ecotourism, and energy-efficient affordable housing—in one of Chicago's most blighted areas.

Carter stresses that framing the environmental debate in terms of opportunities will engage the people who need the most help. It's about investing in the green economy, creating jobs, and building spaces that aren't environmentally challenged. It won't be easy, she says. But it's essential to dream big.

"It's about sacrifice," she says, "for something better and bigger than you could have possibly imagined."

Test-Your-Knowledge Form

We encourage you to photocopy and use this page as a tool to assess how the articles in *Annual Editions* expand on the information in your textbook. By reflecting on the articles you will gain enhanced text information. You can also access this useful form on a product's book support website at *http://www.mhhe.com/cls*.

NAME: DATE:

TITLE AND NUMBER OF ARTICLE:

BRIEFLY STATE THE MAIN IDEA OF THIS ARTICLE:

LIST THREE IMPORTANT FACTS THAT THE AUTHOR USES TO SUPPORT THE MAIN IDEA:

WHAT INFORMATION OR IDEAS DISCUSSED IN THIS ARTICLE ARE ALSO DISCUSSED IN YOUR TEXTBOOK OR OTHER READINGS THAT YOU HAVE DONE? LIST THE TEXTBOOK CHAPTERS AND PAGE NUMBERS:

LIST ANY EXAMPLES OF BIAS OR FAULTY REASONING THAT YOU FOUND IN THE ARTICLE:

LIST ANY NEW TERMS/CONCEPTS THAT WERE DISCUSSED IN THE ARTICLE, AND WRITE A SHORT DEFINITION:

We Want Your Advice

ANNUAL EDITIONS revisions depend on two major opinion sources: one is our Advisory Board, listed in the front of this volume, which works with us in scanning the thousands of articles published in the public press each year; the other is you—the person actually using the book. Please help us and the users of the next edition by completing the prepaid article rating form on this page and returning it to us. Thank you for your help!

ANNUAL EDITIONS: Environment 10/11

ARTICLE RATING FORM

Here is an opportunity for you to have direct input into the next revision of this volume.
We would like you to rate each of the articles listed below, using the following scale:

1. **Excellent: should definitely be retained**
2. **Above average: should probably be retained**
3. **Below average: should probably be deleted**
4. **Poor: should definitely be deleted**

Your ratings will play a vital part in the next revision.
Please mail this prepaid form to us as soon as possible.
Thanks for your help!

RATING	ARTICLE
________	1. IPCC, 2007: Summary for Policymakers
________	2. Global Warming Battlefields: How Climate Change Threatens Security
________	3. China Needs Help with Climate Change
________	4. Where Oil and Water Do Mix: Environmental Scarcity and Future Conflict in the Middle East and North Africa
________	5. Do Global Attitudes and Behaviors Support Sustainable Development?
________	6. Paying for Climate Change
________	7. High-Tech Trash: Will Your Discarded TV or Computer End up in a Ditch in Ghana?
________	8. Down with Carbon: Scientists Work to Put the Greenhouse Gas in Its Place
________	9. Gassing up with Hydrogen
________	10. Wind Power: Obstacles and Opportunities
________	11. A Solar Grand Plan
________	12. The Rise of Renewable Energy
________	13. What Nuclear Renaissance?
________	14. The Biofuel Future: Scientists Seek Ways to Make Green Energy Pay Off
________	15. Putting Your Home on an Energy Diet
________	16. Forest Invades Tundra . . . and the New Tenants Could Aggravate Global Warming
________	17. America's Coral Reefs: Awash with Problems
________	18. Seabird Signals
________	19. Taming the Blue Frontier
________	20. Nature's Revenge
________	21. Tracking U.S. Groundwater: Reserves for the Future?
________	22. How Much Is Clean Water Worth?
________	23. Searching for Sustainability: Forest Policies, Smallholders, and the Trans-Amazon Highway
________	24. Diet, Energy, and Global Warming
________	25. Landfill-on-Sea
________	26. The Truth about Denial
________	27. Swift Boating, Stealth Budgeting, and Unitary Executives
________	28. The Myth of the 1970s Global Cooling Scientific Consensus
________	29. How to Stop Climate Change: The Easy Way
________	30. Environmental Justice for All

NO POSTAGE NECESSARY IF MAILED IN THE UNITED STATES

BUSINESS REPLY MAIL
FIRST CLASS MAIL PERMIT NO. 551 DUBUQUE IA

POSTAGE WILL BE PAID BY ADDRESSEE

McGraw-Hill Contemporary Learning Series
501 BELL STREET
DUBUQUE, IA 52001

ABOUT YOU

Name Date

Are you a teacher? ❐ A student? ❐
Your school's name

Department

Address City State Zip

School telephone #

YOUR COMMENTS ARE IMPORTANT TO US!

Please fill in the following information:
For which course did you use this book?

Did you use a text with this ANNUAL EDITION? ❐ yes ❐ no
What was the title of the text?

What are your general reactions to the Annual Editions concept?

Have you read any pertinent articles recently that you think should be included in the next edition? Explain.

Are there any articles that you feel should be replaced in the next edition? Why?

Are there any World Wide Websites that you feel should be included in the next edition? Please annotate.

May we contact you for editorial input? ❐ yes ❐ no
May we quote your comments? ❐ yes ❐ no

NOTES

NOTES

NOTES

NOTES